Climate Change and
American Policy

Compiled by John R. Burch, Jr.

The Bibliography of Appalachia: More Than 4,700 Books, Articles, Monographs and Dissertations, Topically Arranged and Indexed (McFarland, 2009)

Owsley County, Kentucky, and the Perpetuation of Poverty (McFarland, 2008)

Climate Change and American Policy

Key Documents, 1979–2015

Compiled and edited by
JOHN R. BURCH, JR.

McFarland & Company, Inc., Publishers
Jefferson, North Carolina

LIBRARY OF CONGRESS CATALOGUING-IN-PUBLICATION DATA

Names: Burch, John R., 1968– editor.
Title: Climate change and American policy : key documents, 1979–2015 / compiled and edited by John R. Burch, Jr.
Description: Jefferson, North Carolina : McFarland & Company, Inc., 2016. | Includes bibliographical references and index.
Identifiers: LCCN 2016032514 | ISBN 9781476665276 (softcover : acid free paper) ∞
Subjects: LCSH: Climatic changes—Government policy—United States.
Classification: LCC QC902.92 .C55 2016 | DDC 363.738/745610973—dc23
LC record available at https://lccn.loc.gov/2016032514

BRITISH LIBRARY CATALOGUING DATA ARE AVAILABLE

ISBN (print) 978-1-4766-6527-6
ISBN (ebook) 978-1-4766-2685-7

© 2016 John R. Burch, Jr. All rights reserved

No part of this book may be reproduced or transmitted in any form or by any means, electronic or mechanical, including photocopying or recording, or by any information storage and retrieval system, without permission in writing from the publisher.

Front cover photograph of glacier in Alaska taken during President Obama's 2015 visit to the GLACIER Conference (White House photograph)

Printed in the United States of America

*McFarland & Company, Inc., Publishers
Box 611, Jefferson, North Carolina 28640
www.mcfarlandpub.com*

Table of Contents

Introduction	1
Declaration of the World Climate Conference	7
The Global 2000 Report to the President: Entering the Twenty-First Century (Major Findings and Conclusions)	11
The Montreal Protocol on Substances That Deplete the Ozone Layer (Excerpts)	16
Message to the Senate Transmitting the Montreal Protocol on Ozone-Depleting Substances	33
Statement of Dr. James Hansen, Director, NASA Goddard Institute for Space Studies	35
Clean Air Act Amendments of 1990: Title IV (Excerpt)	39
United Nations Framework Convention on Climate Change (UNFCCC)	45
Senate Resolution 98 [Report No. 105–54] (Byrd-Hagel Resolution)	65
Kyoto Protocol to the United Nations Framework Convention on Climate Change	68
Statement on the Kyoto Protocol on Climate Change	90
Climate Change Impacts on the United States: The Potential Consequences of Climate Variability and Change (Summary)	92
Letter to Members of the Senate on the Kyoto Protocol on Climate Change	97
Global Climate Change Policy Book: Executive Summary	99
Climate Stewardship Act of 2003	104
Hot & Cold Media Spin Cycle: A Challenge to Journalists Who Cover Global Warming	128
Atmosphere of Pressure: Political Interference in Federal Climate Science (2007) (Executive Summary)	141
Massachusetts et al. v. Environmental Protection Agency et al. (Syllabus)	147

vi Table of Contents

H. Res. 593—Congratulating scientists F. Sherwood Rowland, Mario Molina, and Paul Crutzen for their work in atmospheric chemistry, particularly concerning the formation and decomposition of ozone, that led to the development of the Montreal Protocol on Substances that Deplete the Ozone Layer	153
Global Warming Twenty Years Later: Tipping Points Near	156
Secretary Chu's Remarks at the Harvard University Commencement—As Prepared for Delivery	161
Global Climate Change Impacts in the United States (Executive Summary)	167
Safe Climate Act (Excerpt)	173
Remarks by the President at the Morning Plenary Session of the United Nations Climate Change Conference	177
Endangerment and Cause or Contribute Findings for Greenhouse Gases Under Section 202(a) of the Clean Air Act: Final Rule (Introduction)	180
Copenhagen Accord	197
Richard A. Muller: Statement to the Committee on Science, Space and Technology of the United States House of Representatives	201
2013 Highlights of Progress: Responses to Climate Change by the National Water Program (Excerpt)	206
Climate Change Impacts in the United States (Overview and Report Findings)	213
One City Built to Last: Transforming New York City's Buildings for a Low-Carbon Future (Executive Summary)	223
U.S.–China Joint Announcement on Climate Change	234
Senate Bill 66: To Prohibit Any Regulation Regarding Carbon Dioxide or Other Greenhouse Gas Emissions Reduction in the United States until China, India, and Russia Implement Similar Reductions	238
Governor Brown Establishes Most Ambitious Greenhouse Gas Reduction Target in North America	240
National Security Implications of Climate-Related Risks and a Changing Climate	243
Federal Plan Requirements for Greenhouse Gas Emissions from Electric Utility Generating Units Constructed on or Before January 8, 2014; Model Trading Rules; Amendments to Framework Regulations (Executive Summary and Organization and Approach for this Proposed Rule)	254
Remarks by the President at the GLACIER Conference—Anchorage, AK	262
Joint Meeting to Hear an Address by Pope Francis of the Holy See	268
Fact Sheet: The United States and China Issue Joint Presidential Statement on Climate Change with New Domestic Policy Commitments and a Common Vision for an Ambitious Global Climate Agreement in Paris	271

NASA Study: Mass Gains of Antarctic Ice Sheet Greater than Losses   277

Sen. Cruz Confronts the Dogma of Climate Change Alarmism: "Public Policy Should Follow Actual Data, Not Political and Partisan Claims That Run Contrary to Evidence" (Opening Statement)   280

Adoption of the Paris Agreement: Proposal by the President: Draft Decision –/CP.21   282

Bibliography   319

Index   329

Introduction

Svante Arrhenius, a chemist and physicist from Sweden, noted in the late nineteenth century that the combustion of fossil fuels could lead to a rise in global temperatures. His warning, which drew little fanfare during his lifetime, was a response to changes being wrought by the Industrial Revolution. His observations proved prescient as it has been estimated that global temperatures have risen by approximately 35 percent from preindustrial levels. By the mid–1950s, scientists in the United States were conducting research that identified carbon dioxide as a possible culprit in growing global temperatures. Their work spurred research around the world into climate change. Although extensive research into the topic had been under way for decades, it was not until 1979 that climate change was formally brought to the attention of the world community through the World Climate Conference, which was organized by the World Meteorological Association, the International Council of Scientific Unions, and the Intergovernmental Oceanographic Commission. These private organizations subsequently formed the World Climate Program, which was charged with the collection and analysis of climate data from around the globe. Their actions influenced President Jimmy Carter's administration to issue *The Global 2000 Report to the President: Entering the Twenty-First Century*, which contained the United States government's first official acknowledgment of climate change as a significant policy issue. The recommendations included in the report were never acted upon, as it was issued in the last year of Carter's presidency.

This work includes documents that unveil how the United States has approached anthropogenic climate change at the local, state, federal, and international levels since the World Climate Conference. Appropriately, the *Declaration of the World Climate Conference* and an excerpt from *The Global 2000 Report to the President: Entering the Twenty-First Century* are two of the 41 documents included in this reference work. Each of the documents is presented chronologically, to allow users to track the evolving political views that have shaped their content. With few exceptions, the texts of the respective documents are supplemented by an analysis that provides both context and an exploration of their relative influence. Bibliographic references for users desiring to explore the document and its subject matter in-depth are also included. Unintentionally, but inevitably, the arrangement also illuminates how presidential administrations have chosen to view anthropogenic climate change as a policy matter. Some, like President George W. Bush's administration, opted to minimize the threat, while others, most notably President Barack Obama, made it a signature part of their legacy. Due to the influence of the presidential administrations, the documents are contextualized through the timeframe of the respective presidencies.

Ronald Reagan (1981–1989)

Upon assuming the presidency, Ronald Reagan set the tone for his administration's environmental policy by symbolically removing the solar panels that President Jimmy Carter had ordered installed on the White House just a few years earlier. Despite his disdain for the Carter administration's work on environmental issues, Reagan proved an ardent supporter of *The Montreal Protocol on Substances That Deplete the Ozone Layer*. It can be argued that this support occurred because there was no doubt at the time that the Earth's atmospheric ozone layer was being threatened by the use of chlorofluorocarbons. This was evidenced by the existence of a hole in the ozone layer over Antarctica that had been first been observed in 1978. By 1985, the hole had grown to the approximate size of North America. This prompted the international community to draft and implement the Montreal Protocol. Through the *Message to the Senate Transmitting the Montreal Protocol on Ozone-Depleting Substances*, Reagan conveyed the urgency to ratify the document to not the only Congress, but the American people. It should be noted that research conducted in the early 1970s by American scientists Frank Sherwood Rowland and Mario Molina into the manner that chlorofluorocarbons impacted the atmosphere helped shape the Montreal Protocol. The importance of their work was acknowledged in 1995 when they were awarded the Nobel Prize in Chemistry, along with Paul Crutzen. This accomplishment was recognized by the United States House of Representatives in 1997 in *H. Res. 593—Congratulating Scientists F. Sherwood Rowland, Mario Molina, and Paul Crutzen for Their Work in Atmospheric Chemistry, Particularly Concerning the Formation and Decomposition of Ozone, that Led to the Development of the Montreal Protocol on Substances that Deplete the Ozone Layer*.

The Reagan administration was not pleased in 1988 with the *Statement of Dr. James Hansen, Director, NASA Goddard Institute for Space Studies*. In that document, Hansen publicly stated that climate change posed a real threat to the future of mankind. This was notable because scientists either working for the federal government or dependent on that same entity for the financing of their research had been unwilling to stake out such a position knowing the administration's hostility on environmental issues. Hansen's esteemed reputation was a problem because his learned opinion brought instant credibility to climate change as a policy issue. Hansen's stance made him both a celebrity for environmentalists and scientists and a political target in future years, as evidenced by the *Atmosphere of Pressure: Political Interference in Federal Climate Science* and *Global Warming Twenty Years Later: Tipping Points Near*.

George H.W. Bush (1989–1993)

One of President George H.W. Bush's signature environmental policy achievements was the passage of the Clean Air Act Amendments of 1990. While notable on a number of fronts, it included a cap-and-trade plan that was implemented to address one of the leading environmental scourges of the day, namely acid rain. The plan is detailed in the *Clean Air Act Amendments of 1990: Title IV*. The success of the plan made it a model for subsequent carbon cap-and-trade proposals, such as President Barack Obama's *Safe Climate Act*.

Bush also had significant influence on the crafting of the *United Nations Framework*

Convention on Climate Change (UNFCCC). At his behest, negotiators from the United States ensured that the document would not include specific targets or time-tables for the reduction of greenhouse gas (GHG) emissions. Their inclusion, in the eyes of the Bush administration, would have resulted in a severe burden on the nation's economic system. Although the UNFCCC marked an important symbolic commitment by the world community to limit the expulsion of GHGs into the atmosphere, the intentional lack of specific initiatives to be addressed limited its effectiveness.

William J. Clinton (1993–2001)

During his first presidential campaign, William (Bill) Clinton promised to champion efforts to reduce the levels of carbon dioxide emitted into the atmosphere to 1990 levels by the turn of the century. Upon ascending to the presidency, he formed what he called "the most environmentally friendly administration in American history." Although his administration enacted some significant environmental legislation, his overall legacy was primarily shaped by the 1997 *Kyoto Protocol to the United Nations Framework Convention on Climate Change*.

The Kyoto Protocol had its origins in a nonbinding treaty concerning the reduction of greenhouse gas (GHG) emissions that was agreed to at the 1992 Earth Summit, held in Rio de Janeiro, Brazil. After it became obvious in 1995 that the signatories to the agreement, including the United States, could not meet the reductions that they had pledged to meet, the international community began to explore how best to encourage countries to commit to a binding agreement. The solution became the Kyoto Protocol.

During the negotiations, it was already well-known that many politicians in the United States were leery of the intentions of the delegates in Kyoto, as evidenced by the passage of the bipartisan *Senate Resolution 98 [Report No. 105–54] (Byrd-Hagel Resolution)*. U.S. negotiators knew that to obtain ratification of the Kyoto Protocol by the Senate, they had to ensure that developing countries were required to make emission cuts using the same time frame assigned to industrialized countries. The final document called for industrialized countries to cut their GHG emission levels by five percent of 1990 levels by 2012. Developing countries were exempted from such requirements because the majority of delegates agreed that the respective countries needed to use what monies they had available to combat more pressing needs, such as endemic poverty. The disparity between the expectations borne by different countries doomed the agreement's prospects within the United States. Opponents of the Kyoto Protocol charged that the agreement was unfair because it required the United States to hurt itself economically while sparing other countries similar pain. Since it was obvious that the Senate would never ratify the agreement, President Clinton decided to avoid the political embarrassment of having it formally rejected by not submitting it to the Senate.

Although Vice President Albert "Al" Gore, Jr., was one of the primary negotiators for the United States in Kyoto, Japan, his reputation as an environmentalist was not impugned by the failure of the United States to ratify the agreement. His activism on climate change while serving in Clinton's administration and after leaving office resulted in him receiving an Oscar for his film *An Inconvenient Truth* and the Nobel Peace Prize in 2007. Criticisms of Gore's environmental politics are obvious in Senator James Inhofe's *Hot & Cold Media Spin: A Challenge to Journalists Who Cover Global Warming*.

The Clinton administration is credited with producing the first report to the president from the United States Global Research Program (USGRP). Titled *Climate Change Impacts on the United States: The Potential Consequences of Climate Variability and Change*, it posited how the United States would be impacted by changes in the climate for the next 25 years. The USGRP was supposed to produce the report every four years, but the second report, *Global Climate Change Impacts in the United States*, was not produced until 2009. The third report, *Climate Change Impacts in the United States*, was published according to the prescribed schedule.

George W. Bush (2001–2009)

Through his *Letter to Members of the Senate on the Kyoto Protocol on Climate Change*, President George W. Bush openly voiced his disdain for the Kyoto Protocol. In doing so, he signaled that his administration was going to favor business interests over those of environmentalists. As an alternative to the international agreement, President Bush's administration issued the *Global Climate Change Policy Book*. The plan called for an 18 percent reduction in greenhouse gas intensity, which sounded impressive to individuals unacquainted with how the international community measured greenhouse gases in the atmosphere. Instead of pledging to make a percentage decrease in overall greenhouse gas emissions, as called for in the Kyoto Protocol, the Bush administration was calling for a decrease in the ratio of greenhouse gasses found in the air, which was a far lower standard. The Bush administration did not aggressively implement even this watered-down plan because many of its proposals were in direct conflict with its National Energy Strategy. The administration got around this problem by making the apparent dictates contained within the *Global Climate Change Policy Book* voluntary.

While the administration was touting its climate plan, an alternative emerged in the United States Senate in 2003. Senators Joe Lieberman and John McCain sponsored the *Climate Stewardship Act*. Dubbed "Kyoto Lite" by critics, the legislation called for the United States to cut greenhouse gas emissions to a level that was much more modest than what was proposed in the Kyoto Protocol. Even a watered-down version of the international agreement proved unacceptable to the majority of the Senators.

In 2006, the United States Supreme Court heard arguments in the case *Massachusetts et al. v. Environmental Protection Agency et al.* The case resulted from an effort by California, Connecticut, Illinois, Maine, Massachusetts, New Jersey, New Mexico, New York, Oregon, Rhode Island, Vermont, and Washington to force the Bush administration's Environment Protection Agency (EPA) to utilize the Clean Air Act, as amended in 1990, to regulate GHGs. The Supreme Court Justices ruled by a 5–4 vote in April 2007 in favor of the states. In response to the ruling, the EPA issued the *Endangerment and Cause or Contribute Findings for Greenhouse Gases Under Section 202(a) of the Clean Air Act: Final Rule* in 2009.

Barack Obama (2009–2017)

President Barack Obama has eagerly endeavored to make conquering climate change a significant part of his presidential legacy. The issue's importance to the administration was voiced in 2009 by Nobel Prize Winner and United States Secretary of Energy Steven Chu in *Secretary Chu's Remarks at the Harvard University Commencement—As Prepared for Delivery*. Later that year, Obama espoused his personal views in *Remarks by the President at the Morning Plenary Session of the United Nations Climate Change Conference*. The Climate Change Conference in Copenhagen, Denmark, provided him a platform to show environmental leadership on the international stage. The meeting resulted in the passage of the *Copenhagen Accord*. The agreement called for the signatory countries to significantly reduce their greenhouse emissions to ensure that the temperature globally did not rise above 2 degrees Celsius from 1900 levels.

The Copenhagen agreement was an intermediary step leading to the Conference of the Parties (COP) 21, which was held in Paris, France, in December 2016. The lead-up to the conference included efforts at both the city, state, federal, and international levels to shape the document titled *Adoption of the Paris Agreement: Proposal by the President: Draft decision -/CP.21*. Local efforts are exemplified by *One City Built to Last: Transforming New York City's Buildings for a Low-Carbon Future*. California's *Governor Brown Establishes Most Ambitious Greenhouse Gas Reduction Target in North America* shows how the state endeavored to establish the standard for the entire world. It is important to note that both New York Mayor Bill de Blasio and California Governor Jerry Brown endeavored to set higher goals for the reduction in GHGs in their communities than were desired by President Obama. Obama's goals for the meeting were included in his *Remarks by the President at the GLACIER Conference—Anchorage, AK*. Also important were Obama's diplomatic endeavors, which are evident in the documents *U.S.-China Joint Announcement on Climate Change*, *Joint Meeting to Hear an Address by Pope Francis of the Holy See*, and *Fact Sheet: The United States and China Issue Joint Presidential Statement on Climate Change with New Domestic Policy Commitments and a Common Vision for an Ambitious Global Climate Agreement in Paris*. The joint releases with China were particularly important due to the issuance by the United States Senate of *Senate Bill 66: To Prohibit Any Regulation Regarding Carbon Dioxide or Other Greenhouse Gas Emissions Reduction in the United States until China, India, and Russia Implement Similar Reductions*. Like *Senate Resolution 98 [Report No. 105–54] (Byrd-Hagel Resolution)* in advance of the Kyoto negotiations, it was a warning that the United States Senate would not ratify the Paris agreement if it did not treat all countries equally. Like the Kyoto Protocol, the Paris Agreement exempted much of the developing countries from its dictates. Recognizing that the Senate might not be bluffing in their threat not to ratify the agreement, negotiators from the United States carefully parsed words so that they could claim that the resulting document was not a "treaty." Treaties, by law, have to be ratified by the Senate. If it was not a treaty, then President Barack Obama could make it official through Executive Action. Whether the Paris Agreement is a treaty or not will likely be determined by the federal courts sometime in the future. The courts will thus determine if the Paris Agreement will be legally binding within the United States, because it is doubtful that the Senate, as it is comprised in 2016, would ratify the document. If the United States fails to meet its obligations as agreed to by the Obama administration, then a key part of the president's legacy will be greatly diminished.

Declaration of the World Climate Conference

Date: February 23, 1979
Location: Geneva, Switzerland
Significance: The declaration that emerged from the World Climate Conference marked the first official recognition that climate change was a global problem.
Source: World Meteorological Organization. 1979. *Declaration of the World Climate Conference.* http://unesdoc.unesco.org/images/0003/000376/037648eb.pdf (accessed July 30, 2015).

The World Climate Conference, a conference of experts on climate and mankind, held in Geneva, from 12 to 23 February 1979, was sponsored by the World Meteorological Organization in collaboration with other international bodies.

The specialists from many disciplines assembled for the Conference expressed their views concerning climatic variability and change and the implications for the world community. On the basis of their deliberations, they adopted the following.

An Appeal to Nations

Having regard to the all-pervading influence of climate on human society and on many fields of human activity and endeavor, the Conference finds that it is now urgently necessary for the nations of the world:

a. to take full advantage of man's present knowledge of climate;
b. to take steps to improve significantly that knowledge;
c. to foresee and to prevent potential man-made changes in climate that might be adverse to the well-being of humanity.

The problem

The global climate has varied slowly over past millennia, centuries, and decades and will vary in the future. Mankind takes advantage of favourable climate, but is also vulnerable to changes and variations of climate and to the occurrence of extreme events such as droughts and floods. Food, water, energy, shelter, and health are all aspects of human life that depend critically on climate. Recent grain harvest failures and the serious decline in some fisheries emphasize this vulnerability. Even normal variations and modest changes relative to the normal climate have a significant influence upon man's activities.

All countries are vulnerable to climatic variations, and developing countries, especially those in arid, semi-arid, or high rainfall regions, are particularly so. On the other hand, unfavourable impacts may be mitigated and positive benefits may be gained from use of available climate knowledge.

The climates of the countries of the world are interdependent. For this reason, and in view of the increasing demand for resources by the growing world population that strives for improved living conditions, there is an urgent need for the development of a common global strategy for a greater understanding and a rational use of climate.

Man today inadvertently modifies climate on a local scale and to a limited extent on a regional scale. There is serious concern that the continued expansion of man's activities on earth may cause significant extended regional and even global changes of climate. This possibility adds further urgency to the need for global co-operation to explore the possible future course of global climate and to take this new understanding into account in planning for the future development of human society.

Climate and the future

Climate will continue to vary and to change due to natural causes. The slow cooling trend in parts of the northern hemisphere during the last few decades is similar to others of natural origin in the past, and thus whether it will continue or not is unknown.

Research is revealing many basic features of climatic changes of the past and is providing the basis for projections of future climate. The causes of climate variations are becoming better understood, but uncertainty exists about many of them and their relative importance.

Nevertheless, we can say with some confidence that the burning of fossil fuels, deforestation, and changes of land use have increased the amount of carbon dioxide in the atmosphere by about 15 per cent during the last century and it is at present increasing by about 0.4 per cent per year. It is likely that an increase will continue in the future. Carbon dioxide plays a fundamental role in determining the temperature of the earth's atmosphere, and it appears plausible that an increased amount of carbon dioxide in the atmosphere can contribute to a gradual warming of the lower atmosphere, especially at high latitudes. Patterns of change would be likely to affect the distribution of temperature, rainfall and other meteorological parameters, but the details of the changes are still poorly understood.

It is possible that some effects on a regional and global scale may be detectable before the end of this century and become significant before the middle of the next century. This time scale is similar to that required to redirect, if necessary, the operation of many aspects of the world economy, including agriculture and the production of energy. Since changes in climate may prove to be beneficial in some parts of the world and adverse in others, significant social and technological readjustments may be required.

Increasing energy use and thus release of heat have already caused local climatic changes. In the future such heat sources from densely populated and heavily industrialized regions could possibly have some effects on climate on a larger scale. Other human activities such as agriculture, pastoral practices, deforestation, increased use of nitrogen fertilizers and release of chlorofluoromethanes might have climatic consequences and therefore require careful study. Also, a systematic search for still other possible effects on climate of major human efforts is needed.

Some forms of warfare have local climatic effects. World thermonuclear conflict, besides its catastrophic consequences for mankind, would degrade the natural environment and might cause climatic changes on a large scale.

It is conceivable that in the future man may be able to produce limited changes in climate on a large scale by deliberate intervention. It would be irresponsible to consider such actions until we have acquired the essential understanding of the mechanisms governing climate that is needed to predict the consequences. Moreover, international agreement must be reached before such projects are implemented.

Conclusions and Recommendations

The World Climate Programme proposed by the World Meteorological Organization deserves the strongest support of all nations

Its main thrusts are:

- Research into the mechanisms of climate in order to clarify the relative roles of natural and anthropogenic influences. This will require the further development of mathematical models which are the tools for simulating, and assessing the predictability of, the climate system. They will also be used to investigate the sensitivity of climate to possible natural and man-made stimuli such as the release of carbon dioxide and to estimate the climatic response.
- Improving the acquisition and availability of climatic data. The success of the climate programme depends on the development of a vast amount of meteorological, hydrological, oceanographic and other pertinent geophysical data. Furthermore, climatic impact studies and practical application of knowledge of climate by nations in addition requires detailed information about their natural resources and socio-economic structures.
- Application of knowledge of climate in planning, development and management. This effort should include programmes to assist national meteorological and hydrological services to increase the awareness of users of the potential benefits to be gained through the use of climate information, to improve capabilities to provide and disseminate this information, and to facilitate training in nationally significant climate applications. It should include programmes to develop new methodologies for the application of climate data in the food, water, energy and health sectors.
- Study of the impacts of climatic variability and change on human activities and the translation of the findings of such studies in terms of greatest use to governments and people. This will require improvements in our understanding of the relationships between climate and human society, including:
    1. The possible range of societal adjustments to climate variations and change;
    2. The characteristics of human societies at different stages of development and in different environments that make them especially vulnerable or resilient in the face of climate variability and change;
    3. The means by which human societies can protect against adverse consequences of, and take advantage of the opportunities presented by, climate variations and changes.

The overall purposes of the Programme are thus to provide the means to foresee possible future changes of climate and to aid nations in the application of climatic data and knowledge to the planning and management of all aspects of man's activities. This will require an interdisciplinary effort of unprecedented scope at the national and international levels.

The conduct of the World Climate Programme involves a broad range of activities and requires leadership and co-ordination among international bodies and close collaboration among nations

It is fully recognized that the international co-operation which is the prerequisite for any world climate programme can only be successfully pursued under conditions of peace.

There is an immediate need for nations to utilize existing knowledge of climate and climatic variations in the planning for social and economic development

In some parts of the world, there is already sufficient information to provide many applied climate services. However, only a start has been made; data and expertise are generally lacking in developing countries. Programmes must be set up to assist them to participate fully in the World Climate Programme through training and the transfer of appropriate methodologies.

* * *

The long-term survival of mankind depends on achieving a harmony between society and nature. The climate is but one characteristic of our natural environment that needs to be wisely utilized. All elements of the environment interact, both locally and remotely. Degradation of the environment in any national or geographic area must be a major concern of society because it may influence climate elsewhere. The nations of the world must work together to preserve the fertility of the soils; to avoid misuse of the world's water resources, forests and rangelands; to arrest desertification; and to lessen pollution of the atmosphere and the oceans. These actions by nations will require great determination and adequate material resources, and they will be meaningful only in a world at peace.

Analysis

Organized by the World Meteorological Organization, the World Climate Conference brought together approximately 300 scholars from a wide variety of disciplines representing more than fifty countries. After presentations and a number of discussions, the participants jointly declared that climate change was a global problem. They also called for the establishment of an international research program to begin collecting data to determine how climate change was impacting specific locales, as well as what was causing the climate to change so rapidly.

In 1980, the World Meteorological Association, the International Council of Scientific Unions, and the Intergovernmental Oceanographic Commission created the World Climate Research Programme. It is presently part of the World Climate Program, which is charged with the collection and analysis of climate data from around the world.

FURTHER READING

Bierly, Eugene W. 1988. "The World Climate Program: Collaboration and Communication on a Global Scale." *Annals of the American Academy of Political and Social Science* 495: 106–116.

The Global 2000 Report to the President

*Entering the Twenty-First Century
(Major Findings and Conclusions)*

Date: July 24, 1980
Location: Washington, D.C.
Significance: *The Global 2000 Report to the President* marked the first time that climate change emerged as a major policy issue for a presidential administration in the United States.
Source: Council on Environmental Quality and the Department of State. 1980. *The Global 2000 Report to the President: Entering the Twenty-First Century: Volume 1: Summary.* Washington, D.C.: Government Printing Office.

If present trends continue, the world in 2000 will be more crowded, more polluted, less stable ecologically, and more vulnerable to disruption than the world we live in now. Serious stresses involving population, resources, and environment are clearly visible ahead. Despite greater material output, the world's people will be poorer in many ways than they are today.

For hundreds of millions of the desperately poor, the outlook for food and other necessities of life will be no better. For many it will be worse. Barring revolutionary advances in technology, life for the most part on earth will be more precarious in 2000 than it is now-unless the nations of the world act decisively to alter current trends.

This, in essence, is the picture emerging from the U.S. Government's projections of probable changes in world population, resources, and environment by the end of the century, as presented in the global 2000 study. They do not predict what will occur. Rather, they depict conditions that are likely to develop if there are no changes in public policies, institutions, or rates of technological advance, and if there are no wars or other major disruptions. A keener awareness of the nature of current trends, however, may induce changes that will alter these trends and the projected outcome.

Principle Findings

Rapid growth in world population will hardly have altered by 2000. The world's population will grow from 4 billion in 1975 to 6.35 billion in 2000 an increase of more than 50 percent. The rate of growth will slow only marginally, from 1.8 percent a year to 1.7 percent. In terms of sheer numbers, population will be growing faster in 2000 than it is today, 100 million people added each year compared with 75 million 1975. Ninety percent of this growth will occur in the poorest countries.

While the economies of the less developed countries (LDCs) are expected to grow at faster rates than those of the industrialized nations, the gross national product per capita in most LDCs remains low. The average gross national product per capita is projected to rise substantially in some LDCs (especially in Latin America), but in the great populous nations of South Asia it remains below $200 a year (in 1975 dollars). The large existing gap between the rich and the poor nations widens.

World food production is projected to increase 90 percent over the 30 years from 1970 to 2000. This translates into global per capita increase of less than 15 percent over the same period. The bulk of that increase goes to countries that already have relatively high per capita food consumption. Meanwhile per capita consumption in South Asia, the Middle East, and LDCs of Africa, will scarcely improve or will actually decline below present inadequate levels. At the same time, real prices for food are expected to double.

Arable land will increase only 4 percent by 2000, so that most of the increased output of food will have to come from higher yields. Most of the elements that now contribute to higher yields—fertilizer, pesticides, power for irrigation, and fuel for machinery—depend on heavily on oil and gas.

During the 1990s world oil production will approach geological estimates of maximum production capacity, even with rapidly increasing petroleum prices. The study projects that the richer industrialized will be able to command enough and have other commercial energy supplies to meet rising demands through 1990. With the expected price increases, many less developed countries will have increasing difficulties meeting energy needs. For the one-quarter of humankind that depends primarily on wood for fuel, the outlook is bleak. Needs for fuelwood will exceed available supplies by about 25 percent before the turn of the century.

While the world's finite fuel resources—coal, oil, gas, oil shale, tar sands and uranium—are theoretically sufficient for centuries, they are not evenly distributed; they pose difficult economic and environmental problems; and they vary greatly in their amenability to exploitation and use.

Nonfuel mineral resources generally appear sufficient to meet projected demands through 2000, but further discovers and investments will be needed to maintain reserves. In addition, production costs will increase with energy prices and may make some nonfuel mineral resources uneconomic. The quarter of the world's population that inhabits industrial countries will continue to absorb three-fourths of the world's mineral production.

Regional water shortages will become more sever. In the 1970–2000 period population growth alone will cause requirements for water to double in nearly half the world. Still greater increases would be needed to improve standards of living. In many LDCs, water supplies will become increasingly erratic by 2000 as a result of extensive deforestation. Development of new water supplies will become more costly virtually everywhere.

Significant losses of world forests will continue over the next 20 years as demand for forest products and fuelwood increases. Growing stocks of commercial-size timber are projected to decline 50 percent per capita. The world's forests are now disappearing at a rate of 18–20 million hectares a year (an area half the size of California), with most of the loss occurring in the humid tropical forests of Africa, Asia and South America. The projections indicate that by 2000 some 40 percent of the remaining forest cover in LDCs will be gone.

Serious deterioration of agriculture soils will occur worldwide, due to erosion, loss of organic matter, desertification, salinization, alkalization, and waterlogging. Already, an area of cropland and grassland approximately the size of main is becoming a barren wasteland each year, and the spread of desert-like conditions is likely to accelerate.

Atmospheric concentration of carbon dioxide and ozone-depleting chemicals are expected to increase at rates that could alter the world's climate and upper atmosphere significantly by 2050. Acid rain from increased combustion of fossil fuels (especially coal) threatens damage to lakes, soils, and crops. Radioactive and other hazardous materials present health and safety problems in increasing numbers of countries.

Extinctions of plant and animal species will increase dramatically. Hundreds of thousands of species—perhaps as many as 20 percent of all species on earth—will be irretrievably lost as their habitats vanish, especially in tropical forests.

The future depicted by the U.S. Government projections, briefly outlined above, may actually understate the impending problems. The methods available for carrying out the study led to certain gaps and inconsistencies that tend to impart an optimistic bias. For example, most of the individual projections for the various sectors studied—food, minerals, energy, and so on-assume that sufficient capital, energy, water, and land will be available in each of these sectors to meet their needs, regardless of competing needs of the other sectors. More consistent, better-integrated projections would produce a still more emphatic picture of intensifying stresses, as the world enters the twenty-first century.

Conclusions

At present and projected growth rates, the world's population would reach 10 billion by 2030 and would approach 30 billion by the end of the twenty-first century. These levels correspond closely to estimates by the U.S. National Academy of Sciences of the maximum carrying capacity of the entire earth. Already the population in sub–Saharan Africa and in the Himalayan hills of Asia have exceeded the carrying capacity of the immediate area, triggering an erosion of the land's capacity to support life. The resulting poverty and ill health have further complicated efforts to reduce fertility. Unless this circle of interlinked problems is broken soon, population growth in such areas will unfortunately be slowed for reasons other than declining birth rates. Hunger and disease will claim more babies and young children, and more of those surviving will be mentally and physically handicapped by childhood malnutrition.

Indeed, the problems of preserving the carrying capacity of the earth and sustaining the possibility of a decent life for the human beings that inhabit it are enormous and close upon us. Yet there is reason for hope. It must be emphasized that the Global 2000 Study's projections are based on the assumption that national policies regarding population, stabilization, resource conservation, and environmental protection will remain essentially unchanged through the end of the century. But in fact, policies are beginning to change. In some areas, forests are being replanted after cutting. Some nations are taking steps to reduce soil losses and desertification. Interest in energy conservation is growing, and large sums are being invested in exploring alternatives to petroleum dependence. The need for family planning is slowly becoming better understood. Water supplies are being improved and waste treatment systems built. High-yield seeds are widely available and seed banks are being expanded. Some wildlands with their genetic resources are being protected. Natural predators and selective pesticides are being substituted for persistent and destructive pesticides.

Encouraging as these developments are, they are far from adequate to meet the global challenges projected in this Study. Vigorous, determined new initiatives are needed if worsening poverty and human suffering, environmental degradation, and international tension and conflicts are to be prevented. There are no quick fixes. The only solutions to the problems of population, resources, and environment are inextricably linked to some of the most perplexing and persistent problems in the world—poverty, injustice, and social conflict. New and imaginative ideas—and a willingness to act on them—are essential.

The needed changes go far beyond the capability and responsibility of this or any other single nation. An era of unprecedented cooperation and commitment is essential. Yet there are opportunities—and a strong rationale—for the United States to provide leadership among nations. A high priority for this Nation must be a thorough assessment of its foreign and domestic policies relating to population, resources, and environment. The United States, possessing the world's largest economy, can expect its policies to have a significant influence on global trends. An equally important priority for the United States is to cooperate generously and justly with other nations—particularly in the areas of trade, investment, and assistance—in seeking solutions to the many problems that extend beyond our national boundaries. There are so many unfulfilled opportunities to cooperate with other nations in efforts to relieve poverty and hunger, stabilize population, and enhance economic and environmental productivity. Further cooperation among nations is also needed to strengthen international mechanisms for protecting and utilizing the "global commons"—the oceans and atmosphere.

To meet the challenges described in this Study, the United States must improve its ability to identify emerging problems and assess alternative responses. In using and evaluating the Government's present capability for long-term global analysis, the Study found serious inconsistencies in the methods and assumptions employed by various agencies in making their projections. The Study itself made a start toward resolving these inadequacies. It represents the Government's first attempt to produce an interrelated set of population, resource, and environmental projections, and it has brought forth the most consistent set of global projections yet achieved by U.S. agencies. Nevertheless, the projections still contain serious gaps and contradictions that must be corrected if the Government's analytic capability is to be improved. It must be acknowledged that at present the Federal agencies are not always capable of providing projections of the quality needed for long-term policy decisions.

While limited resources may be a contributing factor in some instances, the primary problem is lack of coordination. The U.S. Government needs a mechanism for continuous review of the assumptions and methods the Federal agencies use in their projection models and for assurance that the agencies' models are sound, consistent, and well documented. The improved analyses that could result would provide not only a clearer sense of emerging problems and opportunities, but also a better basis for decisions of worldwide significance that the President, the Congress, and the Federal Government as a whole must make.

With its limitations and rough approximations, the Global 2000 Study may be seen as no more than a reconnaissance of the future; nonetheless its conclusions are reinforced by similar findings of other recent global studies that were examined in the course of the Global 2000 Study (see Appendix). All these studies are in general agreement on the nature of the problems and on the threats they pose to the future welfare of humankind. The available evidence leaves no doubt that the world—including this Nation—faces enormous, urgent, and complex problems in the decades immediately ahead. Prompt and vigorous changes in public policy around the world are needed to avoid or minimize these problems before they become unmanageable. Long lead times are required for effective action. If decisions are delayed until the problems become worse, options for effective action will be severely reduced.

Analysis

In 1977, President Jimmy Carter tasked the Council of Environmental Quality and the Department of State to work with other federal agencies to examine how changes in the world's population, natural resources, and environment would impact the future of humankind. After three years of study, the results were published under the title *The Global 2000 Report to the President: Entering the Twenty-First Century*. The administration praised the researchers, led by Gus Speth, Chairman of the Council of Environmental Quality, for producing the best scientific work on the relationships between the respective topics that had been conducted up to that time. The study was notable for two reasons. It was the first time that a presidential administration acknowledged that global climate change was an important issue that needed to be addressed. Secondly, it revealed the lack of scientific expertise available to the federal government at the time and the consequences that resulted from that shortcoming, for example, not being able to provide the quality data required for long-term forecasting. Future presidents would have many more experts available to them, although that did not mean that they necessarily followed the guidance that was provided.

In response to the report, President Carter created the Presidential Task Force on Global Resources and Environment. Gus Speth was selected to lead the endeavor. The task force never really had the opportunity to accomplish very much as it came into being late into President Jimmy Carter's only term in office.

Further Reading

Carter, Jimmy. 1982. *Public Papers of the Presidents, Jimmy Carter: 1980–1981, Book 2—May 24 to September 26, 1980*. Washington, D.C.: Government Printing Office: 1415–1416.

Council on Environmental Quality and the Department of State. 1980. *The Global 2000 Report to the President: Entering the Twenty-First Century*. 3 vols. Washington, D.C.: Government Printing Office.

United States Congress, Joint Economic Committee, Subcommittee on International Economics. 1980. *The Global 2000 Report: Hearing before the Subcommittee on International Economics of the Joint Economic Committee, Congress of the United States, Ninety-Sixth Congress, Second Session, September 4, 1980*. Washington, D.C.: Government Printing Office.

The Montreal Protocol on Substances That Deplete the Ozone Layer (Excerpts)

Date: September 16, 1987
Location: Montreal, Canada
Significance: The Montreal Protocol was created to protect the Earth's ozone layer. With that issue addressed, the international community turned its attention to other challenges such as the regulation of greenhouse gases.
Source: United Nations Environment Programme. 2012. *Handbook for the Montreal Protocol on Substances that Deplete the Ozone Layer*. http://ozone.unep.org/Publications/MP_Handbook/MP-Handbook-2012.pdf (accessed August 11, 2015).

The Montreal Protocol on Substances that Deplete the Ozone Layer as adjusted and amended by the Second Meeting of the Parties (London, 27–29 June 1990) and by the Fourth Meeting of the Parties (Copenhagen, 23–25 November 1992) and further adjusted by the Seventh Meeting of the Parties (Vienna, 5–7 December 1995) and further adjusted and amended by the Ninth Meeting of the Parties (Montreal, 15–17 September 1997) and by the Eleventh Meeting of the Parties (Beijing, 29 November–3 December 1999) and further adjusted by the Nineteenth Meeting of the Parties (Montreal, 17–21 September 2007)

Preamble

The Parties to this Protocol,
Being Parties to the Vienna Convention for the Protection of the Ozone Layer,
Mindful of their obligation under that Convention to take appropriate measures to protect human health and the environment against adverse effects resulting or likely to result from human activities which modify or are likely to modify the ozone layer,
Recognizing that world-wide emissions of certain substances can significantly deplete and otherwise modify the ozone layer in a manner that is likely to result in adverse effects on human health and the environment,
Conscious of the potential climatic effects of emissions of these substances,
Aware that measures taken to protect the ozone layer from depletion should be based on relevant scientific knowledge, taking into account technical and economic considerations,

Determined to protect the ozone layer by taking precautionary measures to control equitably total global emissions of substances that deplete it, with the ultimate objective of their elimination on the basis of developments in scientific knowledge, taking into account technical and economic considerations and bearing in mind the developmental needs of developing countries,

Acknowledging that special provision is required to meet the needs of developing countries, including the provision of additional financial resources and access to relevant technologies, bearing in mind that the magnitude of funds necessary is predictable, and the funds can be expected to make a substantial difference in the world's ability to address the scientifically established problem of ozone depletion and its harmful effects,

Noting the precautionary measures for controlling emissions of certain chlorofluorocarbons that have already been taken at national and regional levels,

Considering the importance of promoting international co-operation in the research, development and transfer of alternative technologies relating to the control and reduction of emissions of substances that deplete the ozone layer, bearing in mind in particular the needs of developing countries,

HAVE AGREED AS FOLLOWS:

...

Introduction to the Adjustments

The Second, Fourth, Seventh, Ninth, Eleventh and Nineteenth Meetings of the Parties to the Montreal Protocol on Substances that Deplete the Ozone Layer decided, on the basis of assessments made pursuant to Article 6 of the Protocol, to adopt adjustments and reductions of production and consumption of the controlled substances in Annexes A, B, C and E to the Protocol as follows (the text here shows the cumulative effect of all the adjustments):

Article 2A: CFCs

1. Each Party shall ensure that for the twelve-month period commencing on the first day of the seventh month following the date of entry into force of this Protocol, and in each twelve-month period thereafter, its calculated level of consumption of the controlled substances in Group I of Annex A does not exceed its calculated level of consumption in 1986. By the end of the same period, each Party producing one or more of these substances shall ensure that its calculated level of production of the substances does not exceed its calculated level of production in 1986, except that such level may have increased by no more than ten per cent based on the 1986 level. Such increase shall be permitted only so as to satisfy the basic domestic needs of the Parties operating under Article 5 and for the purposes of industrial rationalization between Parties.
2. Each Party shall ensure that for the period from 1 July 1991 to 31 December 1992 its calculated levels of consumption and production of the controlled substances in Group I of Annex A do not exceed 150 per cent of its calculated levels of production and consumption of those substances in 1986; with effect from 1 January 1993, the twelve-month control period for these controlled substances shall run from 1 January to 31 December each year.

3. Each Party shall ensure that for the twelve-month period commencing on 1 January 1994, and in each twelve-month period thereafter, its calculated level of consumption of the controlled substances in Group I of Annex A does not exceed, annually, twenty-five per cent of its calculated level of consumption in 1986. Each Party producing one or more of these substances shall, for the same periods, ensure that its calculated level of production of the substances does not exceed, annually, twenty-five per cent of its calculated level of production in 1986. However, in order to satisfy the basic domestic needs of the Parties operating under paragraph 1 of Article 5, its calculated level of production may exceed that limit by up to ten per cent of its calculated level of production in 1986.
4. Each Party shall ensure that for the twelve-month period commencing on 1 January 1996, and in each twelve-month period thereafter, its calculated level of consumption of the controlled substances in Group I of Annex A does not exceed zero. Each Party producing one or more of these substances shall, for the same periods, ensure that its calculated level of production of the substances does not exceed zero. However, in order to satisfy the basic domestic needs of the Parties operating under paragraph 1 of Article 5, its calculated level of production may exceed that limit by a quantity equal to the annual average of its production of the controlled substances in Group I of Annex A for basic domestic needs for the period 1995 to 1997 inclusive. This paragraph will apply save to the extent that the Parties decide to permit the level of production or consumption that is necessary to satisfy uses agreed by them to be essential.
5. Each Party shall ensure that for the twelve-month period commencing on 1 January 2003 and in each twelve-month period thereafter, its calculated level of production of the controlled substances in Group I of Annex A for the basic domestic needs of the Parties operating under paragraph 1 of Article 5 does not exceed eighty per cent of the annual average of its production of those substances for basic domestic needs for the period 1995 to 1997 inclusive.
6. Each Party shall ensure that for the twelve-month period commencing on 1 January 2005 and in each twelve-month period thereafter, its calculated level of production of the controlled substances in Group I of Annex A for the basic domestic needs of the Parties operating under paragraph 1 of Article 5 does not exceed fifty per cent of the annual average of its production of those substances for basic domestic needs for the period 1995 to 1997 inclusive.
7. Each Party shall ensure that for the twelve-month period commencing on 1 January 2007 and in each twelve-month period thereafter, its calculated level of production of the controlled substances in Group I of Annex A for the basic domestic needs of the Parties operating under paragraph 1 of Article 5 does not exceed fifteen per cent of the annual average of its production of those substances for basic domestic needs for the period 1995 to 1997 inclusive.
8. Each Party shall ensure that for the twelve-month period commencing on 1 January 2010 and in each twelve-month period thereafter, its calculated level of production of the controlled substances in Group I of Annex A for the basic domestic needs of the Parties operating under paragraph 1 of Article 5 does not exceed zero.
9. For the purposes of calculating basic domestic needs under paragraphs 4 to 8 of this Article, the calculation of the annual average of production by a Party includes any production entitlements that it has transferred in accordance with paragraph

5 of Article 2, and excludes any production entitlements that it has acquired in accordance with paragraph 5 of Article 2.

Article 2B: Halons
1. Each Party shall ensure that for the twelve-month period commencing on 1 January 1992, and in each twelve-month period thereafter, its calculated level of consumption of the controlled substances in Group II of Annex A does not exceed, annually, its calculated level of consumption in 1986. Each Party producing one or more of these substances shall, for the same periods, ensure that its calculated level of production of the substances does not exceed, annually, its calculated level of production in 1986. However, in order to satisfy the basic domestic needs of the Parties operating under paragraph 1 of Article 5, its calculated level of production may exceed that limit by up to ten per cent of its calculated level of production in 1986.
2. Each Party shall ensure that for the twelve-month period commencing on 1 January 1994, and in each twelve-month period thereafter, its calculated level of consumption of the controlled substances in Group II of Annex A does not exceed zero. Each Party producing one or more of these substances shall, for the same periods, ensure that its calculated level of production of the substances does not exceed zero. However, in order to satisfy the basic domestic needs of the Parties operating under paragraph 1 of Article 5, its calculated level of production may, until 1 January 2002 exceed that limit by up to fifteen per cent of its calculated level of production in 1986; thereafter, it may exceed that limit by a quantity equal to the annual average of its production of the controlled substances in Group II of Annex A for basic domestic needs for the period 1995 to 1997 inclusive. This paragraph will apply save to the extent that the Parties decide to permit the level of production or consumption that is necessary to satisfy uses agreed by them to be essential.
3. Each Party shall ensure that for the twelve-month period commencing on 1 January 2005 and in each twelve-month period thereafter, its calculated level of production of the controlled substances in Group II of Annex A for the basic domestic needs of the Parties operating under paragraph 1 of Article 5 does not exceed fifty per cent of the annual average of its production of those substances for basic domestic needs for the period 1995 to 1997 inclusive.
4. Each Party shall ensure that for the twelve-month period commencing on 1 January 2010 and in each twelve-month period thereafter, its calculated level of production of the controlled substances in Group II of Annex A for the basic domestic needs of the Parties operating under paragraph 1 of Article 5 does not exceed zero.

Article 2C: Other fully halogenated CFCs
1. Each Party shall ensure that for the twelve-month period commencing on 1 January 1993, its calculated level of consumption of the controlled substances in Group I of Annex B does not exceed, annually, eighty per cent of its calculated level of consumption in 1989. Each Party producing one or more of these substances shall, for the same period, ensure that its calculated level of production of the substances does not exceed, annually, eighty per cent of its calculated level of production in 1989. However, in order to satisfy the basic domestic needs of the Parties operating under paragraph 1 of Article 5, its calculated level of production may exceed that limit by up to ten per cent of its calculated level of production in 1989.

2. Each Party shall ensure that for the twelve-month period commencing on 1 January 1994, and in each twelve-month period thereafter, its calculated level of consumption of the controlled substances in Group I of Annex B does not exceed, annually, twenty-five per cent of its calculated level of consumption in 1989. Each Party producing one or more of these substances shall, for the same periods, ensure that its calculated level of production of the substances does not exceed, annually, twenty-five per cent of its calculated level of production in 1989. However, in order to satisfy the basic domestic needs of the Parties operating under paragraph 1 of Article 5, its calculated level of production may exceed that limit by up to ten per cent of its calculated level of production in 1989.
3. Each Party shall ensure that for the twelve-month period commencing on 1 January 1996, and in each twelve-month period thereafter, its calculated level of consumption of the controlled substances in Group I of Annex B does not exceed zero. Each Party producing one or more of these substances shall, for the same periods, ensure that its calculated level of production of the substances does not exceed zero. However, in order to satisfy the basic domestic needs of the Parties operating under paragraph 1 of Article 5, its calculated level of production may, until 1 January 2003 exceed that limit by up to fifteen per cent of its calculated level of production in 1989; thereafter, it may exceed that limit by a quantity equal to eighty per cent of the annual average of its production of the controlled substances in Group I of Annex B for basic domestic needs for the period 1998 to 2000 inclusive. This paragraph will apply save to the extent that the Parties decide to permit the level of production or consumption that is necessary to satisfy uses agreed by them to be essential.
4. Each Party shall ensure that for the twelve-month period commencing on 1 January 2007 and in each twelve-month period thereafter, its calculated level of production of the controlled substances in Group I of Annex B for the basic domestic needs of the Parties operating under paragraph 1 of Article 5 does not exceed fifteen per cent of the annual average of its production of those substances for basic domestic needs for the period 1998 to 2000 inclusive.
5. Each Party shall ensure that for the twelve-month period commencing on 1 January 2010 and in each twelve-month period thereafter, its calculated level of production of the controlled substances in Group I of Annex B for the basic domestic needs of the Parties operating under paragraph 1 of Article 5 does not exceed zero.

Article 2D: Carbon tetrachloride
1. Each Party shall ensure that for the twelve-month period commencing on 1 January 1995, its calculated level of consumption of the controlled substance in Group II of Annex B does not exceed, annually, fifteen per cent of its calculated level of consumption in 1989. Each Party producing the substance shall, for the same period, ensure that its calculated level of production of the substance does not exceed, annually, fifteen per cent of its calculated level of production in 1989. However, in order to satisfy the basic domestic needs of the Parties operating under paragraph 1 of Article 5, its calculated level of production may exceed that limit by up to ten per cent of its calculated level of production in 1989.
2. Each Party shall ensure that for the twelve-month period commencing on 1 January 1996, and in each twelve-month period thereafter, its calculated level of consumption of the controlled substance in Group II of Annex B does not exceed zero. Each Party

producing the substance shall, for the same periods, ensure that its calculated level of production of the substance does not exceed zero. However, in order to satisfy the basic domestic needs of the Parties operating under paragraph 1 of Article 5, its calculated level of production may exceed that limit by up to fifteen per cent of its calculated level of production in 1989. This paragraph will apply save to the extent that the Parties decide to permit the level of production or consumption that is necessary to satisfy uses agreed by them to be essential.

Article 2E: 1,1,1-Trichloroethane (Methyl chloroform)
1. Each Party shall ensure that for the twelve-month period commencing on 1 January 1993, its calculated level of consumption of the controlled substance in Group III of Annex B does not exceed, annually, its calculated level of consumption in 1989. Each Party producing the substance shall, for the same period, ensure that its calculated level of production of the substance does not exceed, annually, its calculated level of production in 1989. However, in order to satisfy the basic domestic needs of the Parties operating under paragraph 1 of Article 5, its calculated level of production may exceed that limit by up to ten per cent of its calculated level of production in 1989.
2. Each Party shall ensure that for the twelve-month period commencing on 1 January 1994, and in each twelve-month period thereafter, its calculated level of consumption of the controlled substance in Group III of Annex B does not exceed, annually, fifty per cent of its calculated level of consumption in 1989. Each Party producing the substance shall, for the same periods, ensure that its calculated level of production of the substance does not exceed, annually, fifty per cent of its calculated level of production in 1989. However, in order to satisfy the basic domestic needs of the Parties operating under paragraph 1 of Article 5, its calculated level of production may exceed that limit by up to ten per cent of its calculated level of production in 1989.
3. Each Party shall ensure that for the twelve-month period commencing on 1 January 1996, and in each twelve-month period thereafter, its calculated level of consumption of the controlled substance in Group III of Annex B does not exceed zero. Each Party producing the substance shall, for the same periods, ensure that its calculated level of production of the substance does not exceed zero. However, in order to satisfy the basic domestic needs of the Parties operating under paragraph 1 of Article 5, its calculated level of production may exceed that limit by up to fifteen per cent of its calculated level of production for 1989. This paragraph will apply save to the extent that the Parties decide to permit the level of production or consumption that is necessary to satisfy uses agreed by them to be essential.

Article 2F: Hydrochlorofluorocarbons
1. Each Party shall ensure that for the twelve-month period commencing on 1 January 1996, and in each twelve-month period thereafter, its calculated level of consumption of the controlled substances in Group I of Annex C does not exceed, annually, the sum of:
    (a) Two point eight per cent of its calculated level of consumption in 1989 of the controlled substances in Group I of Annex A; and
    (b) Its calculated level of consumption in 1989 of the controlled substances in Group I of Annex C.

2. Each Party producing one or more of these substances shall ensure that for the twelve-month period commencing on 1 January 2004, and in each twelve-month period thereafter, its calculated level of production of the controlled substances in Group I of Annex C does not exceed, annually, the average of:
    (a) The sum of its calculated level of consumption in 1989 of the controlled substances in Group I of Annex C and two point eight per cent of its calculated level of consumption in 1989 of the controlled substances in Group I of Annex A; and
    (b) The sum of its calculated level of production in 1989 of the controlled substances in Group I of Annex C and two point eight per cent of its calculated level of production in 1989 of the controlled substances in Group I of Annex A. However, in order to satisfy the basic domestic needs of the Parties operating under paragraph 1 of Article 5, its calculated level of production may exceed that limit by up to fifteen per cent of its calculated level of production of the controlled substances in Group I of Annex C as defined above.
3. Each Party shall ensure that for the twelve month period commencing on 1 January 2004, and in each twelve-month period thereafter, its calculated level of consumption of the controlled substances in Group I of Annex C does not exceed, annually, sixty-five per cent of the sum referred to in paragraph 1 of this Article.
4. Each Party shall ensure that for the twelve-month period commencing on 1 January 2010, and in each twelve-month period thereafter, its calculated level of consumption of the controlled substances in Group I of Annex C does not exceed, annually, twenty-five per cent of the sum referred to in paragraph 1 of this Article. Each Party producing one or more of these substances shall, for the same periods, ensure that its calculated level of production of the controlled substances in Group I of Annex C does not exceed, annually, twenty-five per cent of the calculated level referred to in paragraph 2 of this Article. However, in order to satisfy the basic domestic needs of the Parties operating under paragraph 1 of Article 5, its calculated level of production may exceed that limit by up to ten per cent of its calculated level of production of the controlled substances in Group I of Annex C as referred to in paragraph 2.
5. Each Party shall ensure that for the twelve-month period commencing on 1 January 2015, and in each twelve-month period thereafter, its calculated level of consumption of the controlled substances in Group I of Annex C does not exceed, annually, ten per cent of the sum referred to in paragraph 1 of this Article. Each Party producing one or more of these substances shall, for the same periods, ensure that its calculated level of production of the controlled substances in Group I of Annex C does not exceed, annually, ten per cent of the calculated level referred to in paragraph 2 of this Article. However, in order to satisfy the basic domestic needs of the Parties operating under paragraph 1 of Article 5, its calculated level of production may exceed that limit by up to ten per cent of its calculated level of production of the controlled substances in Group I of Annex C as referred to in paragraph 2.
6. Each Party shall ensure that for the twelve-month period commencing on 1 January 2020, and in each twelve-month period thereafter, its calculated level of consumption of the controlled substances in Group I of Annex C does not exceed zero. Each Party producing one or more of these substances shall, for the same periods, ensure that its calculated level of production of the controlled substances in Group I of Annex C does not exceed zero. However:

(a) Each Party may exceed that limit on consumption by up to zero point five per cent of the sum referred to in paragraph 1 of this Article in any such twelve-month period ending before 1 January 2030, provided that such consumption shall be restricted to the servicing of refrigeration and air-conditioning equipment existing on 1 January 2020;

(b) Each Party may exceed that limit on production by up to zero point five per cent of the average referred to in paragraph 2 of this Article in any such twelve-month period ending before 1 January 2030, provided that such production shall be restricted to the servicing of refrigeration and air-conditioning equipment existing on 1 January 2020.

7. As of 1 January 1996, each Party shall endeavour to ensure that: (a) The use of controlled substances in Group I of Annex C is limited to those applications where other more environmentally suitable alternative substances or technologies are not available; (b) The use of controlled substances in Group I of Annex C is not outside the areas of application currently met by controlled substances in Annexes A, B and C, except in rare cases for the protection of human life or human health; and (c) Controlled substances in Group I of Annex C are selected for use in a manner that minimizes ozone depletion, in addition to meeting other environmental, safety and economic considerations.

Article 2G: Hydrobromofluorocarbons

Each Party shall ensure that for the twelve-month period commencing on 1 January 1996, and in each twelvemonth period thereafter, its calculated level of consumption of the controlled substances in Group II of Annex C does not exceed zero. Each Party producing the substances shall, for the same periods, ensure that its calculated level of production of the substances does not exceed zero. This paragraph will apply save to the extent that the Parties decide to permit the level of production or consumption that is necessary to satisfy uses agreed by them to be essential.

Article 2H: Methyl bromide

1. Each Party shall ensure that for the twelve-month period commencing on 1 January 1995, and in each twelve-month period thereafter, its calculated level of consumption of the controlled substance in Annex E does not exceed, annually, its calculated level of consumption in 1991. Each Party producing the substance shall, for the same period, ensure that its calculated level of production of the substance does not exceed, annually, its calculated level of production in 1991. However, in order to satisfy the basic domestic needs of the Parties operating under paragraph 1 of Article 5, its calculated level of production may exceed that limit by up to ten per cent of its calculated level of production in 1991.

2. Each Party shall ensure that for the twelve-month period commencing on 1 January 1999, and in the twelve-month period thereafter, its calculated level of consumption of the controlled substance in Annex E does not exceed, annually, seventy-five per cent of its calculated level of consumption in 1991. Each Party producing the substance shall, for the same periods, ensure that its calculated level of production of the substance does not exceed, annually, seventy-five per cent of its calculated level of production in 1991. However, in order to satisfy the basic domestic needs of the Parties operating under paragraph 1 of Article 5, its calculated level of

production may exceed that limit by up to ten per cent of its calculated level of production in 1991.

3. Each Party shall ensure that for the twelve-month period commencing on 1 January 2001, and in the twelve-month period thereafter, its calculated level of consumption of the controlled substance in Annex E does not exceed, annually, fifty per cent of its calculated level of consumption in 1991. Each Party producing the substance shall, for the same periods, ensure that its calculated level of production of the substance does not exceed, annually, fifty per cent of its calculated level of production in 1991. However, in order to satisfy the basic domestic needs of the Parties operating under paragraph 1 of Article 5, its calculated level of production may exceed that limit by up to ten per cent of its calculated level of production in 1991.

4. Each Party shall ensure that for the twelve-month period commencing on 1 January 2003, and in the twelve-month period thereafter, its calculated level of consumption of the controlled substance in Annex E does not exceed, annually, thirty per cent of its calculated level of consumption in 1991. Each Party producing the substance shall, for the same periods, ensure that its calculated level of production of the substance does not exceed, annually, thirty per cent of its calculated level of production in 1991. However, in order to satisfy the basic domestic needs of the Parties operating under paragraph 1 of Article 5, its calculated level of production may exceed that limit by up to ten per cent of its calculated level of production in 1991.

5. Each Party shall ensure that for the twelve-month period commencing on 1 January 2005, and in each twelve-month period thereafter, its calculated level of consumption of the controlled substance in Annex E does not exceed zero. Each Party producing the substance shall, for the same periods, ensure that its calculated level of production of the substance does not exceed zero. However, in order to satisfy the basic domestic needs of the Parties operating under paragraph 1 of Article 5, its calculated level of production may, until 1 January 2002 exceed that limit by up to fifteen per cent of its calculated level of production in 1991; thereafter, it may exceed that limit by a quantity equal to the annual average of its production of the controlled substance in Annex E for basic domestic needs for the period 1995 to 1998 inclusive. This paragraph will apply save to the extent that the Parties decide to permit the level of production or consumption that is necessary to satisfy uses agreed by them to be critical uses.

5 *bis*. Each Party shall ensure that for the twelve-month period commencing on 1 January 2005 and in each twelve-month period thereafter, its calculated level of production of the controlled substance in Annex E for the basic domestic needs of the Parties operating under paragraph 1 of Article 5 does not exceed eighty per cent of the annual average of its production of the substance for basic domestic needs for the period 1995 to 1998 inclusive.

5 *ter*. Each Party shall ensure that for the twelve-month period commencing on 1 January 2015 and in each twelve-month period thereafter, its calculated level of production of the controlled substance in Annex E for the basic domestic needs of the Parties operating under paragraph 1 of Article 5 does not exceed zero.

6. The calculated levels of consumption and production under this Article shall not include the amounts used by the Party for quarantine and pre-shipment applications.

Article 2I: Bromochloromethane
 Each Party shall ensure that for the twelve-month period commencing on 1 January 2002, and in each twelvemonth period thereafter, its calculated level of consumption and production of the controlled substances in Group III of Annex C does not exceed zero. This paragraph will apply save to the extent that the Parties decide to permit the level of production or consumption that is necessary to satisfy uses agreed by them to be essential.
 ...

Article 5: Special situation of developing countries
1. Any Party that is a developing country and whose annual calculated level of consumption of the controlled substances in Annex A is less than 0.3 kilograms per capita on the date of the entry into force of the Protocol for it, or any time thereafter until 1 January 1999, shall, in order to meet its basic domestic needs, be entitled to delay for ten years its compliance with the control measures set out in Articles 2A to 2E, provided that any further amendments to the adjustments or Amendment adopted at the Second Meeting of the Parties in London, 29 June 1990, shall apply to the Parties operating under this paragraph after the review provided for in paragraph 8 of this Article has taken place and shall be based on the conclusions of that review.
1 *bis*. The Parties shall, taking into account the review referred to in paragraph 8 of this Article, the assessments made pursuant to Article 6 and any other relevant information, decide by 1 January 1996, through the procedure set forth in paragraph 9 of Article 2: (a) With respect to paragraphs 1 to 6 of Article 2F, what base year, initial levels, control schedules and phase-out date for consumption of the controlled substances in Group I of Annex C will apply to Parties operating under paragraph 1 of this Article; (b) With respect to Article 2G, what phase-out date for production and consumption of the controlled substances in Group II of Annex C will apply to Parties operating under paragraph 1 of this Article; and (c) With respect to Article 2H, what base year, initial levels and control schedules for consumption and production of the controlled substance in Annex E will apply to Parties operating under paragraph 1 of this Article.
2. However, any Party operating under paragraph 1 of this Article shall exceed neither an annual calculated level of consumption of the controlled substances in Annex A of 0.3 kilograms per capita nor an annual calculated level of consumption of controlled substances of Annex B of 0.2 kilograms per capita.
3. When implementing the control measures set out in Articles 2A to 2E, any Party operating under paragraph 1 of this Article shall be entitled to use:
    (a) For controlled substances under Annex A, either the average of its annual calculated level of consumption for the period 1995 to 1997 inclusive or a calculated level of consumption of 0.3 kilograms per capita, whichever is the lower, as the basis for determining its compliance with the control measures relating to consumption.
    (b) For controlled substances under Annex B, the average of its annual calculated level of consumption for the period 1998 to 2000 inclusive or a calculated level of consumption of 0.2 kilograms per capita, whichever is the lower, as the basis for determining its compliance with the control measures relating to consumption.

(c) For controlled substances under Annex A, either the average of its annual calculated level of production for the period 1995 to 1997 inclusive or a calculated level of production of 0.3 kilograms per capita, whichever is the lower, as the basis for determining its compliance with the control measures relating to production.

(d) For controlled substances under Annex B, either the average of its annual calculated level of production for the period 1998 to 2000 inclusive or a calculated level of production of 0.2 kilograms per capita, whichever is the lower, as the basis for determining its compliance with the control measures relating to production.

4. If a Party operating under paragraph 1 of this Article, at any time before the control measures obligations in Articles 2A to 2I become applicable to it, finds itself unable to obtain an adequate supply of controlled substances, it may notify this to the Secretariat. The Secretariat shall forthwith transmit a copy of such notification to the Parties, which shall consider the matter at their next Meeting, and decide upon appropriate action to be taken.

5. Developing the capacity to fulfil the obligations of the Parties operating under paragraph 1 of this Article to comply with the control measures set out in Articles 2A to 2E and Article 2I, and any control measures in Articles 2F to 2H that are decided pursuant to paragraph 1 bis of this Article, and their implementation by those same Parties will depend upon the effective implementation of the financial co-operation as provided by Article 10 and the transfer of technology as provided by Article 10A.

6. Any Party operating under paragraph 1 of this Article may, at any time, notify the Secretariat in writing that, having taken all practicable steps it is unable to implement any or all of the obligations laid down in Articles 2A to 2E and Article 2I, or any or all obligations in Articles 2F to 2H that are decided pursuant to paragraph 1 bis of this Article, due to the inadequate implementation of Articles 10 and 10A. The Secretariat shall forthwith transmit a copy of the notification to the Parties, which shall consider the matter at their next Meeting, giving due recognition to paragraph 5 of this Article and shall decide upon appropriate action to be taken.

7. During the period between notification and the Meeting of the Parties at which the appropriate action referred to in paragraph 6 above is to be decided, or for a further period if the Meeting of the Parties so decides, the non-compliance procedures referred to in Article 8 shall not be invoked against the notifying Party.

8. A Meeting of the Parties shall review, not later than 1995, the situation of the Parties operating under paragraph 1 of this Article, including the effective implementation of financial co-operation and transfer of technology to them, and adopt such revisions that may be deemed necessary regarding the schedule of control measures applicable to those Parties.

8 *bis*. Based on the conclusions of the review referred to in paragraph 8 above:

(a) With respect to the controlled substances in Annex A, a Party operating under paragraph 1 of this Article shall, in order to meet its basic domestic needs, be entitled to delay for ten years its compliance with the control measures adopted by the Second Meeting of the Parties in London, 29 June 1990, and reference by the Protocol to Articles 2A and 2B shall be read accordingly;

(b) With respect to the controlled substances in Annex B, a Party operating under paragraph 1 of this Article shall, in order to meet its basic domestic needs, be entitled to delay for ten years its compliance with the control measures adopted by the Second Meeting of the Parties in London, 29 June 1990, and reference by the Protocol to Articles 2C to 2E shall be read accordingly.

8 *ter.* Pursuant to paragraph 1 bis above:
(a) Each Party operating under paragraph 1 of this Article shall ensure that for the twelve-month period commencing on 1 January 2013, and in each twelve-month period thereafter, its calculated level of consumption of the controlled substances in Group I of Annex C does not exceed, annually, the average of its calculated levels of consumption in 2009 and 2010. Each Party operating under paragraph 1 of this Article shall ensure that for the twelve-month period commencing on 1 January 2013 and in each twelve-month period thereafter, its calculated level of production of the controlled substances in Group I of Annex C does not exceed, annually, the average of its calculated levels of production in 2009 and 2010;
(b) Each Party operating under paragraph 1 of this Article shall ensure that for the twelve-month period commencing on 1 January 2015, and in each twelve-month period thereafter, its calculated level of consumption of the controlled substances in Group I of Annex C does not exceed, annually, ninety per cent of the average of its calculated levels of consumption in 2009 and 2010. Each such Party producing one or more of these substances shall, for the same periods, ensure that its calculated level of production of the controlled substances in Group I of Annex C does not exceed, annually, ninety per cent of the average of its calculated levels of production in 2009 and 2010;
(c) Each Party operating under paragraph 1 of this Article shall ensure that for the twelve-month period commencing on 1 January 2020, and in each twelve-month period thereafter, its calculated level of consumption of the controlled substances in Group I of Annex C does not exceed, annually, sixty-five per cent of the average of its calculated levels of consumption in 2009 and 2010. Each such Party producing one or more of these substances shall, for the same periods, ensure that its calculated level of production of the controlled substances in Group I of Annex C does not exceed, annually, sixty-five per cent of the average of its calculated levels of production in 2009 and 2010;
(d) Each Party operating under paragraph 1 of this Article shall ensure that for the twelve-month period commencing on 1 January 2025, and in each twelve-month period thereafter, its calculated level of consumption of the controlled substances in Group I of Annex C does not exceed, annually, thirty two point five per cent of the average of its calculated levels of consumption in 2009 and 2010. Each such Party producing one or more of these substances shall, for the same periods, ensure that its calculated level of production of the controlled substances in Group I of Annex C does not exceed, annually, thirty-two point five per cent of the average of its calculated levels of production in 2009 and 2010;
(e) Each Party operating under paragraph 1 of this Article shall ensure that for the twelve-month period commencing on 1 January 2030, and in each twelve-month period thereafter, its calculated level of consumption of the controlled

substances in Group I of Annex C does not exceed zero. Each such Party producing one or more of these substances shall, for the same periods, ensure that its calculated level of production of the controlled substances in Group I of Annex C does not exceed zero. However:
  (i) Each such Party may exceed that limit on consumption in any such twelve-month period so long as the sum of its calculated levels of consumption over the ten-year period from 1 January 2030 to 1 January 2040, divided by ten, does not exceed two point five per cent of the average of its calculated levels of consumption in 2009 and 2010, and provided that such consumption shall be restricted to the servicing of refrigeration and air-conditioning equipment existing on 1 January 2030;
  (ii) Each such Party may exceed that limit on production in any such twelve-month period so long as the sum of its calculated levels of production over the ten-year period from 1 January 2030 to 1 January 2040, divided by ten, does not exceed two point five per cent of the average of its calculated levels of production in 2009 and 2010, and provided that such production shall be restricted to the servicing of refrigeration and air-conditioning equipment existing on 1 January 2030.
(f) Each Party operating under paragraph 1 of this Article shall comply with Article 2G;
(g) With regard to the controlled substance contained in Annex E:
  (i) As of 1 January 2002 each Party operating under paragraph 1 of this Article shall comply with the control measures set out in paragraph 1 of Article 2H and, as the basis for its compliance with these control measures, it shall use the average of its annual calculated level of consumption and production, respectively, for the period of 1995 to 1998 inclusive;
  (ii) Each Party operating under paragraph 1 of this Article shall ensure that for the twelve-month period commencing on 1 January 2005, and in each twelve-month period thereafter, its calculated levels of consumption and production of the controlled substance in Annex E do not exceed, annually, eighty per cent of the average of its annual calculated levels of consumption and production, respectively, for the period of 1995 to 1998 inclusive;
  (iii) Each Party operating under paragraph 1 of this Article shall ensure that for the twelve-month period commencing on 1 January 2015 and in each twelve-month period thereafter, its calculated levels of consumption and production of the controlled substance in Annex E do not exceed zero. This paragraph will apply save to the extent that the Parties decide to permit the level of production or consumption that is necessary to satisfy uses agreed by them to be critical uses;
  (iv) The calculated levels of consumption and production under this sub-paragraph shall not include the amounts used by the Party for quarantine and pre-shipment applications.
9. Decisions of the Parties referred to in paragraph 4, 6 and 7 of this Article shall be taken according to the same procedure applied to decision-making under Article 10.
  …

Article 9: Research, development, public awareness and exchange of information
1. The Parties shall co-operate, consistent with their national laws, regulations and practices and taking into account in particular the needs of developing countries, in promoting, directly or through competent international bodies, research, development and exchange of information on:
    (a) best technologies for improving the containment, recovery, recycling, or destruction of controlled substances or otherwise reducing their emissions;
    (b) possible alternatives to controlled substances, to products containing such substances, and to products manufactured with them; and
    (c) costs and benefits of relevant control strategies.
2. The Parties, individually, jointly or through competent international bodies, shall co-operate in promoting public awareness of the environmental effects of the emissions of controlled substances and other substances that deplete the ozone layer.
3. Within two years of the entry into force of this Protocol and every two years thereafter, each Party shall submit to the Secretariat a summary of the activities it has conducted pursuant to this Article.

Article 10: Financial mechanism
1. The Parties shall establish a mechanism for the purposes of providing financial and technical cooperation, including the transfer of technologies, to Parties operating under paragraph 1 of Article 5 of this Protocol to enable their compliance with the control measures set out in Articles 2A to 2E and Article 2I, and any control measures in Articles 2F to 2H that are decided pursuant to paragraph 1 bis of Article 5 of the Protocol. The mechanism, contributions to which shall be additional to other financial transfers to Parties operating under that paragraph, shall meet all agreed incremental costs of such Parties in order to enable their compliance with the control measures of the Protocol. An indicative list of the categories of incremental costs shall be decided by the meeting of the Parties.
2. The mechanism established under paragraph 1 shall include a Multilateral Fund. It may also include other means of multilateral, regional and bilateral co-operation.
3. The Multilateral Fund shall:
    (a) Meet, on a grant or concessional basis as appropriate, and according to criteria to be decided upon by the Parties, the agreed incremental costs;
    (b) Finance clearing-house functions to:
        (i) Assist Parties operating under paragraph 1 of Article 5, through country specific studies and other technical co-operation, to identify their needs for co-operation;
        (ii) Facilitate technical co-operation to meet these identified needs;
        (iii) Distribute, as provided for in Article 9, information and relevant materials, and hold workshops, training sessions, and other related activities, for the benefit of Parties that are developing countries; and
        (iv) Facilitate and monitor other multilateral, regional and bilateral co-operation available to Parties that are developing countries;
    (c) Finance the secretarial services of the Multilateral Fund and related support costs.
4. The Multilateral Fund shall operate under the authority of the Parties who shall decide on its overall policies.

5. The Parties shall establish an Executive Committee to develop and monitor the implementation of specific operational policies, guidelines and administrative arrangements, including the disbursement of resources, for the purpose of achieving the objectives of the Multilateral Fund. The Executive Committee shall discharge its tasks and responsibilities, specified in its terms of reference as agreed by the Parties, with the co-operation and assistance of the International Bank for Reconstruction and Development (World Bank), the United Nations Environment Programme, the United Nations Development Programme or other appropriate agencies depending on their respective areas of expertise. The members of the Executive Committee, which shall be selected on the basis of a balanced representation of the Parties operating under paragraph 1 of Article 5 and of the Parties not so operating, shall be endorsed by the Parties.
6. The Multilateral Fund shall be financed by contributions from Parties not operating under paragraph 1 of Article 5 in convertible currency or, in certain circumstances, in kind and/or in national currency, on the basis of the United Nations scale of assessments. Contributions by other Parties shall be encouraged. Bilateral and, in particular cases agreed by a decision of the Parties, regional co-operation may, up to a percentage and consistent with any criteria to be specified by decision of the Parties, be considered as a contribution to the Multilateral Fund, provided that such co-operation, as a minimum:
    (a) Strictly relates to compliance with the provisions of this Protocol;
    (b) Provides additional resources; and
    (c) Meets agreed incremental costs.
7. The Parties shall decide upon the programme budget of the Multilateral Fund for each fiscal period and upon the percentage of contributions of the individual Parties thereto.
8. Resources under the Multilateral Fund shall be disbursed with the concurrence of the beneficiary Party.
9. Decisions by the Parties under this Article shall be taken by consensus whenever possible. If all efforts at consensus have been exhausted and no agreement reached, decisions shall be adopted by a two-thirds majority vote of the Parties present and voting, representing a majority of the Parties operating under paragraph 1 of Article 5 present and voting and a majority of the Parties not so operating present and voting.
10. The financial mechanism set out in this Article is without prejudice to any future arrangements that may be developed with respect to other environmental issues. Article

10A: Transfer of technology

Each Party shall take every practicable step, consistent with the programmes supported by the financial mechanism, to ensure:
    (a) that the best available, environmentally safe substitutes and related technologies are expeditiously transferred to Parties operating under paragraph 1 of Article 5; and
    (b) that the transfers referred to in subparagraph (a) occur under fair and most favourable conditions.

...

Article 16: Entry into force
1. This Protocol shall enter into force on 1 January 1989, provided that at least eleven instruments of ratification, acceptance, approval of the Protocol or accession thereto have been deposited by States or regional economic integration organizations representing at least two-thirds of 1986 estimated global consumption of the controlled substances, and the provisions of paragraph 1 of Article 17 of the Convention have been fulfilled. In the event that these conditions have not been fulfilled by that date, the Protocol shall enter into force on the ninetieth day following the date on which the conditions have been fulfilled.
2. For the purposes of paragraph 1, any such instrument deposited by a regional economic integration organization shall not be counted as additional to those deposited by member States of such organization.
3. After the entry into force of this Protocol, any State or regional economic integration organization shall become a Party to it on the ninetieth day following the date of deposit of its instrument of ratification, acceptance, approval or accession.

Article 17: Parties joining after entry into force
Subject to Article 5, any State or regional economic integration organization which becomes a Party to this Protocol after the date of its entry into force, shall fulfil forthwith the sum of the obligations under Article 2, as well as under Articles 2A to 2I and Article 4, that apply at that date to the States and regional economic integration organizations that became Parties on the date the Protocol entered into force.

Article 18: Reservations
No reservations may be made to this Protocol.

Article 19: Withdrawal
Any Party may withdraw from this Protocol by giving written notification to the Depositary at any time after four years of assuming the obligations specified in paragraph 1 of Article 2A. Any such withdrawal shall take effect upon expiry of one year after the date of its receipt by the Depositary, or on such later date as may be specified in the notification of the withdrawal.

Article 20: Authentic texts
The original of this Protocol, of which the Arabic, Chinese, English, French, Russian and Spanish texts are equally authentic, shall be deposited with the Secretary-General of the United Nations.
IN WITNESS WHEREOF THE UNDERSIGNED, BEING DULY AUTHORIZED TO THAT EFFECT, HAVE SIGNED THIS PROTOCOL. DONE AT MONTREAL THIS SIXTEENTH DAY OF SEPTEMBER, ONE THOUSAND NINE HUNDRED AND EIGHTY SEVEN.

Analysis

In the early 1970s, Frank Sherwood Rowland and Mario Molina, who were chemists at the University of California, Irvine, began studying the manner that chlorofluorocarbons (CFCs) impacted the atmosphere. The scientists discovered that when the CFCs reached the stratosphere, where they were exposed to ultraviolet radiation, they released chlorine atoms. Those atoms over time broke down the ozone layer. The depletion of the ozone layer meant that more ultraviolet radiation would reach the surface, which would have numerous negative impacts on humans, such as a rise in skin cancers. They espoused their ozone depletion hypothesis in a 1974 publication. Two years later, the United States Academy of Sciences released a report supporting Rowland and Molina's theory. This led to the United States banning CFCs as propellants in aerosol cans in 1978. Despite the potential implications, the threat posed was largely ignored internationally outside of the scientific community as it was viewed as a theoretical problem. Ironically, in 1978, researchers began to take note of a doughnut-shaped hole that emerged in October of that year over Antarctica. It closed, only to emerge larger the following year. By 1985, the annual recurrence of that pattern resulted in a massive hole in the ozone approximately the size of North America. Its existence spurred the drafting and passage of the Montreal Protocol. The work of Sherwood and Molina resulted in them receiving the Nobel Prize in Chemistry in 1995.

The international effort to protect the ozone layer began with twenty-eight countries becoming signatories to the Vienna Convention for the Protection of the Ozone Layer in March 1985. In November 1986, the United Nations' Environment Programme convened a meeting in Geneva, Switzerland, which saw representatives of forty-five countries discuss how to address the threat posed by the hole in the ozone layer. Those discussions proved divisive, as the European Community and Japan rejected calls by other countries, led by the United States, to ban CFCs and halons, which had similar effects on ozone. Japan and the European Community exported many products containing CFCs and halons so any bans would hurt their economies. Japan and the European Community finally relented at a February 27, 1985, meeting in Vienna, Austria after being presented overwhelming scientific evidence that CFCs and halons were the primary culprits in the depletion of the ozone layer. With all parties in agreement, the stage was set for a final agreement to be crafted in Montreal, Canada.

The Montreal Protocol called for signatories of the agreement to cease production of chlorofluorocarbon 11, 12, 113, 114, and 115. It called for the same for halons 1211, 1301, and 2402. In order to encourage extensive international participation, special dispensation was made for "developing" countries. They were allowed to keep using the banned CFCs and halons if there was not a suitable alternative available. To help those countries find another option, the protocol created the Multilateral Fund for the Implementation of the Montreal Protocol, which had the distinction of being the first international fund ever created to address any aspect of climate change. Amended or adjusted seven times since passage, the Montreal Protocol has proven to be an effective mechanism for the protection of the ozone layer.

Further Reading

Benedick, Richard Elliott. 1998. *Ozone Diplomacy: New Directions in Safeguarding the Planet.* Cambridge, MA: Harvard University Press.

Gareau, Brian J. 2013. *From Precaution to Profit: Contemporary Challenges to Environmental Protection in the Montreal Protocol.* New Haven, CT: Yale University Press.

Peloso, Chris. 2010. "Crafting an International Climate Change Protocol: Applying the Lessons Learned From the Success of the Montreal Protocol and the Ozone Depletion Problem." *Journal of Land Use & Environmental Law* 25: 305–329.

Wu, Yutian, Lorenzo M. Polvani, and Richard Seager. 2013. "The Importance of the Montreal Protocol in Protecting Earth's Hydroclimate." *Journal of Climate* 26: 4049–4068.

Message to the Senate Transmitting the Montreal Protocol on Ozone-Depleting Substances

Date: December 21, 1987
Location: Washington, D.C.
Significance: President Ronald Reagan expressed his support to the United States Senate for the ratification of the Montreal Protocol on Substances that Deplete the Ozone Layer.
Source: Reagan, Ronald. 1989. *Public Papers of the Presidents, Ronald Reagan: 1987, Book 2—July 4 to December 31, 1987.* Washington, D.C.: Government Printing Office.

To the Senate of the United States:
I transmit herewith, for the advice and consent of the Senate to ratification, the Montreal Protocol on Substances that deplete the Ozone Layer, done at Montreal on September 16, 1987. The report of the Department of State is also enclosed for the information of the Senate.

The Montreal Protocol provides for internationally coordinated control of ozone-depleting substances in order to protect public health and the environment from potential adverse effects of depletion of stratospheric ozone. The Protocol was negotiated under the auspices of the United Nations Environment Program, pursuant to the Vienna Convention for the Protection of the Ozone Layer, which was ratified by the United States in August 1986.

In this historic agreement, the international community undertakes cooperative measures to protect a vital global resource. The United States played a leading role in the negotiation of the Protocol. United States ratification is necessary for entry into force and effective implementation of the Protocol. Early ratification by the United States will encourage similar action by other nations whose participation is also essential.

I recommend that the Senate give early and favorable consideration to the Protocol and give its advice and consent to ratification.
RONALD REAGAN

Analysis

President Ronald Reagan strongly supported the Montreal Protocol on Substances that Deplete the Ozone Layer and actively worked to get the agreement ratified by the Senate. One factor that helped ensure widespread support in the Senate was that the agreement did not seriously impact the country's chemical industry because it did not produce a significant amount of products containing chlorofluorocarbons.

Soon thereafter, discussions began at the international level to draft an agreement modeled on the Montreal Protocol that would reduce pollution resulting from the use of fossil fuels such as coal and natural gas. Since the United States' economy was heavily dependent on fossil fuels, the Reagan administration made it abundantly clear that it would not support the initiative.

Further Reading

Gareau, Brian J. 2013. *From Precaution to Profit: Contemporary Challenges to Environmental Protection in the Montreal Protocol.* New Haven, CT: Yale University Press.

Statement of Dr. James Hansen, Director, NASA Goddard Institute for Space Studies

Date: June 23, 1988
Location: Washington, D.C.
Significance: In his influential and controversial presentation before the United States Senate's Commission on Energy and Natural Resources, Dr. James Hansen made the case that global temperatures were rising significantly due to the emission of greenhouse gases.
Source: United States Congress. Senate. Commission on Energy and Natural Resources. 1988. *Greenhouse Effect and Global Climate Change: Hearings Before the Committee on Energy and Natural Resources, United States Senate, One Hundredth Congress, 2nd session.* Washington, D.C.: Government Printing Office.

Dr. Hansen: Mr. Chairman and committee members, thank you for the opportunity to present the results of my research on the greenhouse effect which has been carried out with my colleagues at the NASA Goddard Institute for Space Studies.

I would like to draw three main conclusions. Number one, the earth is warmer in 1988 than at any time in the history of instrumental measurements. Number two, the global warming is now large enough that we can ascribe with a high degree of confidence a cause and effect relationship to the greenhouse effect. And number three, our computer climate simulations indicate that the greenhouse effect is already large enough to begin to effect the probability of extreme events such as summer heat waves.

My first viewgraph, which I would like to ask Suki to put up if he would, shows the global temperature over the period of instrumental records which is about 100 years. The present temperature is the highest in the period of record. The rate of warming in the past 25 years, as you can see on the right, is the highest on record. The four warmest years, as the Senator mentioned, have all been in the 1980s. And 1988 so far is so much warmer than in 1987, that barring a remarkable and improbable cooling, 1988 will be the warmest year on the record.

Now let me turn to my second point which is causal association of the greenhouse effect and the global warming. Causal association requires first that the warming be larger than natural climate variability and second, that the magnitude and nature of the warming be consistent with the greenhouse mechanism. These points are both addressed on my second viewgraph. The observed warming during the past 30 years, which is the period when we have accurate measurements of atmospheric composition, is shown by the heavy black line in this graph. The warming is almost 0.4 degrees Centigrade by 1987 relative to climatology, which is defined as the 30 year mean, 1950 to 1980 and, in fact, the warming is more than 0.4 degrees Centigrade in 1988. The probability of a chance warming of this magnitude is about 1 percent. So, with 99 percent confidence we can state that the warming during this time period is a real warming trend.

The other curves in this figure are the results of global climate model calculations for three scenarios of atmospheric trace gas growth. We have considered several scenarios because there are uncertainties in the exact trace gas growth in the past and especially in the future. We have considered cases ranging from business as usual, which is scenario A, to draconian emission cuts, scenario C, which would totally eliminate net trace gas growth by year 2000.

The main point to be made here is that the expected global warming is of the same magnitude as the observed warming. Since there is only a 1 percent chance of an accidental warming of this magnitude, the agreement with the expected greenhouse effect is of considerable significance. Moreover, if you look at the next level of detail in the global temperature change, there are clear signs of the greenhouse effect. Observational data suggests a cooling in the stratosphere while the ground is warming. The data suggest somewhat more warming over land and sea ice regions than over open ocean, more warming at high latitudes than at low latitudes, and more warming in the winter than in the summer. In all of these cases, the signal is at best just starting to emerge, and we need more data. Some of these details, such as the northern hemisphere high latitude temperature trends, do not look exactly like the greenhouse effect, but that is expected. There are certainly other climate change factors involved in addition to the greenhouse effect.

Altogether the evidence that the earth is warming by an amount which is too large to be a chance fluctuation and the similarity of the warming to that expected from the greenhouse effect represents a very strong case. In my opinion, that the greenhouse effect has been detected, and it is changing our climate now.

Then my third point. Finally, I would like to address the question of whether the greenhouse effect is already large enough to affect the probability of extreme events, such as summer heatwaves. As shown in my next viewgraph, we have used the temperature changes computed in our global climate model to estimate the impact of the greenhouse effect on the frequency of hot summers in Washington, D.C., and Omaha, Nebraska. A hot summer is defined as the hottest one-third of the summers in the 1950 to 1980 period, which is the period the Weather Bureau uses for defining climatology. So, in that period the probability of having a hot summer was 33 percent, but by the 1990s, you can see that the greenhouse effect has increased the probability of a hot summer to somewhere between 55 and 70 percent in Washington according to our climate model simulations. In the late 1980s, the probability of a hot summer would be somewhat less than that. You can interpolate to a value of something like 40 to 60 percent.

I believe that this change in the frequency of hot summers is large enough to be noticeable to the average person. So, we have already reached a point that the greenhouse effect is important. It may also have important implication other than for creature comforts.

My last viewgraph shows global maps of temperature anomalies for a particular month, July, for several different years between 1986 and 2029, as computed with our global climate model for the immediate trace gas scenario B. As shown by the graphs on the left where yellow and red colors represent areas that are warmer than climatology and blue areas represent areas that are colder than climatology, at the present time in the 1980s the greenhouse warming is smaller than the natural variability of the local temperature. So, in any given month, there is almost as much area that is cooler than normal as there is area warmer than normal. A few decades in the future, as shown on the right, it is warm almost everywhere.

However, the point that I would like to make is that in the late 1980's and 1990's we notice a clear tendency in our model for greater than average warming in the southeast United States and the Midwest. In our model this result seems to arise because the Atlantic Ocean off the coast of the United States warms more slowly than the land. This leads to high pressure along the east coast and circulation of warm air north into the midwest or the southeast. There is only a tendency for this phenomenon. It is certainly not going to happen every year and climate models are certainly an imperfect tool at this time. However, we conclude that there is evidence that the greenhouse effect increases the likelihood of heat wave drought situations in the southeast and midwest United States even though we cannot blame a specific drought on the greenhouse effect.

Therefore, I believe that it is not a good idea to use the period 1950 to 1980 for which climatology is normally defined as an indication of how frequently droughts will occur in the future. If our model is approximately correct, such situations may be more common in the next 10 to 15 years than they were in the period 1950 to 1980.

Finally, I would like to stress that there is a need for improving these global climate models, and there is a need for global observations if we're going to obtain a full understanding of these phenomena.

That conclude my statement, and I'd be glad to answer questions if you'd like.

Analysis

At the time of his testimony, Dr. James Hansen was the head of the NASA Goddard Institute for Space Studies. In that role, he led one of the federal agencies actively pursuing scientific research on the causes and consequences of global climate change. Unlike many scientists dependent on the federal government for employment or research funding, Hansen was unafraid to honestly express his learned opinions publicly about climate change to officials in the legislative and executive branches of the government. This was an admirable quality considering many of those individuals were predisposed, due to their official political positions on climate change, to reject the results of the research conducted by Hansen and his colleagues at NASA.

During his presentation, Hansen used meteorological data and climate models to argue that there was a growing warming trend that was impacting global temperatures. He blamed greenhouse gas emissions (GHG) for causing most of the changes in the climate. Hansen warned that global warming was accelerating thus immediate federal action was required to cut emission levels before the damage to the environment became irreversible.

Hansen's testimony was problematic to climate change deniers because he was a well-known and reputable scientist. His one vulnerability was that he was a federal employee. Over time, political operatives began censoring Hansen's work and that of other scientists dependent on the federal government for employment or research grants. These activities drew the ire of scientists around the country and their professional organizations who saw any interference into the conduct of research and the sharing of data as a threat to scientific integrity.

Further Reading

Besel, Richard D. 2013. "Accommodating Climate Change Science: James Hansen and the Rhetorical / Political Emergence of Global Warming." *Science in Context* 26: 137–152.

Bowen, Mark. 2008. *Censoring Science: Inside the Political Attack on Dr. James Hansen and the Truth of Global Warming*. New York: Dutton.

Clean Air Act Amendments of 1990
Title IV (Excerpt)

Date: November 15, 1990
Location: Washington, D.C.
Significance: Title IV of the Clean Air Act Amendments included a cap-and-trade program to address the pollutants that were contributing to the creation of acid rain. It became the model for both national and international proposals to mediate the emission of greenhouse gases into the atmosphere.
Source: 101st Congress. 1990. *Public Law 101–549-Nov. 15, 1990.* https://history.nih.gov/research/downloads/PL101–549.pdf (accessed April 23, 2016).

An Act

To amend the Clean Air Act to provide for attainment and maintenance of health protective national ambient air quality standards, and for other purposes.

Be it enacted by the Senate and House of Representatives of the United States of America in Congress assembled

...

Title IV—Acid Deposition Control

Sec. 401. Acid deposition control.
Sec. 402. Fossil fuel use.
Sec. 403. Repeal of percent reduction.
Sec. 404. Acid deposition standards.
Sec. 405. National acid lakes registry.
Sec. 406. Industrial SO2 Emissions.
Sec. 407. Sense of the Congress on emission reductions costs.
Sec. 408. Monitor acid rain program in Canada.
Sec. 409. Report on clean coals technologies export programs.
Sec. 410. Acid deposition research by the United States Fish and Wildlife Service.
Sec. 411. Study of buffering and neutralizing agents.
Sec. 412. Canforming amendment.
Sec. 413. Special clean coal technology project.

SEC. 401. Acid Deposition Control
The Clean Air Act is amended by adding the following new title after title III:

"Title IV—Acid Deposition Control

"Sec. 401. Findings and purpose.
"Sec. 402. Definitions.
"Sec. 403. Sulfur dioxide allowance program for existing and new units.
"Sec. 404. Phase I sulfur dioxide requirements.
"Sec. 405. Phase II sulfur dioxide requirements.
"Sec. 406. Allowances for States with emissions rates at or below 0.80 lbs/mmBtu.
"Sec. 407. Nitrogen oxides emission reduction program.
"Sec. 408. Permits and compliance plans.
"Sec. 409. Repowered sources.
"Sec. 410. Election for additional sources.
"Sec. 411. Excess emissions penalty.
"Sec. 412. Monitoring, reporting, and recordkeeping requirements.
"Sec. 413. General compliance with other provisions.
"Sec. 414. Enforcement.
"Sec. 415. Clean coal technology regulatory incentives.
"Sec. 416. Contingency guarantee; auctions, reserve.
...

"SEC. 403. Sulfur Dioxide Allowance Program for Existing and New Units.
"(a) ALLOCATIONS OF ANNUAL ALLOWANCES FOR EXISTING AND NEW UNITS.-(1) For the emission limitation programs under this title, the Administrator shall allocate annual allowances for the unit, to be held or distributed by the designated representative of the owner or operator of each affected unit at an affected source in accordance with this title, in an amount equal to the annual tonnage emission limitation calculated under section 404, 405, 406, 409, or 410 except as otherwise specifically provided elsewhere in this title. Except as provided in sections 405(a)(2), 405(a)(3), 409 and 410, beginning January 1, 2000, the Administrator shall not allocate annual allowances to emit sulfur dioxide pursuant to section 405 in such an amount as would result in total annual emissions of sulfur dioxide from utility units in excess of 8.90 million tons except that the Administrator shall not take into account unused allowances carried forward by owners and operators of affected units or by other persons holding such allowances, following the year for which they were allocated. If necessary to meeting the restrictions imposed in the preceding sentence, the Administrator shall reduce, pro rata, the basic Phase II allowance allocations for each unit subject to the requirements of section 405. Subject to the provisions of section 416, the Administrator shall allocate allowances for each affected unit at an affected source annually, as provided in paragraphs (2) and (3) and section 408. Except as provided in sections 409 and 410, the removal of an existing affected unit or source from commercial operation at any time after the date of the enactment of the Clean Air Act Amendments of 1990 (whether before or after January 1, 1995, or January 1, 2000) shall not terminate or otherwise affect the allocation of allowances pursuant to section 404 or 405 to which the unit is entitled. Allowances shall be allocated

by the Administrator without cost to the recipient, except for allowances sold by the Administrator pursuant to section 416. Not later than December 31, 1991, the Administrator shall publish a proposed list of the basic Phase ll allowance allocations, the Phase ll bonus allowance allocations and, if applicable, allocations pursuant to section 405(a)(3) for each unit subject to the emissions limitation requirements of section 405 for the year 2000 and the year 2010. After notice and opportunity for public comment, but not later than December 31, 1992, the Administrator shall publish a final list of such allocations, subject to the provisions of section 405(a)(2). Any owner or operator of an existing unit subject to the requirements of section 405(b) or (c) who is considering applying for an extension of the emission limitation requirement compliance deadline for that unit from January 1, 2000, until not later than December 31, 2000, pursuant to section 409, shall notify the Administrator no later than March 31, 1991. Such notification shall be used as the basis for estimating the basic Phase Il allowances under this subsection. Prior to June 1,1998, the Administrator shall publish a revised final statement of allowance allocations, subject to the provisions of section 405(a)(2) and taking into account the effect of any compliance date extensions granted pursuant to section 409 on such allocations. Any person who may make an election concerning the amount of allowances to be allocated to a unit or units shall make such election and so inform the Administrator not later than March 31, 1991, in the case of an election under section 405 (or June 30, 1991, in the case of an election under section 406). If such person fails to make such election, the Administrator shall set forth for each unit owned or operated by such person, the amount of allowances reflecting the election that would, in the judgment of the Administrator, provide the greatest benefit for the owner or operator of the unit. If such person is a Governor who may make an election under section 406 and the Governor fails to make an election, the Administrator shall set forth for each unit in the State the amount of allowances reflecting the election that would, in the judgment of the Administrator, provide the greatest benefit for units in the State.

"(b) ALLOWANCE TRANSFER SYSTEM.-Allowances allocated under this title may be transferred among designated representatives of the owners or operators of affected sources under this title and any other person who holds such allowances, as provided by the allowance system regulations to be promulgated by the Administrator not later than eighteen months after the date of enactment of the Clean Air Act Amendments of 1990. Such regulations shall establish the allowance system prescribed under this section, including, but not limited to, requirements for the allocation, transfer, and use of allowances under this title. Such regulations shall prohibit the use of any allowance prior to the calendar year for which the allowance was allocated, and shall provide, consistent with the purposes of this title, for the identification of unused allowances, and for such unused allowances to be carried forward and added to allowances allocated in subsequent years, including allowances allocated to units subject to Phase I requirements (as described in section 404) which are applied to emissions limitations requirements in Phase II (as described in section 405). Transfers of allowances shall not be effective until written certification of the transfer, signed by a responsible official of each party to the transfer, is received and recorded by the Administrator. Such regulations shall permit the transfer of allowances prior to the issuance of such allowances. Recorded pre-allocation transfers shall be deducted by the Administrator from the number of allowances which would otherwise be allocated to the transferor, and added to those allowances allocated to the transferee. Pre-allocation transfers shall not affect the prohibition contained in this subsection against the use of allowances prior to the year for which they are allocated.

"(c) INTERPOLLUTANT TRADING.-Not later than January 1, 1994, the Administrator shall furnish to the Congress a study evaluating the environmental and economic consequences of amending this title to permit trading sulfur dioxide allowances for nitrogen oxides allowances.

"(d) ALLOWANCE TRACKING SYSTEM.-(1) The Administrator shall promulgate, not later than 18 months after the date of enactment of the Clean Air Act Amendments of 1990, a system for issuing, recording, and tracking allowances, which shall specify all necessary procedures and requirements for an orderly and competitive functioning of the allowance system. All allowance allocations and transfers shall, upon recordation by the Administrator, be deemed a part of each unit's permit requirements pursuant to section 408, without any further permit review and revision.

"(2) In order to insure electric reliability, such regulations shall not prohibit or affect temporary increases and decreases in emissions within utility systems, power pools, or utilities entering into allowance pool agreements, that result from their operations, including emergencies and central dispatch, and such temporary emissions increases and decreases shall not require transfer of allowances among units nor shall it require recordation. The owners or operators of such units shall act through a designated representative. Notwithstanding the preceding sentence, the total tonnage of emissions in any calendar year (calculated at the end thereof) from all units in such a utility system, power pool, or allowance pool agreements shall not exceed the total allowances for such units for the calendar year concerned.

"(e) NEW UTILITY UNITS.-After January 1, 2000, it shall be unlawful for a new utility unit to emit an annual tonnage of sulfur dioxide in excess of the number of allowances to emit held for the unit by the unit's owner or operator. Such new utility units shall not be eligible for an allocation of sulfur dioxide allowances under subsection (a)(l), unless the unit is subject to the provisions of subsection (g)(2) or (3) of section 405. New utility units may obtain allowances from any person, in accordance with this title. The owner or operator of any new utility unit in violation of this subsection shall be liable for fulfilling the obligations specified in section 411 of this title.

"(f) NATURE OF ALLOWANCES.-An allowance allocated under this title is a limited authorization to emit sulfur dioxide in accordance with the provisions of this title. Such allowance does not constitute a property right. Nothing in this title or in any other provision of law shall be construed to limit the authority of the United States to terminate or limit such authorization. Nothing in this section relating to allowances shall be construed as affecting the application of, or compliance with, any other revision of this Act to an affected unit or source, including the provisions related to applicable National Ambient Air Quality Standards and State implementation plans. Nothing in this section shall be construed as requiring a change of any kind in any State law regulating electric utility rates and charges or affecting any State law regarding such State regulation or as limiting State regulation (including any prudency review) under such a State law. Nothing in this section shall be construed as modifying the Federal Power Act or as affecting the authority of the Federal Energy Regulatory Commission under that Act. Nothing in this title shall be construed to interfere with or impair any program for competitive bidding for power supply in a State in which such program is established. Allowances, once allocated to a person by the Administrator, may be received, held, and temporarily or permanently transferred in accordance with this title and the regulations of the Administrator without regard to whether or not a permit is in effect under title V or section 408 with respect to the unit for which such allowance was originally allocated and recorded. Each permit under this title and each permit issued under title V for any affected unit shall provide that the affected unit may not emit an annual tonnage of sulfur dioxide in excess of the allowances held for that unit.

"(g) PROHIBITION.-It shall be unlawful for any person to hold, use, or transfer any allowance allocated under this title, except in accordance with regulations promulgated by the Administrator. It shall be unlawful for any affected unit to emit sulfur dioxide in excess of the number of allowances held for that unit for that year by the owner or operator of the unit. Upon the allocation of allowances under this title, the prohibition contained in the preceding sentence shall supersede any other emission limitation applicable under this title to the units for which such allowances are allocated. Allowances may not be used prior to the calendar year for which they are allocated. Nothing in this section or in the allowance system regulations shall relieve the Administrator of the Administrator's permitting, monitoring and enforcement obligations under this Act, nor relieve affected sources of their requirements and liabilities under this Act.

"(h) COMPETITIVE BIDDING FOR POWER SUPPLY.-Nothing in this title shall be construed to interfere with or impair any program for competitive bidding for power supply in a State in which such program is established.

"(i) APPLICABILITY OF THE ANTITRUST LAWS.-

"(1) Nothing in this section affects-

"(A) the applicability of the antitrust laws to the transfer, use, or sale of allowances, or

"(B) the authority of the Federal Energy Regulatory Commission under any provision of law respecting unfair methods of competition or anticompetitive acts or practices.

"(2) As used in this section, 'antitrust laws' means those Acts set forth in section 1 of the Clayton Act (15 U.S.C. 12), as amended.

(j) PUBLIC UTILITY HOLDING COMPANY ACT.-The acquisition or disposition of allowances pursuant to this title including the issuance of securities or the undertaking of any other financing transaction in connection with such allowances shall not be subject to the provisions of the Public Utility Holding Company Act of 1935.

Analysis

While campaigning for the presidency, Vice President George H.W. Bush promised to update the Clean Air Act of 1970. Upon assuming the presidency, his administration began work on fulfilling his promise. The administration's efforts were aided by congressmen from the Democratic Party, including Senate Majority Leader George Mitchell of Maine and Representative Henry Waxman from California. The resulting Clean Air Amendments (CAA) of 1990 passed the House of Representatives by a vote of 401–21 and the Senate 89–11.

One of the leading environmental problems of the day that was addressed within the CAA was "acid rain." It was water, in the form of rain or snow, which contained nitrogen oxides and/or sulfur dioxides. Once it fell to the ground, it poisoned ecosystems. To remedy the issue, a cap-and-trade regulatory system was created that established a maximum amount of SO2 that could be emitted by electric-generating facilities over the course of a year. Each facility was assigned a specific number of allowances (an allowance equals a ton of SO2). If a facility did not use all of its allowances, it could trade its surplus to another entity that exceeded its quota. Any facility that exceeded its share and could not obtain the extra unused allowance from another provider was fined by the Environmental Protection Agency and also lost an equal amount from their quota the following year. Over time, the cap-and-trade regime proved extremely successful in alleviating the scourge of acid rain.

Many environmental activists and politicians have sought to use the acid rain cap-and-trade program as a model to address the emission of greenhouse gases (GHG) at both the national and international levels. Despite support from Presidents William Clinton and Barack Obama, the United States Congress has continually refused, on a bipartisan basis, to support such an initiative.

FURTHER READING

Chan, Gabriel, et al. 2012. "The SO2 Allowance-Trading System and the Clean Air Act Amendments of 1990: Reflections on 20 Years of Policy Innovation." *National Tax Journal* 65: 419–452.

Hausker, Karl. 1992. "The Politics and Economics of Auction Design in the Market for Sulfur Dioxide Pollution." *Journal of Policy Analysis and Management* 11: 553–572.

Hays, Samuel P. 1998. *Explorations in Environmental History: Essays by Samuel P. Hays*. Pittsburgh: University of Pittsburgh Press.

Schmalensee, Richard, and Robert N. Stavins. 2013. "The SO2 Allowance Trading System: The Ironic History of a Grand Policy Experiment." *Journal of Economic Perspectives* 27: 103–121.

United Nations Framework Convention on Climate Change (UNFCCC)

Date: May 9, 1992
Location: New York, New York
Significance: The UNFCCC marked a starting point for the international community to address global warming through the regulation of greenhouse gases (GHGs).
Source: United Nations. 1992. *United Nations Framework Convention on Climate Change.* http://unfccc.int/resource/docs/convkp/conveng.pdf (accessed July 25, 2015).

The Parties to this Convention,

Acknowledging that change in the Earth's climate and its adverse effects are a common concern of humankind,

Concerned that human activities have been substantially increasing the atmospheric concentrations of greenhouse gases, that these increases enhance the natural greenhouse effect, and that this will result on average in an additional warming of the Earth's surface and atmosphere and may adversely affect natural ecosystems and humankind,

Noting that the largest share of historical and current global emissions of greenhouse gases has originated in developed countries, that per capita emissions in developing countries are still relatively low and that the share of global emissions originating in developing countries will grow to meet their social and development needs,

Aware of the role and importance in terrestrial and marine ecosystems of sinks and reservoirs of greenhouse gases,

Noting that there are many uncertainties in predictions of climate change, particularly with regard to the timing, magnitude and regional patterns thereof,

Acknowledging that the global nature of climate change calls for the widest possible cooperation by all countries and their participation in an effective and appropriate international response, in accordance with their common but differentiated responsibilities and respective capabilities and their social and economic conditions,

Recalling the pertinent provisions of the Declaration of the United Nations Conference on the Human Environment, adopted at Stockholm on 16 June 1972,

Recalling also that States have, in accordance with the Charter of the United Nations and the principles of international law, the sovereign right to exploit their own resources pursuant to their own environmental and developmental policies, and the responsibility to ensure that activities within their jurisdiction or control do not cause damage to the environment of other States or of areas beyond the limits of national jurisdiction,

Reaffirming the principle of sovereignty of States in international cooperation to address climate change,

Recognizing that States should enact effective environmental legislation, that environmental standards, management objectives and priorities should reflect the environmental and developmental context to which they apply, and that standards applied by some countries may be inappropriate and of unwarranted economic and social cost to other countries, in particular developing countries,

Recalling the provisions of General Assembly resolution 44/228 of 22 December 1989 on the United Nations Conference on Environment and Development, and resolutions 43/53 of 6 December 1988, 44/207 of 22 December 1989, 45/212 of 21 December 1990 and 46/169 of 19 December 1991 on protection of global climate for present and future generations of mankind,

Recalling also the provisions of General Assembly resolution 44/206 of 22 December 1989 on the possible adverse effects of sea-level rise on islands and coastal areas, particularly low-lying coastal areas and the pertinent provisions of General Assembly resolution 44/172 of 19 December 1989 on the implementation of the Plan of Action to Combat Desertification,

Recalling further the Vienna Convention for the Protection of the Ozone Layer, 1985, and the Montreal Protocol on Substances that Deplete the Ozone Layer, 1987, as adjusted and amended on 29 June 1990,

Noting the Ministerial Declaration of the Second World Climate Conference adopted on 7 November 1990,

Conscious of the valuable analytical work being conducted by many States on climate change and of the important contributions of the World Meteorological Organization, the United Nations Environment Programme and other organs, organizations and bodies of the United Nations system, as well as other international and intergovernmental bodies, to the exchange of results of scientific research and the coordination of research,

Recognizing that steps required to understand and address climate change will be environmentally, socially and economically most effective if they are based on relevant scientific, technical and economic considerations and continually re-evaluated in the light of new findings in these areas,

Recognizing that various actions to address climate change can be justified economically in their own right and can also help in solving other environmental problems,

Recognizing also the need for developed countries to take immediate action in a flexible manner on the basis of clear priorities, as a first step towards comprehensive response strategies at the global, national and, where agreed, regional levels that take into account all greenhouse gases, with due consideration of their relative contributions to the enhancement of the greenhouse effect,

Recognizing further that low-lying and other small island countries, countries with low-lying coastal, arid and semi-arid areas or areas liable to floods, drought and desertification, and developing countries with fragile mountainous ecosystems are particularly vulnerable to the adverse effects of climate change,

Recognizing the special difficulties of those countries, especially developing countries, whose economies are particularly dependent on fossil fuel production, use and exportation, as a consequence of action taken on limiting greenhouse gas emissions,

Affirming that responses to climate change should be coordinated with social and economic development in an integrated manner with a view to avoiding adverse impacts on the latter, taking into full account the legitimate priority needs of developing countries for the achievement of sustained economic growth and the eradication of poverty,

Recognizing that all countries, especially developing countries, need access to resources required to achieve sustainable social and economic development and that, in order for developing countries to progress towards that goal, their energy consumption will need to grow taking into account the possibilities for achieving greater energy efficiency and for controlling greenhouse gas emissions in general, including through the application of new technologies on terms which make such an application economically and socially beneficial,

Determined to protect the climate system for present and future generations,

Have agreed as follows:

Article 1: Definitions

For the purposes of this Convention:

1. "Adverse effects of climate change" means changes in the physical environment or biota resulting from climate change which have significant deleterious effects on the composition, resilience or productivity of natural and managed ecosystems or on the operation of socio-economic systems or on human health and welfare.
2. "Climate change" means a change of climate which is attributed directly or indirectly to human activity that alters the composition of the global atmosphere and which is in addition to natural climate variability observed over comparable time periods.
3. "Climate system" means the totality of the atmosphere, hydrosphere, biosphere and geosphere and their interactions.
4. "Emissions" means the release of greenhouse gases and/or their precursors into the atmosphere over a specified area and period of time.
5. "Greenhouse gases" means those gaseous constituents of the atmosphere, both natural and anthropogenic, that absorb and re-emit infrared radiation.
6. "Regional economic integration organization" means an organization constituted by sovereign States of a given region which has competence in respect of matters governed by this Convention or its protocols and has been duly authorized, in accordance with its internal procedures, to sign, ratify, accept, approve or accede to the instruments concerned
7. "Reservoir" means a component or components of the climate system where a greenhouse gas or a precursor of a greenhouse gas is stored.
8. "Sink" means any process, activity or mechanism which removes a greenhouse gas, an aerosol or a precursor of a greenhouse gas from the atmosphere.
9. "Source" means any process or activity which releases a greenhouse gas, an aerosol or a precursor of a greenhouse gas into the atmosphere.

Article 2: Objective

The ultimate objective of this Convention and any related legal instruments that the Conference of the Parties may adopt is to achieve, in accordance with the relevant provisions of the Convention, stabilization of greenhouse gas concentrations in the atmosphere at a level that would prevent dangerous anthropogenic interference with the climate system. Such a level should be achieved within a time frame sufficient to allow ecosystems to adapt naturally to climate change, to ensure that food production is not threatened and to enable economic development to proceed in a sustainable manner.

Article 3: Principles

In their actions to achieve the objective of the Convention and to implement its provisions, the Parties shall be guided, inter alia, by the following:

1. The Parties should protect the climate system for the benefit of present and future generations of humankind, on the basis of equity and in accordance with their common but differentiated responsibilities and respective capabilities. Accordingly, the developed country Parties should take the lead in combating climate change and the adverse effects thereof.
2. The specific needs and special circumstances of developing country Parties, especially those that are particularly vulnerable to the adverse effects of climate change, and of those Parties, especially developing country Parties, that would have to bear a disproportionate or abnormal burden under the Convention, should be given full consideration.
3. The Parties should take precautionary measures to anticipate, prevent or minimize the causes of climate change and mitigate its adverse effects. Where there are threats of serious or irreversible damage, lack of full scientific certainty should not be used as a reason for postponing such measures, taking into account that policies and measures to deal with climate change should be cost-effective so as to ensure global benefits at the lowest possible cost. To achieve this, such policies and measures should take into account different socio-economic contexts, be comprehensive, cover all relevant sources, sinks and reservoirs of greenhouse gases and adaptation, and comprise all economic sectors. Efforts to address climate change may be carried out cooperatively by interested Parties.
4. The Parties have a right to, and should, promote sustainable development. Policies and measures to protect the climate system against human-induced change should be appropriate for the specific conditions of each Party and should be integrated with national development programmes, taking into account that economic development is essential for adopting measures to address climate change.
5. The Parties should cooperate to promote a supportive and open international economic system that would lead to sustainable economic growth and development in all Parties, particularly developing country Parties, thus enabling them better to address the problems of climate change. Measures taken to combat climate change, including unilateral ones, should not constitute a means of arbitrary or unjustifiable discrimination or a disguised restriction on international trade.

Article 4: Commitments

1. All Parties, taking into account their common but differentiated responsibilities and their specific national and regional development priorities, objectives and circumstances, shall:
    (a) Develop, periodically update, publish and make available to the Conference of the Parties, in accordance with Article 12, national inventories of anthropogenic emissions by sources and removals by sinks of all greenhouse gases not controlled by the Montreal Protocol, using comparable methodologies to be agreed upon by the Conference of the Parties;
    (b) Formulate, implement, publish and regularly update national and, where appropriate, regional programmes containing measures to mitigate climate change by addressing anthropogenic emissions by sources and removals by sinks of all greenhouse gases not controlled by the Montreal Protocol, and measures to facilitate adequate adaptation to climate change;
    (c) Promote and cooperate in the development, application and diffusion, including transfer, of technologies, practices and processes that control, reduce or prevent anthropogenic emissions of greenhouse gases not controlled by the Montreal Protocol in all relevant sectors, including the energy, transport, industry, agriculture, forestry and waste management sectors;
    (d) Promote sustainable management, and promote and cooperate in the conservation and enhancement, as appropriate, of sinks and reservoirs of all greenhouse gases not controlled by the Montreal Protocol, including biomass, forests and oceans as well as other terrestrial, coastal and marine ecosystems;
    (e) Cooperate in preparing for adaptation to the impacts of climate change; develop and elaborate appropriate and integrated plans for coastal zone management, water resources and agriculture, and for the protection and rehabilitation of areas, particularly in Africa, affected by drought and desertification, as well as floods;
    (f) Take climate change considerations into account, to the extent feasible, in their relevant social, economic and environmental policies and actions, and employ appropriate methods, for example impact assessments, formulated and determined nationally, with a view to minimizing adverse effects on the economy, on public health and on the quality of the environment, of projects or measures undertaken by them to mitigate or adapt to climate change;
    (g) Promote and cooperate in scientific, technological, technical, socio-economic and other research, systematic observation and development of data archives related to the climate system and intended to further the understanding and to reduce or eliminate the remaining uncertainties regarding the causes, effects, magnitude and timing of climate change and the economic and social consequences of various response strategies;
    (h) Promote and cooperate in the full, open and prompt exchange of relevant scientific, technological, technical, socio-economic and legal information related to the climate system and climate change, and to the economic and social consequences of various response strategies;
    (i) Promote and cooperate in education, training and public awareness related

to climate change and encourage the widest participation in this process, including that of non-governmental organizations; and

(j) Communicate to the Conference of the Parties information related to implementation, in accordance with Article 12.

2. The developed country Parties and other Parties included in Annex I commit themselves specifically as provided for in the following:

(a) Each of these Parties shall adopt national1 policies and take corresponding measures on the mitigation of climate change, by limiting its anthropogenic emissions of greenhouse gases and protecting and enhancing its greenhouse gas sinks and reservoirs. These policies and measures will demonstrate that developed countries are taking the lead in modifying longer-term trends in anthropogenic emissions consistent with the objective of the Convention, recognizing that the return by the end of the present decade to earlier levels of anthropogenic emissions of carbon dioxide and other greenhouse gases not controlled by the Montreal Protocol would contribute to such modification, and taking into account the differences in these Parties' starting points and approaches, economic structures and resource bases, the need to maintain strong and sustainable economic growth, available technologies and other individual circumstances, as well as the need for equitable and appropriate contributions by each of these Parties to the global effort regarding that objective. These Parties may implement such policies and measures jointly with other Parties and may assist other Parties in contributing to the achievement of the objective of the Convention and, in particular, that of this subparagraph; This includes policies and measures adopted by regional economic integration organizations.

(b) In order to promote progress to this end, each of these Parties shall communicate, within six months of the entry into force of the Convention for it and periodically thereafter, and in accordance with Article 12, detailed information on its policies and measures referred to in subparagraph (a) above, as well as on its resulting projected anthropogenic emissions by sources and removals by sinks of greenhouse gases not controlled by the Montreal Protocol for the period referred to in subparagraph (a), with the aim of returning individually or jointly to their 1990 levels these anthropogenic emissions of carbon dioxide and other greenhouse gases not controlled by the Montreal Protocol. This information will be reviewed by the Conference of the Parties, at its first session and periodically thereafter, in accordance with Article 7;

(c) Calculations of emissions by sources and removals by sinks of greenhouse gases for the purposes of subparagraph (b) above should take into account the best available scientific knowledge, including of the effective capacity of sinks and the respective contributions of such gases to climate change. The Conference of the Parties shall consider and agree on methodologies for these calculations at its first session and review them regularly thereafter;

(d) The Conference of the Parties shall, at its first session, review the adequacy of subparagraphs (a) and (b) above. Such review shall be carried out in the light of the best available scientific information and assessment on climate change and its impacts, as well as relevant technical, social and economic information. Based on this review, the Conference of the Parties shall take appropriate action, which may include the adoption of amendments to the commitments in subparagraphs (a) and (b) above. The Conference of the Parties, at its first session, shall also take decisions regarding criteria for joint implementation as indicated in subparagraph (a) above. A second review of subparagraphs (a) and (b) shall take place not later than 31 December 1998, and thereafter at regular intervals determined by the Conference of the Parties, until the objective of the Convention is met;

(e) Each of these Parties shall: (i) coordinate as appropriate with other such Parties, relevant economic and administrative instruments developed to achieve the objective of the Convention; and (ii) identify and periodically review its own policies and practices which encourage activities that lead to greater levels of anthropogenic emissions of greenhouse gases not controlled by the Montreal Protocol than would otherwise occur;

(f) The Conference of the Parties shall review, not later than 31 December 1998, available information with a view to taking decisions regarding such amendments to the lists in Annexes I and II as may be appropriate, with the approval of the Party concerned;

(g) Any Party not included in Annex I may, in its instrument of ratification, acceptance, approval or accession, or at any time thereafter, notify the Depositary that it intends to be bound by subparagraphs (a) and (b) above. The Depositary shall inform the other signatories and Parties of any such notification.

3. The developed country Parties and other developed Parties included in Annex II shall provide new and additional financial resources to meet the agreed full costs incurred by developing country Parties in complying with their obligations under Article 12, paragraph 1. They shall also provide such financial resources, including for the transfer of technology, needed by the developing country Parties to meet the agreed full incremental costs of implementing measures that are covered by paragraph 1 of this Article and that are agreed between a developing country Party and the international entity or entities referred to in Article 11, in accordance with that Article. The implementation of these commitments shall take into account the need for adequacy and predictability in the flow of funds and the importance of appropriate burden sharing among the developed country Parties.

4. The developed country Parties and other developed Parties included in Annex II shall also assist the developing country Parties that are particularly vulnerable to the adverse effects of climate change in meeting costs of adaptation to those adverse effects.

5. The developed country Parties and other developed Parties included in Annex II shall take all practicable steps to promote, facilitate and finance, as appropriate, the transfer of, or access to, environmentally sound technologies and know-how to other Parties, particularly developing country Parties, to enable them to implement the provisions of the Convention. In this process, the developed country Parties shall support the development and enhancement of endogenous capacities and technologies of developing country Parties. Other Parties and organizations in a position to do so may also assist in facilitating the transfer of such technologies.

6. In the implementation of their commitments under paragraph 2 above, a certain degree of flexibility shall be allowed by the Conference of the Parties to the Parties included in Annex I undergoing the process of transition to a market economy, in

order to enhance the ability of these Parties to address climate change, including with regard to the historical level of anthropogenic emissions of greenhouse gases not controlled by the Montreal Protocol chosen as a reference.
7. The extent to which developing country Parties will effectively implement their commitments under the Convention will depend on the effective implementation by developed country Parties of their commitments under the Convention related to financial resources and transfer of technology and will take fully into account that economic and social development and poverty eradication are the first and overriding priorities of the developing country Parties.
8. In the implementation of the commitments in this Article, the Parties shall give full consideration to what actions are necessary under the Convention, including actions related to funding, insurance and the transfer of technology, to meet the specific needs and concerns of developing country Parties arising from the adverse effects of climate change and/or the impact of the implementation of response measures, especially on:
    (a) Small island countries;
    (b) Countries with low-lying coastal areas;
    (c) Countries with arid and semi-arid areas, forested areas and areas liable to forest decay;
    (d) Countries with areas prone to natural disasters;
    (e) Countries with areas liable to drought and desertification;
    (f) Countries with areas of high urban atmospheric pollution;
    (g) Countries with areas with fragile ecosystems, including mountainous ecosystems;
    (h) Countries whose economies are highly dependent on income generated from the production, processing and export, and/or on consumption of fossil fuels and associated energy-intensive products; and
    (i) Landlocked and transit countries.
    Further, the Conference of the Parties may take actions, as appropriate, with respect to this paragraph.
9. The Parties shall take full account of the specific needs and special situations of the least developed countries in their actions with regard to funding and transfer of technology.
10. The Parties shall, in accordance with Article 10, take into consideration in the implementation of the commitments of the Convention the situation of Parties, particularly developing country Parties, with economies that are vulnerable to the adverse effects of the implementation of measures to respond to climate change. This applies notably to Parties with economies that are highly dependent on income generated from the production, processing and export, and/or consumption of fossil fuels and associated energy-intensive products and/or the use of fossil fuels for which such Parties have serious difficulties in switching to alternatives.

Article 5: Research and Systematic Observation

In carrying out their commitments under Article 4, paragraph 1 (g), the Parties shall:

(a) Support and further develop, as appropriate, international and intergovernmental programmes and networks or organizations aimed at defining, conducting, assessing and financing research, data collection and systematic observation, taking into account the need to minimize duplication of effort;
(b) Support international and intergovernmental efforts to strengthen systematic observation and national scientific and technical research capacities and capabilities, particularly in developing countries, and to promote access to, and the exchange of, data and analyses thereof obtained from areas beyond national jurisdiction; and
(c) Take into account the particular concerns and needs of developing countries and cooperate in improving their endogenous capacities and capabilities to participate in the efforts referred to in subparagraphs (a) and (b) above.

Article 6: Education, Training and Public Awareness

In carrying out their commitments under Article 4, paragraph 1 (i), the Parties shall:

(a) Promote and facilitate at the national and, as appropriate, subregional and regional levels, and in accordance with national laws and regulations, and within their respective capacities:
   (i) the development and implementation of educational and public awareness programmes on climate change and its effects;
   (ii) public access to information on climate change and its effects;
   (iii) public participation in addressing climate change and its effects and developing adequate responses; and
   (iv) training of scientific, technical and managerial personnel;
(b) Cooperate in and promote, at the international level, and, where appropriate, using existing bodies:
   (i) the development and exchange of educational and public awareness material on climate change and its effects; and
   (ii) the development and implementation of education and training programmes, including the strengthening of national institutions and the exchange or secondment of personnel to train experts in this field, in particular for developing countries.

Article 7: Conference of the Parties

1. A Conference of the Parties is hereby established.
2. The Conference of the Parties, as the supreme body of this Convention, shall keep under regular review the implementation of the Convention and any related legal instruments that the Conference of the Parties may adopt, and shall make, within its mandate, the decisions necessary to promote the effective implementation of the Convention. To this end, it shall:
   (a) Periodically examine the obligations of the Parties and the institutional

arrangements under the Convention, in the light of the objective of the Convention, the experience gained in its implementation and the evolution of scientific and technological knowledge;

(b) Promote and facilitate the exchange of information on measures adopted by the Parties to address climate change and its effects, taking into account the differing circumstances, responsibilities and capabilities of the Parties and their respective commitments under the Convention;

(c) Facilitate, at the request of two or more Parties, the coordination of measures adopted by them to address climate change and its effects, taking into account the differing circumstances, responsibilities and capabilities of the Parties and their respective commitments under the Convention;

(d) Promote and guide, in accordance with the objective and provisions of the Convention, the development and periodic refinement of comparable methodologies, to be agreed on by the Conference of the Parties, inter alia, for preparing inventories of greenhouse gas emissions by sources and removals by sinks, and for evaluating the effectiveness of measures to limit the emissions and enhance the removals of these gases;

(e) Assess, on the basis of all information made available to it in accordance with the provisions of the Convention, the implementation of the Convention by the Parties, the overall effects of the measures taken pursuant to the Convention, in particular environmental, economic and social effects as well as their cumulative impacts and the extent to which progress towards the objective of the Convention is being achieved;

(f) Consider and adopt regular reports on the implementation of the Convention and ensure their publication;

(g) Make recommendations on any matters necessary for the implementation of the Convention;

(h) Seek to mobilize financial resources in accordance with Article 4, paragraphs 3, 4 and 5, and Article 11;

(i) Establish such subsidiary bodies as are deemed necessary for the implementation of the Convention;

(j) Review reports submitted by its subsidiary bodies and provide guidance to them;

(k) Agree upon and adopt, by consensus, rules of procedure and financial rules for itself and for any subsidiary bodies;

(l) Seek and utilize, where appropriate, the services and cooperation of, and information provided by, competent international organizations and inter-governmental and non-governmental bodies; and

(m) Exercise such other functions as are required for the achievement of the objective of the Convention as well as all other functions assigned to it under the Convention.

3. The Conference of the Parties shall, at its first session, adopt its own rules of procedure as well as those of the subsidiary bodies established by the Convention, which shall include decision-making procedures for matters not already covered by decision-making procedures stipulated in the Convention. Such procedures may include specified majorities required for the adoption of particular decisions.

4. The first session of the Conference of the Parties shall be convened by the interim

secretariat referred to in Article 21 and shall take place not later than one year after the date of entry into force of the Convention. Thereafter, ordinary sessions of the Conference of the Parties shall be held every year unless otherwise decided by the Conference of the Parties.

5. Extraordinary sessions of the Conference of the Parties shall be held at such other times as may be deemed necessary by the Conference, or at the written request of any Party, provided that, within six months of the request being communicated to the Parties by the secretariat, it is supported by at least one third of the Parties.

6. The United Nations, its specialized agencies and the International Atomic Energy Agency, as well as any State member thereof or observers thereto not Party to the Convention, may be represented at sessions of the Conference of the Parties as observers. Any body or agency, whether national or international, governmental or non-governmental, which is qualified in matters covered by the Convention, and which has informed the secretariat of its wish to be represented at a session of the Conference of the Parties as an observer, may be so admitted unless at least one third of the Parties present object. The admission and participation of observers shall be subject to the rules of procedure adopted by the Conference of the Parties.

Article 8: Secretariat

1. A secretariat is hereby established.
2. The functions of the secretariat shall be:
    (a) To make arrangements for sessions of the Conference of the Parties and its subsidiary bodies established under the Convention and to provide them with services as required;
    (b) To compile and transmit reports submitted to it;
    (c) To facilitate assistance to the Parties, particularly developing country Parties, on request, in the compilation and communication of information required in accordance with the provisions of the Convention;
    (d) To prepare reports on its activities and present them to the Conference of the Parties;
    (e) To ensure the necessary coordination with the secretariats of other relevant international bodies;
    (f) To enter, under the overall guidance of the Conference of the Parties, into such administrative and contractual arrangements as may be required for the effective discharge of its functions; and
    (g) To perform the other secretariat functions specified in the Convention and in any of its protocols and such other functions as may be determined by the Conference of the Parties.
3. The Conference of the Parties, at its first session, shall designate a permanent secretariat and make arrangements for its functioning.

Article 9: Subsidiary Body for Scientific and Technological Advice

1. A subsidiary body for scientific and technological advice is hereby established to provide the Conference of the Parties and, as appropriate, its other subsidiary bodies with timely information and advice on scientific and technological matters relating to the Convention. This body shall be open to participation by all Parties and shall be multidisciplinary. It shall comprise government representatives competent in the relevant field of expertise. It shall report regularly to the Conference of the Parties on all aspects of its work.
2. Under the guidance of the Conference of the Parties, and drawing upon existing competent international bodies, this body shall:
   (a) Provide assessments of the state of scientific knowledge relating to climate change and its effects;
   (b) Prepare scientific assessments on the effects of measures taken in the implementation of the Convention;
   (c) Identify innovative, efficient and state-of-the-art technologies and know-how and advise on the ways and means of promoting development and/or transferring such technologies;
   (d) Provide advice on scientific programmes, international cooperation in research and development related to climate change, as well as on ways and means of supporting endogenous capacity-building in developing countries; and
   (e) Respond to scientific, technological and methodological questions that the Conference of the Parties and its subsidiary bodies may put to the body.
3. The functions and terms of reference of this body may be further elaborated by the Conference of the Parties.

Article 10: Subsidiary Body for Implementation

1. A subsidiary body for implementation is hereby established to assist the Conference of the Parties in the assessment and review of the effective implementation of the Convention. This body shall be open to participation by all Parties and comprise government representatives who are experts on matters related to climate change. It shall report regularly to the Conference of the Parties on all aspects of its work.
2. Under the guidance of the Conference of the Parties, this body shall:
   (a) Consider the information communicated in accordance with Article 12, paragraph 1, to assess the overall aggregated effect of the steps taken by the Parties in the light of the latest scientific assessments concerning climate change;
   (b) Consider the information communicated in accordance with Article 12, paragraph 2, in order to assist the Conference of the Parties in carrying out the reviews required by Article 4, paragraph 2 (d); and
   (c) Assist the Conference of the Parties, as appropriate, in the preparation and implementation of its decisions.

Article 11: Financial Mechanism

1. A mechanism for the provision of financial resources on a grant or concessional basis, including for the transfer of technology, is hereby defined. It shall function under the guidance of and be accountable to the Conference of the Parties, which shall decide on its policies, programme priorities and eligibility criteria related to this Convention. Its operation shall be entrusted to one or more existing international entities.
2. The financial mechanism shall have an equitable and balanced representation of all Parties within a transparent system of governance.
3. The Conference of the Parties and the entity or entities entrusted with the operation of the financial mechanism shall agree upon arrangements to give effect to the above paragraphs, which shall include the following:
    (a) Modalities to ensure that the funded projects to address climate change are in conformity with the policies, programme priorities and eligibility criteria established by the Conference of the Parties;
    (b) Modalities by which a particular funding decision may be reconsidered in light of these policies, programme priorities and eligibility criteria;
    (c) Provision by the entity or entities of regular reports to the Conference of the Parties on its funding operations, which is consistent with the requirement for accountability set out in paragraph 1 above; and
    (d) Determination in a predictable and identifiable manner of the amount of funding necessary and available for the implementation of this Convention and the conditions under which that amount shall be periodically reviewed.
4. The Conference of the Parties shall make arrangements to implement the above-mentioned provisions at its first session, reviewing and taking into account the interim arrangements referred to in Article 21, paragraph 3, and shall decide whether these interim arrangements shall be maintained. Within four years thereafter, the Conference of the Parties shall review the financial mechanism and take appropriate measures.
5. The developed country Parties may also provide and developing country Parties avail themselves of, financial resources related to the implementation of the Convention through bilateral, regional and other multilateral channels.

Article 12: Communication of Information Related to Implementation

1. In accordance with Article 4, paragraph 1, each Party shall communicate to the Conference of the Parties, through the secretariat, the following elements of information:
    (a) A national inventory of anthropogenic emissions by sources and removals by sinks of all greenhouse gases not controlled by the Montreal Protocol, to the extent its capacities permit, using comparable methodologies to be promoted and agreed upon by the Conference of the Parties;
    (b) A general description of steps taken or envisaged by the Party to implement the Convention; and
    (c) Any other information that the Party considers relevant to the achievement of the objective of the Convention and suitable for inclusion in its commu-

nication, including, if feasible, material relevant for calculations of global emission trends.
2. Each developed country Party and each other Party included in Annex I shall incorporate in its communication the following elements of information:
    (a) A detailed description of the policies and measures that it has adopted to implement its commitment under Article 4, paragraphs 2 (a) and 2 (b); and
    (b) A specific estimate of the effects that the policies and measures referred to in subparagraph (a) immediately above will have on anthropogenic emissions by its sources and removals by its sinks of greenhouse gases during the period referred to in Article 4, paragraph 2 (a).
3. In addition, each developed country Party and each other developed Party included in Annex II shall incorporate details of measures taken in accordance with Article 4, paragraphs 3, 4 and 5. 4. Developing country Parties may, on a voluntary basis, propose projects for financing, including specific technologies, materials, equipment, techniques or practices that would be needed to implement such projects, along with, if possible, an estimate of all incremental costs, of the reductions of emissions and increments of removals of greenhouse gases, as well as an estimate of the consequent benefits.
5. Each developed country Party and each other Party included in Annex I shall make its initial communication within six months of the entry into force of the Convention for that Party. Each Party not so listed shall make its initial communication within three years of the entry into force of the Convention for that Party, or of the availability of financial resources in accordance with Article 4, paragraph 3. Parties that are least developed countries may make their initial communication at their discretion. The frequency of subsequent communications by all Parties shall be determined by the Conference of the Parties, taking into account the differentiated timetable set by this paragraph.
6. Information communicated by Parties under this Article shall be transmitted by the secretariat as soon as possible to the Conference of the Parties and to any subsidiary bodies concerned. If necessary, the procedures for the communication of information may be further considered by the Conference of the Parties.
7. From its first session, the Conference of the Parties shall arrange for the provision to developing country Parties of technical and financial support, on request, in compiling and communicating information under this Article, as well as in identifying the technical and financial needs associated with proposed projects and response measures under Article 4. Such support may be provided by other Parties, by competent international organizations and by the secretariat, as appropriate.
8. Any group of Parties may, subject to guidelines adopted by the Conference of the Parties, and to prior notification to the Conference of the Parties, make a joint communication in fulfilment of their obligations under this Article, provided that such a communication includes information on the fulfilment by each of these Parties of its individual obligations under the Convention.
9. Information received by the secretariat that is designated by a Party as confidential, in accordance with criteria to be established by the Conference of the Parties, shall be aggregated by the secretariat to protect its confidentiality before being made available to any of the bodies involved in the communication and review of information.
10. Subject to paragraph 9 above, and without prejudice to the ability of any Party to

make public its communication at any time, the secretariat shall make communications by Parties under this Article publicly available at the time they are submitted to the Conference of the Parties.

Article 13: Resolution of Questions Regarding Implementation

The Conference of the Parties shall, at its first session, consider the establishment of a multilateral consultative process, available to Parties on their request, for the resolution of questions regarding the implementation of the Convention.

Article 14: Settlement of Disputes

1. In the event of a dispute between any two or more Parties concerning the interpretation or application of the Convention, the Parties concerned shall seek a settlement of the dispute through negotiation or any other peaceful means of their own choice.
2. When ratifying, accepting, approving or acceding to the Convention, or at any time thereafter, a Party which is not a regional economic integration organization may declare in a written instrument submitted to the Depositary that, in respect of any dispute concerning the interpretation or application of the Convention, it recognizes as compulsory ipso facto and without special agreement, in relation to any Party accepting the same obligation:
    (a) Submission of the dispute to the International Court of Justice; and/or
    (b) Arbitration in accordance with procedures to be adopted by the Conference of the Parties as soon as practicable, in an annex on arbitration. A Party which is a regional economic integration organization may make a declaration with like effect in relation to arbitration in accordance with the procedures referred to in subparagraph (b) above.
3. A declaration made under paragraph 2 above shall remain in force until it expires in accordance with its terms or until three months after written notice of its revocation has been deposited with the Depositary.
4. A new declaration, a notice of revocation or the expiry of a declaration shall not in any way affect proceedings pending before the International Court of Justice or the arbitral tribunal, unless the parties to the dispute otherwise agree.
5. Subject to the operation of paragraph 2 above, if after twelve months following notification by one Party to another that a dispute exists between them, the Parties concerned have not been able to settle their dispute through the means mentioned in paragraph 1 above, the dispute shall be submitted, at the request of any of the parties to the dispute, to conciliation.
6. A conciliation commission shall be created upon the request of one of the parties to the dispute. The commission shall be composed of an equal number of members appointed by each party concerned and a chairman chosen jointly by the members appointed by each party. The commission shall render a recommendatory award, which the parties shall consider in good faith.

7. Additional procedures relating to conciliation shall be adopted by the Conference of the Parties, as soon as practicable, in an annex on conciliation.
8. The provisions of this Article shall apply to any related legal instrument which the Conference of the Parties may adopt, unless the instrument provides otherwise.

Article 15: Amendments to the Convention

1. Any Party may propose amendments to the Convention.
2. Amendments to the Convention shall be adopted at an ordinary session of the Conference of the Parties. The text of any proposed amendment to the Convention shall be communicated to the Parties by the secretariat at least six months before the meeting at which it is proposed for adoption. The secretariat shall also communicate proposed amendments to the signatories to the Convention and, for information, to the Depositary.
3. The Parties shall make every effort to reach agreement on any proposed amendment to the Convention by consensus. If all efforts at consensus have been exhausted, and no agreement reached, the amendment shall as a last resort be adopted by a three-fourths majority vote of the Parties present and voting at the meeting. The adopted amendment shall be communicated by the secretariat to the Depositary, who shall circulate it to all Parties for their acceptance.
4. Instruments of acceptance in respect of an amendment shall be deposited with the Depositary. An amendment adopted in accordance with paragraph 3 above shall enter into force for those Parties having accepted it on the ninetieth day after the date of receipt by the Depositary of an instrument of acceptance by at least three fourths of the Parties to the Convention.
5. The amendment shall enter into force for any other Party on the ninetieth day after the date on which that Party deposits with the Depositary its instrument of acceptance of the said amendment.
6. For the purposes of this Article, "Parties present and voting" means Parties present and casting an affirmative or negative vote.

Article 16: Adoption and Amendment of Annexes to the Convention

1. Annexes to the Convention shall form an integral part thereof and, unless otherwise expressly provided, a reference to the Convention constitutes at the same time a reference to any annexes thereto. Without prejudice to the provisions of Article 14, paragraphs 2 (b) and 7, such annexes shall be restricted to lists, forms and any other material of a descriptive nature that is of a scientific, technical, procedural or administrative character.
2. Annexes to the Convention shall be proposed and adopted in accordance with the procedure set forth in Article 15, paragraphs 2, 3 and 4.
3. An annex that has been adopted in accordance with paragraph 2 above shall enter into force for all Parties to the Convention six months after the date of the communication

by the Depositary to such Parties of the adoption of the annex, except for those Parties that have notified the Depositary, in writing, within that period of their non-acceptance of the annex. The annex shall enter into force for Parties which withdraw their notification of non-acceptance on the ninetieth day after the date on which withdrawal of such notification has been received by the Depositary.
4. The proposal, adoption and entry into force of amendments to annexes to the Convention shall be subject to the same procedure as that for the proposal, adoption and entry into force of annexes to the Convention in accordance with paragraphs 2 and 3 above.
5. If the adoption of an annex or an amendment to an annex involves an amendment to the Convention, that annex or amendment to an annex shall not enter into force until such time as the amendment to the Convention enters into force.

Article 17: Protocols

1. The Conference of the Parties may, at any ordinary session, adopt protocols to the Convention.
2. The text of any proposed protocol shall be communicated to the Parties by the secretariat at least six months before such a session.
3. The requirements for the entry into force of any protocol shall be established by that instrument.
4. Only Parties to the Convention may be Parties to a protocol.
5. Decisions under any protocol shall be taken only by the Parties to the protocol concerned.

Article 18: Right to Vote

1. Each Party to the Convention shall have one vote, except as provided for in paragraph 2 below.
2. Regional economic integration organizations, in matters within their competence, shall exercise their right to vote with a number of votes equal to the number of their member States that are Parties to the Convention. Such an organization shall not exercise its right to vote if any of its member States exercises its right, and vice versa.

Article 19: Depositary

The Secretary-General of the United Nations shall be the Depositary of the Convention and of protocols adopted in accordance with Article 17.

Article 20: Signature

This Convention shall be open for signature by States Members of the United Nations or of any of its specialized agencies or that are Parties to the Statute of the International

Court of Justice and by regional economic integration organizations at Rio de Janeiro, during the United Nations Conference on Environment and Development, and thereafter at United Nations Headquarters in New York from 20 June 1992 to 19 June 1993.

Article 21: Interim Arrangements

1. The secretariat functions referred to in Article 8 will be carried out on an interim basis by the secretariat established by the General Assembly of the United Nations in its resolution 45/212 of 21 December 1990, until the completion of the first session of the Conference of the Parties.
2. The head of the interim secretariat referred to in paragraph 1 above will cooperate closely with the Intergovernmental Panel on Climate Change to ensure that the Panel can respond to the need for objective scientific and technical advice. Other relevant scientific bodies could also be consulted.
3. The Global Environment Facility of the United Nations Development Programme, the United Nations Environment Programme and the International Bank for Reconstruction and Development shall be the international entity entrusted with the operation of the financial mechanism referred to in Article 11 on an interim basis. In this connection, the Global Environment Facility should be appropriately restructured and its membership made universal to enable it to fulfil the requirements of Article 11.

Article 22: Ratification, Acceptance, Approval or Accession

1. The Convention shall be subject to ratification, acceptance, approval or accession by States and by regional economic integration organizations. It shall be open for accession from the day after the date on which the Convention is closed for signature. Instruments of ratification, acceptance, approval or accession shall be deposited with the Depositary.
2. Any regional economic integration organization which becomes a Party to the Convention without any of its member States being a Party shall be bound by all the obligations under the Convention. In the case of such organizations, one or more of whose member States is a Party to the Convention, the organization and its member States shall decide on their respective responsibilities for the performance of their obligations under the Convention. In such cases, the organization and the member States shall not be entitled to exercise rights under the Convention concurrently.
3. In their instruments of ratification, acceptance, approval or accession, regional economic integration organizations shall declare the extent of their competence with respect to the matters governed by the Convention. These organizations shall also inform the Depositary, who shall in turn inform the Parties, of any substantial modification in the extent of their competence.

Article 23: Entry Into Force

1. The Convention shall enter into force on the ninetieth day after the date of deposit of the fiftieth instrument of ratification, acceptance, approval or accession.
2. For each State or regional economic integration organization that ratifies, accepts or approves the Convention or accedes thereto after the deposit of the fiftieth instrument of ratification, acceptance, approval or accession, the Convention shall enter into force on the ninetieth day after the date of deposit by such State or regional economic integration organization of its instrument of ratification, acceptance, approval or accession.
3. For the purposes of paragraphs 1 and 2 above, any instrument deposited by a regional economic integration organization shall not be counted as additional to those deposited by States members of the organization.

Article 24: Reservations

No reservations may be made to the Convention.

Article 25: Withdrawal

1. At any time after three years from the date on which the Convention has entered into force for a Party, that Party may withdraw from the Convention by giving written notification to the Depositary.
2. Any such withdrawal shall take effect upon expiry of one year from the date of receipt by the Depositary of the notification of withdrawal, or on such later date as may be specified in the notification of withdrawal.
3. Any Party that withdraws from the Convention shall be considered as also having withdrawn from any protocol to which it is a Party.

Article 26: Authentic Texts

The original of this Convention, of which the Arabic, Chinese, English, French, Russian and Spanish texts are equally authentic, shall be deposited with the Secretary-General of the United Nations. IN WITNESS WHEREOF the undersigned, being duly authorized to that effect, have signed this Convention. DONE at New York this ninth day of May one thousand nine hundred and ninety-two.

Analysis

The UNFCCC began the process of regulating the emission of GHGs that culminated with the passage of the Kyoto Protocol. As such, it did not include any legally binding commitments. It did not even define what constituted GHGs, although it did make specific mention of carbon dioxide, hydrofluorocarbons, methane, nitrogen oxide, perfluorocarbons, and sulfur hexafluoride. This was intentional so that any GHGs identified in future years were automatically included under the terms of the agreement. The major benefit provided by the UNFCCC was that it provided a framework for negotiations so that the international community could reach the consensus required to draft and pass a legally binding agreement to reduce the amount of GHGs emitted into the atmosphere.

Further Reading

Afionis, Stavros. 2016. *The European Union in International Climate Change Negotiations*. New York: Routledge.

Eastwood, Lauren. 2016. *Negotiating the Environment: Civil Society, Globalization, and the UN*. New York: Routledge.

Victor, David G. 2001. *The Collapse of the Kyoto Protocol and the Struggle to Slow Global Warming*. Princeton, NJ: Princeton University Press.

Weart, Spencer R. 2008. *The Discovery of Global Warming*. Rev. and expanded ed. Cambridge, MA: Harvard University Press.

Senate Resolution 98
[Report No. 105–54]
(Byrd-Hagel Resolution)

Date: July 25, 1997
Location: Washington, D.C.
Significance: This resolution, supported by a bipartisan group of Senators, warned President William Clinton's administration and the world community that the agreement being negotiated in Kyoto, Japan concerning greenhouse gas emissions was not going to be ratified by the Senate if the terms did not apply equally to all countries.
Source: 105th Congress, 1st Session, United States Senate. 1997. *Senate Resolution 98 [Report No. 105–54].* http://www.gpo.gov/fdsys/pkg/BILLS-105sres98ats/pdf/BILLS-105sres98ats.pdf (accessed July 18, 2015).

IN THE SENATE OF THE UNITED STATES
Mr. BYRD (for himself, Mr. HAGEL, Mr. HOLLINGS, Mr. CRAIG, Mr. INOUYE, Mr. WARNER, Mr. FORD, Mr. THOMAS, Mr. DORGAN, Mr. HELMS, Mr. LEVIN, Mr. ROBERTS, Mr. ABRAHAM, Mr. MCCONNELL, Mr. ASHCROFT, Mr. BROWNBACK, Mr. KEMPTHORNE, Mr. THURMOND, Mr. BURNS, Mr. CONRAD, Mr. GLENN, Mr. ENZI, Mr. INHOFE, Mr. BOND, Mr. COVERDELL, Mr. DEWINE, Mrs. HUTCHISON, Mr. GORTON, Mr. HATCH, Mr. BREAUX, Mr. CLELAND, Mr. DURBIN, Mr. HUTCHINSON, Mr. JOHNSON, Ms. LANDRIEU, Ms. MIKULSKI, Mr. NICKLES, Mr. SANTORUM, Mr. SHELBY, Mr. SMITH of Oregon, Mr. BENNETT, Mr. FAIRCLOTH, Mr. FRIST, Mr. GRASSLEY, Mr. ALLARD, Mr. MURKOWSKI, Mr. AKAKA, Mr. COATS, Mr. COCHRAN, Mr. DOMENICI, Mr. GRAMM, Mr. GRAMS, Mr. LOTT, Ms. MOSELEY-BRAUN, Mr. ROBB, Mr. ROCKEFELLER, Mr. SESSIONS, Mr. SMITH of New Hampshire, Mr. SPECTER, Mr. STEVENS, Mr. LUGAR, Mr. REID, Mr. BRYAN, Mr. THOMPSON, and Mr. CAMPBELL) submitted the following resolution; which was referred to the Committee on Foreign Relations
JULY 21, 1997 Reported by Mr. HELMS, without amendment
JULY 25, 1997 Considered and agreed to

Resolution

Expressing the sense of the Senate regarding the conditions for the United States becoming a signatory to any international agreement on greenhouse gas emissions under the United Nations Framework Convention on Climate Change.

Whereas the United Nations Framework Convention on Climate Change (in this resolution referred to as the "Convention"), adopted in May 1992, entered into force in 1994 and is not yet fully implemented;

Whereas the Convention, intended to address climate change on a global basis, identifies the former Soviet Union and the countries of Eastern Europe and the Organization For Economic Co-operation and Development (OECD), including the United States, as "Annex I Parties", and the remaining 129 countries, including China, Mexico, India, Brazil, and South Korea, as "Developing Country Parties"; Whereas in April 1995, the Convention's "Conference of the Parties" adopted the so-called "Berlin Mandate";

Whereas the "Berlin Mandate" calls for the adoption, as soon as December 1997, in Kyoto, Japan, of a protocol or another legal instrument that strengthens commitments to limit greenhouse gas emissions by Annex I Parties for the post–2000 period and establishes a negotiation process called the "Ad Hoc Group on the Berlin Mandate";

Whereas the "Berlin Mandate" specifically exempts all Developing Country Parties from any new commitments in such negotiation process for the post–2000 period;

Whereas although the Convention, approved by the United States Senate, called on all signatory parties to adopt policies and programs aimed at limiting their greenhouse gas (GHG) emissions, in July 1996 the Undersecretary of State for Global Affairs called for the first time for "legally binding" emission limitation targets and timetables for Annex I Parties, a position reiterated by the Secretary of State in testimony before the Committee on Foreign Relations of the Senate on January 8, 1997;

Whereas greenhouse gas emissions of Developing Country Parties are rapidly increasing and are expected to surpass emissions of the United States and other OECD countries as early as 2015;

Whereas the Department of State has declared that it is critical for the Parties to the Convention to include Developing Country Parties in the next steps for global action and, therefore, has proposed that consideration of additional steps to include limitations on Developing Country Parties' greenhouse gas emissions would not begin until after a protocol or other legal instrument is adopted in Kyoto, Japan in December 1997;

Whereas the exemption for Developing Country Parties is inconsistent with the need for global action on climate change and is environmentally flawed;

Whereas the Senate strongly believes that the proposals under negotiation, because of the disparity of treatment between Annex I Parties and Developing Countries and the level of required emission reductions, could result in serious harm to the United States economy, including significant job loss, trade disadvantages, increased energy and consumer costs, or any combination thereof; and

Whereas it is desirable that a bipartisan group of Senators be appointed by the Majority and Minority Leaders of the Senate for the purpose of monitoring the status of negotiations on Global Climate Change and reporting periodically to the Senate on those negotiations: Now, therefore, be it

Resolved, That it is the sense of the Senate that—

(1) the United States should not be a signatory to any protocol to, or other agreement regarding, the United Nations Framework Convention on Climate Change of 1992, at negotiations in Kyoto in December 1997, or thereafter, which would—

(A) mandate new commitments to limit or 8 reduce greenhouse gas emissions for the Annex I Parties, unless the protocol or other agreement also mandates new specific scheduled commitments to limit or reduce greenhouse gas emissions for Developing Country Parties within the same compliance period, or

(B) would result in serious harm to the economy of the United States; and

(2) any such protocol or other agreement which would require the advice and consent of the Senate to ratification should be accompanied by a detailed explanation of any legislation or regulatory actions that may be required to implement the protocol or other agreement and should also be accompanied by an analysis of the detailed financial costs and other impacts on the economy of the United States which would be incurred by the implementation of the protocol or other agreement.

SEC. 2. The Secretary of the Senate shall transmit a copy of this resolution to the President.

Analysis

At the initial Conference of the Parties, held March 28–April 7, 1995, the Berlin Mandate (BM) was enacted as part of the United Nations Framework Convention on Climate Change. Although it was an international agreement to stabilize the amount of greenhouse gas (GHG) emitted globally, the BM was little more than a joint decision to take action. It did not call on specific countries to reduce their output of GHGs. The agreement also did not determine what level of GHG emissions were required to prevent damage to the global climate system. The one thing that was notable in the BM was that it called for legally binding standards to be set at the 1997 Conference of the Parties to be held in Kyoto, Japan. From the rhetoric at the Berlin meeting, it was apparent that the developing world believed that it should be exempt from limitations on GHG output since they had not been responsible for the excessive emissions since the Industrial Revolution that had resulted in global climate change. In their eyes, countries like the Australia, Great Britain, and the United States were responsible for the problem and should thus bear the brunt of the reduction in GHGs. Understandably, the United States Senate did not agree with that reasoning.

The Byrd-Hagel Resolution, which was introduced by Senator Robert Byrd (Democrat, West Virginia) and Senator Chuck Hagel (Republican, Nebraska), was a bipartisan repudiation of the BM. Passed by a vote of 95–0, it signaled to both President William Clinton's administration and the international community that the United States Senate would not ratify the document produced in Kyoto, Japan, unless all countries were treated equally. The United States was not going to suffer economic damage unless the pain was shared globally. Despite the clear warning, the Clinton administration worked with its counterparts at Kyoto to produce the very agreements that the Senate had already deemed unacceptable. If they believed the Byrd-Hagel resolution was a bluff, they soon learned that the will of the Senate had been accurately represented. Rather than suffer a humiliating defeat, the Clinton administration ultimately decided not to submit the Kyoto Protocol to the Senate for ratification.

Further Reading

Fisher, Dana R. 2006. "Bringing the Material Back In: Understanding the U.S. Position on Climate Change." *Sociological Forum* 21: 467–494.

Hagel, Chuck. 1997. "Kyoto: The Political Realities." *Institute of Public Affairs Review* 50: 15–17.

Hovi, John, Detlef F. Sprinz, and Guri Bang. 2012. "Why the United States Did Not Become a Party to the Kyoto Protocol: German, Norwegian, and US Perspectives." *European Journal of International Relations* 18: 129–150.

Kyoto Protocol to the United Nations Framework Convention on Climate Change

Date: December 11, 1997
Location: Kyoto, Japan
Significance: Although strongly supported by President William Clinton's administration, the United States did not become a party to the Kyoto Protocol.
Source: United Nations. 2008. *Kyoto Protocol to the United Nations Framework Convention on Climate Change.* http://unfccc.int/resource/docs/convkp/kpeng.pdf (accessed July 26, 2015).

The Parties to this Protocol,
Being Parties to the United Nations Framework Convention on Climate Change, hereinafter referred to as "the Convention,"
In pursuit of the ultimate objective of the Convention as stated in its Article 2,
Recalling the provisions of the Convention,
Being guided by Article 3 of the Convention,
Pursuant to the Berlin Mandate adopted by decision 1/CP.1 of the Conference of the Parties to the Convention at its first session,
Have agreed as follows:

Article 1

For the purposes of this Protocol, the definitions contained in Article 1 of the Convention shall apply. In addition:

1. "Conference of the Parties" means the Conference of the Parties to the Convention.
2. "Convention" means the United Nations Framework Convention on Climate Change, adopted in New York on 9 May 1992.
3. "Intergovernmental Panel on Climate Change" means the Intergovernmental Panel on Climate Change established in 1988 jointly by the World Meteorological Organization and the United Nations Environment Programme.
4. "Montreal Protocol" means the Montreal Protocol on Substances that Deplete the Ozone Layer, adopted in Montreal on 16 September 1987 and as subsequently adjusted and amended.
5. "Parties present and voting" means Parties present and casting an affirmative or negative vote.

6. "Party" means, unless the context otherwise indicates, a Party to this Protocol.
7. "Party included in Annex I" means a Party included in Annex I to the Convention, as may be amended, or a Party which has made a notification under Article 4, paragraph 2 (g), of the Convention.

Article 2

1. Each Party included in Annex I, in achieving its quantified emission limitation and reduction commitments under Article 3, in order to promote sustainable development, shall:
    (a) Implement and/or further elaborate policies and measures in accordance with its national circumstances, such as:
        (i) Enhancement of energy efficiency in relevant sectors of the national economy;
        (ii) Protection and enhancement of sinks and reservoirs of greenhouse gases not controlled by the Montreal Protocol, taking into account its commitments under relevant international environmental agreements; promotion of sustainable forest management practices, afforestation and reforestation;
        (iii) Promotion of sustainable forms of agriculture in light of climate change considerations;
        (iv) Research on, and promotion, development and increased use of, new and renewable forms of energy, of carbon dioxide sequestration technologies and of advanced and innovative environmentally sound technologies;
        (v) Progressive reduction or phasing out of market imperfections, fiscal incentives, tax and duty exemptions and subsidies in all greenhouse gas emitting sectors that run counter to the objective of the Convention and application of market instruments;
        (vi) Encouragement of appropriate reforms in relevant sectors aimed at promoting policies and measures which limit or reduce emissions of greenhouse gases not controlled by the Montreal Protocol;
        (vii) Measures to limit and/or reduce emissions of greenhouse gases not controlled by the Montreal Protocol in the transport sector;
        (viii) Limitation and/or reduction of methane emissions through recovery and use in waste management, as well as in the production, transport and distribution of energy;
    (b) Cooperate with other such Parties to enhance the individual and combined effectiveness of their policies and measures adopted under this Article, pursuant to Article 4, paragraph 2 (e) (i), of the Convention. To this end, these Parties shall take steps to share their experience and exchange information on such policies and measures, including developing ways of improving their comparability, transparency and effectiveness. The Conference of the Parties serving as the meeting of the Parties to this Protocol shall, at its first session or as soon as practicable thereafter, consider ways to facilitate such cooperation, taking into account all relevant information.

2. The Parties included in Annex I shall pursue limitation or reduction of emissions of greenhouse gases not controlled by the Montreal Protocol from aviation and marine bunker fuels, working through the International Civil Aviation Organization and the International Maritime Organization, respectively.
3. The Parties included in Annex I shall strive to implement policies and measures under this Article in such a way as to minimize adverse effects, including the adverse effects of climate change, effects on international trade, and social, environmental and economic impacts on other Parties, especially developing country Parties and in particular those identified in Article 4, paragraphs 8 and 9, of the Convention, taking into account Article 3 of the Convention. The Conference of the Parties serving as the meeting of the Parties to this Protocol may take further action, as appropriate, to promote the implementation of the provisions of this paragraph.
4. The Conference of the Parties serving as the meeting of the Parties to this Protocol, if it decides that it would be beneficial to coordinate any of the policies and measures in paragraph 1 (a) above, taking into account different national circumstances and potential effects, shall consider ways and means to elaborate the coordination of such policies and measures.

Article 3

1. The Parties included in Annex I shall, individually or jointly, ensure that their aggregate anthropogenic carbon dioxide equivalent emissions of the greenhouse gases listed in Annex A do not exceed their assigned amounts, calculated pursuant to their quantified emission limitation and reduction commitments inscribed in Annex B and in accordance with the provisions of this Article, with a view to reducing their overall emissions of such gases by at least 5 per cent below 1990 levels in the commitment period 2008 to 2012.
2. Each Party included in Annex I shall, by 2005, have made demonstrable progress in achieving its commitments under this Protocol.
3. The net changes in greenhouse gas emissions by sources and removals by sinks resulting from direct human-induced land-use change and forestry activities, limited to afforestation, reforestation and deforestation since 1990, measured as verifiable changes in carbon stocks in each commitment period, shall be used to meet the commitments under this Article of each Party included in Annex I. The greenhouse gas emissions by sources and removals by sinks associated with those activities shall be reported in a transparent and verifiable manner and reviewed in accordance with Articles 7 and 8.
4. Prior to the first session of the Conference of the Parties serving as the meeting of the Parties to this Protocol, each Party included in Annex I shall provide, for consideration by the Subsidiary Body for Scientific and Technological Advice, data to establish its level of carbon stocks in 1990 and to enable an estimate to be made of its changes in carbon stocks in subsequent years. The Conference of the Parties serving as the meeting of the Parties to this Protocol shall, at its first session or as soon as practicable thereafter, decide upon modalities, rules and guidelines as to how, and which, additional human-induced activities related to changes in

greenhouse gas emissions by sources and removals by sinks in the agricultural soils and the land-use change and forestry categories shall be added to, or subtracted from, the assigned amounts for Parties included in Annex I, taking into account uncertainties, transparency in reporting, verifiability, the methodological work of the Intergovernmental Panel on Climate Change, the advice provided by the Subsidiary Body for Scientific and Technological Advice in accordance with Article 5 and the decisions of the Conference of the Parties. Such a decision shall apply in the second and subsequent commitment periods. A Party may choose to apply such a decision on these additional human-induced activities for its first commitment period, provided that these activities have taken place since 1990.

5. The Parties included in Annex I undergoing the process of transition to a market economy whose base year or period was established pursuant to decision 9/CP.2 of the Conference of the Parties at its second session shall use that base year or period for the implementation of their commitments under this Article. Any other Party included in Annex I undergoing the process of transition to a market economy which has not yet submitted its first national communication under Article 12 of the Convention may also notify the Conference of the Parties serving as the meeting of the Parties to this Protocol that it intends to use an historical base year or period other than 1990 for the implementation of its commitments under this Article. The Conference of the Parties serving as the meeting of the Parties to this Protocol shall decide on the acceptance of such notification.

6. Taking into account Article 4, paragraph 6, of the Convention, in the implementation of their commitments under this Protocol other than those under this Article, a certain degree of flexibility shall be allowed by the Conference of the Parties serving as the meeting of the Parties to this Protocol to the Parties included in Annex I undergoing the process of transition to a market economy.

7. In the first quantified emission limitation and reduction commitment period, from 2008 to 2012, the assigned amount for each Party included in Annex I shall be equal to the percentage inscribed for it in Annex B of its aggregate anthropogenic carbon dioxide equivalent emissions of the greenhouse gases listed in Annex A in 1990, or the base year or period determined in accordance with paragraph 5 above, multiplied by five. Those Parties included in Annex I for whom land-use change and forestry constituted a net source of greenhouse gas emissions in 1990 shall include in their 1990 emissions base year or period the aggregate anthropogenic carbon dioxide equivalent emissions by sources minus removals by sinks in 1990 from land-use change for the purposes of calculating their assigned amount.

8. Any Party included in Annex I may use 1995 as its base year for hydrofluorocarbons, perfluorocarbons and sulphur hexafluoride, for the purposes of the calculation referred to in paragraph 7 above.

9. Commitments for subsequent periods for Parties included in Annex I shall be established in amendments to Annex B to this Protocol, which shall be adopted in accordance with the provisions of Article 21, paragraph 7. The Conference of the Parties serving as the meeting of the Parties to this Protocol shall initiate the consideration of such commitments at least seven years before the end of the first commitment period referred to in paragraph 1 above.

10. Any emission reduction units, or any part of an assigned amount, which a Party

acquires from another Party in accordance with the provisions of Article 6 or of Article 17 shall be added to the assigned amount for the acquiring Party.

11. Any emission reduction units, or any part of an assigned amount, which a Party transfers to another Party in accordance with the provisions of Article 6 or of Article 17 shall be subtracted from the assigned amount for the transferring Party.
12. Any certified emission reductions which a Party acquires from another Party in accordance with the provisions of Article 12 shall be added to the assigned amount for the acquiring Party.
13. If the emissions of a Party included in Annex I in a commitment period are less than its assigned amount under this Article, this difference shall, on request of that Party, be added to the assigned amount for that Party for subsequent commitment periods.
14. Each Party included in Annex I shall strive to implement the commitments mentioned in paragraph 1 above in such a way as to minimize adverse social, environmental and economic impacts on developing country Parties, particularly those identified in Article 4, paragraphs 8 and 9, of the Convention. In line with relevant decisions of the Conference of the Parties on the implementation of those paragraphs, the Conference of the Parties serving as the meeting of the Parties to this Protocol shall, at its first session, consider what actions are necessary to minimize the adverse effects of climate change and/or the impacts of response measures on Parties referred to in those paragraphs. Among the issues to be considered shall be the establishment of funding, insurance and transfer of technology.

Article 4

1. Any Parties included in Annex I that have reached an agreement to fulfil their commitments under Article 3 jointly, shall be deemed to have met those commitments provided that their total combined aggregate anthropogenic carbon dioxide equivalent emissions of the greenhouse gases listed in Annex A do not exceed their assigned amounts calculated pursuant to their quantified emission limitation and reduction commitments inscribed in Annex B and in accordance with the provisions of Article 3. The respective emission level allocated to each of the Parties to the agreement shall be set out in that agreement.
2. The Parties to any such agreement shall notify the secretariat of the terms of the agreement on the date of deposit of their instruments of ratification, acceptance or approval of this Protocol, or accession thereto. The secretariat shall in turn inform the Parties and signatories to the Convention of the terms of the agreement.
3. Any such agreement shall remain in operation for the duration of the commitment period specified in Article 3, paragraph 7.
4. If Parties acting jointly do so in the framework of, and together with, a regional economic integration organization, any alteration in the composition of the organization after adoption of this Protocol shall not affect existing commitments under this Protocol. Any alteration in the composition of the organization shall only apply for the purposes of those commitments under Article 3 that are adopted subsequent to that alteration.
5. In the event of failure by the Parties to such an agreement to achieve their total

combined level of emission reductions, each Party to that agreement shall be responsible for its own level of emissions set out in the agreement.
6. If Parties acting jointly do so in the framework of, and together with, a regional economic integration organization which is itself a Party to this Protocol, each member State of that regional economic integration organization individually, and together with the regional economic integration organization acting in accordance with Article 24, shall, in the event of failure to achieve the total combined level of emission reductions, be responsible for its level of emissions as notified in accordance with this Article.

Article 5

1. Each Party included in Annex I shall have in place, no later than one year prior to the start of the first commitment period, a national system for the estimation of anthropogenic emissions by sources and removals by sinks of all greenhouse gases not controlled by the Montreal Protocol. Guidelines for such national systems, which shall incorporate the methodologies specified in paragraph 2 below, shall be decided upon by the Conference of the Parties serving as the meeting of the Parties to this Protocol at its first session.
2. Methodologies for estimating anthropogenic emissions by sources and removals by sinks of all greenhouse gases not controlled by the Montreal Protocol shall be those accepted by the Intergovernmental Panel on Climate Change and agreed upon by the Conference of the Parties at its third session. Where such methodologies are not used, appropriate adjustments shall be applied according to methodologies agreed upon by the Conference of the Parties serving as the meeting of the Parties to this Protocol at its first session. Based on the work of, inter alia, the Intergovernmental Panel on Climate Change and advice provided by the Subsidiary Body for Scientific and Technological Advice, the Conference of the Parties serving as the meeting of the Parties to this Protocol shall regularly review and, as appropriate, revise such methodologies and adjustments, taking fully into account any relevant decisions by the Conference of the Parties. Any revision to methodologies or adjustments shall be used only for the purposes of ascertaining compliance with commitments under Article 3 in respect of any commitment period adopted subsequent to that revision.
3. The global warming potentials used to calculate the carbon dioxide equivalence of anthropogenic emissions by sources and removals by sinks of greenhouse gases listed in Annex A shall be those accepted by the Intergovernmental Panel on Climate Change and agreed upon by the Conference of the Parties at its third session. Based on the work of, inter alia, the Intergovernmental Panel on Climate Change and advice provided by the Subsidiary Body for Scientific and Technological Advice, the Conference of the Parties serving as the meeting of the Parties to this Protocol shall regularly review and, as appropriate, revise the global warming potential of each such greenhouse gas, taking fully into account any relevant decisions by the Conference of the Parties. Any revision to a global warming potential shall apply only to commitments under Article 3 in respect of any commitment period adopted subsequent to that revision.

Article 6

1. For the purpose of meeting its commitments under Article 3, any Party included in Annex I may transfer to, or acquire from, any other such Party emission reduction units resulting from projects aimed at reducing anthropogenic emissions by sources or enhancing anthropogenic removals by sinks of greenhouse gases in any sector of the economy, provided that:
    (a) Any such project has the approval of the Parties involved;
    (b) Any such project provides a reduction in emissions by sources, or an enhancement of removals by sinks, that is additional to any that would otherwise occur;
    (c) It does not acquire any emission reduction units if it is not in compliance with its obligations under Articles 5 and 7; and
    (d) The acquisition of emission reduction units shall be supplemental to domestic actions for the purposes of meeting commitments under Article 3.
2. The Conference of the Parties serving as the meeting of the Parties to this Protocol may, at its first session or as soon as practicable thereafter, further elaborate guidelines for the implementation of this Article, including for verification and reporting.
3. A Party included in Annex I may authorize legal entities to participate, under its responsibility, in actions leading to the generation, transfer or acquisition under this Article of emission reduction units.
4. If a question of implementation by a Party included in Annex I of the requirements referred to in this Article is identified in accordance with the relevant provisions of Article 8, transfers and acquisitions of emission reduction units may continue to be made after the question has been identified, provided that any such units may not be used by a Party to meet its commitments under Article 3 until any issue of compliance is resolved.

Article 7

1. Each Party included in Annex I shall incorporate in its annual inventory of anthropogenic emissions by sources and removals by sinks of greenhouse gases not controlled by the Montreal Protocol, submitted in accordance with the relevant decisions of the Conference of the Parties, the necessary supplementary information for the purposes of ensuring compliance with Article 3, to be determined in accordance with paragraph 4 below.
2. Each Party included in Annex I shall incorporate in its national communication, submitted under Article 12 of the Convention, the supplementary information necessary to demonstrate compliance with its commitments under this Protocol, to be determined in accordance with paragraph 4 below.
3. Each Party included in Annex I shall submit the information required under paragraph 1 above annually, beginning with the first inventory due under the Convention for the first year of the commitment period after this Protocol has entered into force for that Party. Each such Party shall submit the information required under paragraph 2 above as part of the first national communication due

under the Convention after this Protocol has entered into force for it and after the adoption of guidelines as provided for in paragraph 4 below. The frequency of subsequent submission of information required under this Article shall be determined by the Conference of the Parties serving as the meeting of the Parties to this Protocol, taking into account any timetable for the submission of national communications decided upon by the Conference of the Parties.

4. The Conference of the Parties serving as the meeting of the Parties to this Protocol shall adopt at its first session, and review periodically thereafter, guidelines for the preparation of the information required under this Article, taking into account guidelines for the preparation of national communications by Parties included in Annex I adopted by the Conference of the Parties. The Conference of the Parties serving as the meeting of the Parties to this Protocol shall also, prior to the first commitment period, decide upon modalities for the accounting of assigned amounts.

Article 8

1. The information submitted under Article 7 by each Party included in Annex I shall be reviewed by expert review teams pursuant to the relevant decisions of the Conference of the Parties and in accordance with guidelines adopted for this purpose by the Conference of the Parties serving as the meeting of the Parties to this Protocol under paragraph 4 below. The information submitted under Article 7, paragraph 1, by each Party included in Annex I shall be reviewed as part of the annual compilation and accounting of emissions inventories and assigned amounts. Additionally, the information submitted under Article 7, paragraph 2, by each Party included in Annex I shall be reviewed as part of the review of communications.
2. Expert review teams shall be coordinated by the secretariat and shall be composed of experts selected from those nominated by Parties to the Convention and, as appropriate, by intergovernmental organizations, in accordance with guidance provided for this purpose by the Conference of the Parties.
3. The review process shall provide a thorough and comprehensive technical assessment of all aspects of the implementation by a Party of this Protocol. The expert review teams shall prepare a report to the Conference of the Parties serving as the meeting of the Parties to this Protocol, assessing the implementation of the commitments of the Party and identifying any potential problems in, and factors influencing, the fulfilment of commitments. Such reports shall be circulated by the secretariat to all Parties to the Convention. The secretariat shall list those questions of implementation indicated in such reports for further consideration by the Conference of the Parties serving as the meeting of the Parties to this Protocol.
4. The Conference of the Parties serving as the meeting of the Parties to this Protocol shall adopt at its first session, and review periodically thereafter, guidelines for the review of implementation of this Protocol by expert review teams taking into account the relevant decisions of the Conference of the Parties.
5. The Conference of the Parties serving as the meeting of the Parties to this Protocol shall, with the assistance of the Subsidiary Body for Implementation and, as appropriate, the Subsidiary Body for Scientific and Technological Advice, consider:

(a) The information submitted by Parties under Article 7 and the reports of the expert reviews thereon conducted under this Article; and
(b) Those questions of implementation listed by the secretariat under paragraph 3 above, as well as any questions raised by Parties.
6. Pursuant to its consideration of the information referred to in paragraph 5 above, the Conference of the Parties serving as the meeting of the Parties to this Protocol shall take decisions on any matter required for the implementation of this Protocol.

Article 9

1. The Conference of the Parties serving as the meeting of the Parties to this Protocol shall periodically review this Protocol in the light of the best available scientific information and assessments on climate change and its impacts, as well as relevant technical, social and economic information. Such reviews shall be coordinated with pertinent reviews under the Convention, in particular those required by Article 4, paragraph 2 (d), and Article 7, paragraph 2 (a), of the Convention. Based on these reviews, the Conference of the Parties serving as the meeting of the Parties to this Protocol shall take appropriate action.
2. The first review shall take place at the second session of the Conference of the Parties serving as the meeting of the Parties to this Protocol. Further reviews shall take place at regular intervals and in a timely manner.

Article 10

All Parties, taking into account their common but differentiated responsibilities and their specific national and regional development priorities, objectives and circumstances, without introducing any new commitments for Parties not included in Annex I, but reaffirming existing commitments under Article 4, paragraph 1, of the Convention, and continuing to advance the implementation of these commitments in order to achieve sustainable development, taking into account Article 4, paragraphs 3, 5 and 7, of the Convention, shall:

(a) Formulate, where relevant and to the extent possible, cost-effective national and, where appropriate, regional programmes to improve the quality of local emission factors, activity data and/or models which reflect the socio-economic conditions of each Party for the preparation and periodic updating of national inventories of anthropogenic emissions by sources and removals by sinks of all greenhouse gases not controlled by the Montreal Protocol, using comparable methodologies to be agreed upon by the Conference of the Parties, and consistent with the guidelines for the preparation of national communications adopted by the Conference of the Parties;
(b) Formulate, implement, publish and regularly update national and, where appropriate, regional programmes containing measures to mitigate climate change and measures to facilitate adequate adaptation to climate change:
   (i) Such programmes would, inter alia, concern the energy, transport and industry sectors as well as agriculture, forestry and waste management. Further-

more, adaptation technologies and methods for improving spatial planning would improve adaptation to climate change; and

(ii) Parties included in Annex I shall submit information on action under this Protocol, including national programmes, in accordance with Article 7; and other Parties shall seek to include in their national communications, as appropriate, information on programmes which contain measures that the Party believes contribute to addressing climate change and its adverse impacts, including the abatement of increases in greenhouse gas emissions, and enhancement of and removals by sinks, capacity building and adaptation measures;

(c) Cooperate in the promotion of effective modalities for the development, application and diffusion of, and take all practicable steps to promote, facilitate and finance, as appropriate, the transfer of, or access to, environmentally sound technologies, know-how, practices and processes pertinent to climate change, in particular to developing countries, including the formulation of policies and programmes for the effective transfer of environmentally sound technologies that are publicly owned or in the public domain and the creation of an enabling environment for the private sector, to promote and enhance the transfer of, and access to, environmentally sound technologies;

(d) Cooperate in scientific and technical research and promote the maintenance and the development of systematic observation systems and development of data archives to reduce uncertainties related to the climate system, the adverse impacts of climate change and the economic and social consequences of various response strategies, and promote the development and strengthening of endogenous capacities and capabilities to participate in international and intergovernmental efforts, programmes and networks on research and systematic observation, taking into account Article 5 of the Convention;

(e) Cooperate in and promote at the international level, and, where appropriate, using existing bodies, the development and implementation of education and training programmes, including the strengthening of national capacity building, in particular human and institutional capacities and the exchange or secondment of personnel to train experts in this field, in particular for developing countries, and facilitate at the national level public awareness of, and public access to information on, climate change. Suitable modalities should be developed to implement these activities through the relevant bodies of the Convention, taking into account Article 6 of the Convention;

(f) Include in their national communications information on programmes and activities undertaken pursuant to this Article in accordance with relevant decisions of the Conference of the Parties; and

(g) Give full consideration, in implementing the commitments under this Article, to Article 4, paragraph 8, of the Convention.

Article 11

1. In the implementation of Article 10, Parties shall take into account the provisions of Article 4, paragraphs 4, 5, 7, 8 and 9, of the Convention.
2. In the context of the implementation of Article 4, paragraph 1, of the Convention, in accordance with the provisions of Article 4, paragraph 3, and Article 11 of the

Convention, and through the entity or entities entrusted with the operation of the financial mechanism of the Convention, the developed country Parties and other developed Parties included in Annex II to the Convention shall:

 (a) Provide new and additional financial resources to meet the agreed full costs incurred by developing country Parties in advancing the implementation of existing commitments under Article 4, paragraph 1 (a), of the Convention that are covered in Article 10, subparagraph (a); and

 (b) Also provide such financial resources, including for the transfer of technology, needed by the developing country Parties to meet the agreed full incremental costs of advancing the implementation of existing commitments under Article 4, paragraph 1, of the Convention that are covered by Article 10 and that are agreed between a developing country Party and the international entity or entities referred to in Article 11 of the Convention, in accordance with that Article.

The implementation of these existing commitments shall take into account the need for adequacy and predictability in the flow of funds and the importance of appropriate burden sharing among developed country Parties. The guidance to the entity or entities entrusted with the operation of the financial mechanism of the Convention in relevant decisions of the Conference of the Parties, including those agreed before the adoption of this Protocol, shall apply mutatis mutandis to the provisions of this paragraph.

3. The developed country Parties and other developed Parties in Annex II to the Convention may also provide, and developing country Parties avail themselves of, financial resources for the implementation of Article 10, through bilateral, regional and other multilateral channels.

Article 12

1. A clean development mechanism is hereby defined.
2. The purpose of the clean development mechanism shall be to assist Parties not included in Annex I in achieving sustainable development and in contributing to the ultimate objective of the Convention, and to assist Parties included in Annex I in achieving compliance with their quantified emission limitation and reduction commitments under Article 3.
3. Under the clean development mechanism:
    (a) Parties not included in Annex I will benefit from project activities resulting in certified emission reductions; and
    (b) Parties included in Annex I may use the certified emission reductions accruing from such project activities to contribute to compliance with part of their quantified emission limitation and reduction commitments under Article 3, as determined by the Conference of the Parties serving as the meeting of the Parties to this Protocol.
4. The clean development mechanism shall be subject to the authority and guidance of the Conference of the Parties serving as the meeting of the Parties to this Protocol and be supervised by an executive board of the clean development mechanism.
5. Emission reductions resulting from each project activity shall be certified by

operational entities to be designated by the Conference of the Parties serving as the meeting of the Parties to this Protocol, on the basis of:
   (a) Voluntary participation approved by each Party involved;
   (b) Real, measurable, and long-term benefits related to the mitigation of climate change; and
   (c) Reductions in emissions that are additional to any that would occur in the absence of the certified project activity.
6. The clean development mechanism shall assist in arranging funding of certified project activities as necessary.
7. The Conference of the Parties serving as the meeting of the Parties to this Protocol shall, at its first session, elaborate modalities and procedures with the objective of ensuring transparency, efficiency and accountability through independent auditing and verification of project activities.
8. The Conference of the Parties serving as the meeting of the Parties to this Protocol shall ensure that a share of the proceeds from certified project activities is used to cover administrative expenses as well as to assist developing country Parties that are particularly vulnerable to the adverse effects of climate change to meet the costs of adaptation.
9. Participation under the clean development mechanism, including in activities mentioned in paragraph 3 (a) above and in the acquisition of certified emission reductions, may involve private and/or public entities, and is to be subject to whatever guidance may be provided by the executive board of the clean development mechanism.
10. Certified emission reductions obtained during the period from the year 2000 up to the beginning of the first commitment period can be used to assist in achieving compliance in the first commitment period.

Article 13

1. The Conference of the Parties, the supreme body of the Convention, shall serve as the meeting of the Parties to this Protocol.
2. Parties to the Convention that are not Parties to this Protocol may participate as observers in the proceedings of any session of the Conference of the Parties serving as the meeting of the Parties to this Protocol. When the Conference of the Parties serves as the meeting of the Parties to this Protocol, decisions under this Protocol shall be taken only by those that are Parties to this Protocol.
3. When the Conference of the Parties serves as the meeting of the Parties to this Protocol, any member of the Bureau of the Conference of the Parties representing a Party to the Convention but, at that time, not a Party to this Protocol, shall be replaced by an additional member to be elected by and from amongst the Parties to this Protocol.
4. The Conference of the Parties serving as the meeting of the Parties to this Protocol shall keep under regular review the implementation of this Protocol and shall make, within its mandate, the decisions necessary to promote its effective implementation. It shall perform the functions assigned to it by this Protocol and shall:
   (a) Assess, on the basis of all information made available to it in accordance

with the provisions of this Protocol, the implementation of this Protocol by the Parties, the overall effects of the measures taken pursuant to this Protocol, in particular environmental, economic and social effects as well as their cumulative impacts and the extent to which progress towards the objective of the Convention is being achieved;

(b) Periodically examine the obligations of the Parties under this Protocol, giving due consideration to any reviews required by Article 4, paragraph 2 (d), and Article 7, paragraph 2, of the Convention, in the light of the objective of the Convention, the experience gained in its implementation and the evolution of scientific and technological knowledge, and in this respect consider and adopt regular reports on the implementation of this Protocol;

(c) Promote and facilitate the exchange of information on measures adopted by the Parties to address climate change and its effects, taking into account the differing circumstances, responsibilities and capabilities of the Parties and their respective commitments under this Protocol;

(d) Facilitate, at the request of two or more Parties, the coordination of measures adopted by them to address climate change and its effects, taking into account the differing circumstances, responsibilities and capabilities of the Parties and their respective commitments under this Protocol;

(e) Promote and guide, in accordance with the objective of the Convention and the provisions of this Protocol, and taking fully into account the relevant decisions by the Conference of the Parties, the development and periodic refinement of comparable methodologies for the effective implementation of this Protocol, to be agreed on by the Conference of the Parties serving as the meeting of the Parties to this Protocol;

(f) Make recommendations on any matters necessary for the implementation of this Protocol;

(g) Seek to mobilize additional financial resources in accordance with Article 11, paragraph 2;

(h) Establish such subsidiary bodies as are deemed necessary for the implementation of this Protocol;

(i) Seek and utilize, where appropriate, the services and cooperation of, and information provided by, competent international organizations and inter-governmental and non-governmental bodies; and

(j) Exercise such other functions as may be required for the implementation of this Protocol, and consider any assignment resulting from a decision by the Conference of the Parties.

5. The rules of procedure of the Conference of the Parties and financial procedures applied under the Convention shall be applied mutatis mutandis under this Protocol, except as may be otherwise decided by consensus by the Conference of the Parties serving as the meeting of the Parties to this Protocol.

6. The first session of the Conference of the Parties serving as the meeting of the Parties to this Protocol shall be convened by the secretariat in conjunction with the first session of the Conference of the Parties that is scheduled after the date of the entry into force of this Protocol. Subsequent ordinary sessions of the Conference of the Parties serving as the meeting of the Parties to this Protocol shall be held every year and in conjunction with ordinary sessions of the Conference of the

Parties, unless otherwise decided by the Conference of the Parties serving as the meeting of the Parties to this Protocol.
7. Extraordinary sessions of the Conference of the Parties serving as the meeting of the Parties to this Protocol shall be held at such other times as may be deemed necessary by the Conference of the Parties serving as the meeting of the Parties to this Protocol, or at the written request of any Party, provided that, within six months of the request being communicated to the Parties by the secretariat, it is supported by at least one third of the Parties.
8. The United Nations, its specialized agencies and the International Atomic Energy Agency, as well as any State member thereof or observers thereto not party to the Convention, may be represented at sessions of the Conference of the Parties serving as the meeting of the Parties to this Protocol as observers. Anybody or agency, whether national or international, governmental or non-governmental, which is qualified in matters covered by this Protocol and which has informed the secretariat of its wish to be represented at a session of the Conference of the Parties serving as the meeting of the Parties to this Protocol as an observer, may be so admitted unless at least one third of the Parties present object. The admission and participation of observers shall be subject to the rules of procedure, as referred to in paragraph 5 above.

Article 14

1. The secretariat established by Article 8 of the Convention shall serve as the secretariat of this Protocol.
2. Article 8, paragraph 2, of the Convention on the functions of the secretariat, and Article 8, paragraph 3, of the Convention on arrangements made for the functioning of the secretariat, shall apply mutatis mutandis to this Protocol. The secretariat shall, in addition, exercise the functions assigned to it under this Protocol.

Article 15

1. The Subsidiary Body for Scientific and Technological Advice and the Subsidiary Body for Implementation established by Articles 9 and 10 of the Convention shall serve as, respectively, the Subsidiary Body for Scientific and Technological Advice and the Subsidiary Body for Implementation of this Protocol. The provisions relating to the functioning of these two bodies under the Convention shall apply mutatis mutandis to this Protocol. Sessions of the meetings of the Subsidiary Body for Scientific and Technological Advice and the Subsidiary Body for Implementation of this Protocol shall be held in conjunction with the meetings of, respectively, the Subsidiary Body for Scientific and Technological Advice and the Subsidiary Body for Implementation of the Convention.
2. Parties to the Convention that are not Parties to this Protocol may participate as observers in the proceedings of any session of the subsidiary bodies. When the subsidiary bodies serve as the subsidiary bodies of this Protocol, decisions under this Protocol shall be taken only by those that are Parties to this Protocol.
3. When the subsidiary bodies established by Articles 9 and 10 of the Convention

exercise their functions with regard to matters concerning this Protocol, any member of the Bureaux of those subsidiary bodies representing a Party to the Convention but, at that time, not a party to this Protocol, shall be replaced by an additional member to be elected by and from amongst the Parties to this Protocol.

Article 16

The Conference of the Parties serving as the meeting of the Parties to this Protocol shall, as soon as practicable, consider the application to this Protocol of, and modify as appropriate, the multilateral consultative process referred to in Article 13 of the Convention, in the light of any relevant decisions that may be taken by the Conference of the Parties. Any multilateral consultative process that may be applied to this Protocol shall operate without prejudice to the procedures and mechanisms established in accordance with Article 18.

Article 17

The Conference of the Parties shall define the relevant principles, modalities, rules and guidelines, in particular for verification, reporting and accountability for emissions trading. The Parties included in Annex B may participate in emissions trading for the purposes of fulfilling their commitments under Article 3. Any such trading shall be supplemental to domestic actions for the purpose of meeting quantified emission limitation and reduction commitments under that Article.

Article 18

The Conference of the Parties serving as the meeting of the Parties to this Protocol shall, at its first session, approve appropriate and effective procedures and mechanisms to determine and to address cases of non-compliance with the provisions of this Protocol, including through the development of an indicative list of consequences, taking into account the cause, type, degree and frequency of non-compliance. Any procedures and mechanisms under this Article entailing binding consequences shall be adopted by means of an amendment to this Protocol.

Article 19

The provisions of Article 14 of the Convention on settlement of disputes shall apply mutatis mutandis to this Protocol.

Article 20

1. Any Party may propose amendments to this Protocol.
2. Amendments to this Protocol shall be adopted at an ordinary session of the Conference of the Parties serving as the meeting of the Parties to this Protocol. The text of any

proposed amendment to this Protocol shall be communicated to the Parties by the secretariat at least six months before the meeting at which it is proposed for adoption. The secretariat shall also communicate the text of any proposed amendments to the Parties and signatories to the Convention and, for information, to the Depositary.
3. The Parties shall make every effort to reach agreement on any proposed amendment to this Protocol by consensus. If all efforts at consensus have been exhausted, and no agreement reached, the amendment shall as a last resort be adopted by a three-fourths majority vote of the Parties present and voting at the meeting. The adopted amendment shall be communicated by the secretariat to the Depositary, who shall circulate it to all Parties for their acceptance.
4. Instruments of acceptance in respect of an amendment shall be deposited with the Depositary. An amendment adopted in accordance with paragraph 3 above shall enter into force for those Parties having accepted it on the ninetieth day after the date of receipt by the Depositary of an instrument of acceptance by at least three fourths of the Parties to this Protocol.
5. The amendment shall enter into force for any other Party on the ninetieth day after the date on which that Party deposits with the Depositary its instrument of acceptance of the said amendment.

Article 21

1. Annexes to this Protocol shall form an integral part thereof and, unless otherwise expressly provided, a reference to this Protocol constitutes at the same time a reference to any annexes thereto. Any annexes adopted after the entry into force of this Protocol shall be restricted to lists, forms and any other material of a descriptive nature that is of a scientific, technical, procedural or administrative character.
2. Any Party may make proposals for an annex to this Protocol and may propose amendments to annexes to this Protocol.
3. Annexes to this Protocol and amendments to annexes to this Protocol shall be adopted at an ordinary session of the Conference of the Parties serving as the meeting of the Parties to this Protocol. The text of any proposed annex or amendment to an annex shall be communicated to the Parties by the secretariat at least six months before the meeting at which it is proposed for adoption. The secretariat shall also communicate the text of any proposed annex or amendment to an annex to the Parties and signatories to the Convention and, for information, to the Depositary.
4. The Parties shall make every effort to reach agreement on any proposed annex or amendment to an annex by consensus. If all efforts at consensus have been exhausted, and no agreement reached, the annex or amendment to an annex shall as a last resort be adopted by a three-fourths majority vote of the Parties present and voting at the meeting. The adopted annex or amendment to an annex shall be communicated by the secretariat to the Depositary, who shall circulate it to all Parties for their acceptance.
5. An annex, or amendment to an annex other than Annex A or B, that has been adopted in accordance with paragraphs 3 and 4 above shall enter into force for all Parties to this Protocol six months after the date of the communication by the Depositary to such Parties of the adoption of the annex or adoption of the amendment to the annex, except for those Parties that have notified the Depositary, in writing, within

that period of their non-acceptance of the annex or amendment to the annex. The annex or amendment to an annex shall enter into force for Parties which withdraw their notification of non-acceptance on the ninetieth day after the date on which withdrawal of such notification has been received by the Depositary.
6. If the adoption of an annex or an amendment to an annex involves an amendment to this Protocol, that annex or amendment to an annex shall not enter into force until such time as the amendment to this Protocol enters into force.
7. Amendments to Annexes A and B to this Protocol shall be adopted and enter into force in accordance with the procedure set out in Article 20, provided that any amendment to Annex B shall be adopted only with the written consent of the Party concerned.

Article 22

1. Each Party shall have one vote, except as provided for in paragraph 2 below.
2. Regional economic integration organizations, in matters within their competence, shall exercise their right to vote with a number of votes equal to the number of their member States that are Parties to this Protocol. Such an organization shall not exercise its right to vote if any of its member States exercises its right, and vice versa.

Article 23

The Secretary-General of the United Nations shall be the Depositary of this Protocol.

Article 24

1. This Protocol shall be open for signature and subject to ratification, acceptance or approval by States and regional economic integration organizations which are Parties to the Convention. It shall be open for signature at United Nations Headquarters in New York from 16 March 1998 to 15 March 1999. This Protocol shall be open for accession from the day after the date on which it is closed for signature. Instruments of ratification, acceptance, approval or accession shall be deposited with the Depositary.
2. Any regional economic integration organization which becomes a Party to this Protocol without any of its member States being a Party shall be bound by all the obligations under this Protocol. In the case of such organizations, one or more of whose member States is a Party to this Protocol, the organization and its member States shall decide on their respective responsibilities for the performance of their obligations under this Protocol. In such cases, the organization and the member States shall not be entitled to exercise rights under this Protocol concurrently.
3. In their instruments of ratification, acceptance, approval or accession, regional economic integration organizations shall declare the extent of their competence with respect to the matters governed by this Protocol. These organizations shall also inform the Depositary, who shall in turn inform the Parties, of any substantial modification in the extent of their competence.

Article 25

1. This Protocol shall enter into force on the ninetieth day after the date on which not less than 55 Parties to the Convention, incorporating Parties included in Annex I which accounted in total for at least 55 per cent of the total carbon dioxide emissions for 1990 of the Parties included in Annex I, have deposited their instruments of ratification, acceptance, approval or accession.
2. For the purposes of this Article, "the total carbon dioxide emissions for 1990 of the Parties included in Annex I" means the amount communicated on or before the date of adoption of this Protocol by the Parties included in Annex I in their first national communications submitted in accordance with Article 12 of the Convention.
3. For each State or regional economic integration organization that ratifies, accepts or approves this Protocol or accedes thereto after the conditions set out in paragraph 1 above for entry into force have been fulfilled, this Protocol shall enter into force on the ninetieth day following the date of deposit of its instrument of ratification, acceptance, approval or accession.
4. For the purposes of this Article, any instrument deposited by a regional economic integration organization shall not be counted as additional to those deposited by States members of the organization.

Article 26

No reservations may be made to this Protocol.

Article 27

1. At any time after three years from the date on which this Protocol has entered into force for a Party, that Party may withdraw from this Protocol by giving written notification to the Depositary.
2. Any such withdrawal shall take effect upon expiry of one year from the date of receipt by the Depositary of the notification of withdrawal, or on such later date as may be specified in the notification of withdrawal.
3. Any Party that withdraws from the Convention shall be considered as also having withdrawn from this Protocol.

Article 28

The original of this Protocol, of which the Arabic, Chinese, English, French, Russian and Spanish texts are equally authentic, shall be deposited with the Secretary-General of the United Nations.

DONE at Kyoto this eleventh day of December one thousand nine hundred and ninety-seven.

IN WITNESS WHEREOF the undersigned, being duly authorized to that effect, have affixed their signatures to this Protocol on the dates indicated.

Annex A

Greenhouse gases
Carbon dioxide (CO_2)
Methane (CH_4)
Nitrous oxide (N_2O)
Hydrofluorocarbons (HFCs)
Perfluorocarbons (PFCs)
Sulphur hexafluoride (SF_6)

Sectors/source categories
Energy
    Fuel combustion
        Energy industries
        Manufacturing industries and construction
        Transport
        Other sectors
        Other
    Fugitive emissions from fuels
        Solid fuels
        Oil and natural gas
        Other
Industrial processes
    Mineral products
    Chemical industry
    Metal production
    Other production
    Production of halocarbons and sulphur hexafluoride
    Consumption of halocarbons and sulphur hexafluoride
    Other
Solvent and other product use
Agriculture
    Enteric fermentation
    Manure management
    Rice cultivation
    Agricultural soils
    Prescribed burning of savannas
    Field burning of agricultural residues
    Other
Waste
    Solid waste disposal on land
    Wastewater handling
    Waste incineration
    Other

Annex B

Party	Quantified emission limitation or reduction commitment (percentage of base year or period)
Australia	108
Austria	92
Belgium	92
Bulgaria*	92
Canada	94
Croatia*	95
Czech Republic*	92
Denmark	92
Estonia*	92
European Community	92
Finland	92
France	92
Germany	92
Greece	92
Hungary*	94
Iceland	110
Ireland	92
Italy	92
Japan	94
Latvia*	92
Liechtenstein	92
Lithuania*	92
Luxembourg	92
Monaco	92
Netherlands	92
New Zealand	100
Norway	101
Poland*	94
Portugal	92
Romania*	92
Russian Federation*	100
Slovakia*	92
Slovenia*	92
Spain	92
Sweden	92
Switzerland	92
Ukraine*	100
United Kingdom of Great Britain and Northern Ireland	92
United States of America	93

* Countries that are undergoing the process of transition to a market economy.

Analysis

The Kyoto Protocol had its origins in a nonbinding treaty concerning the reduction of greenhouse gas emissions (GHG) that was agreed to by 160 nations at the 1992 Earth Summit, held in Rio de Janeiro, Brazil. After it became obvious in 1995 that countries, including the United States, could not meet the reductions that they had pledged to meet, the international community began to explore how best to encourage countries to commit to a binding agreement. Initially, it was thought that the United Nations Framework Convention on Climate Change might provide the solution, but it was determined that it did not contain the necessary cuts in GHGs to effectively mitigate the consequences of climate change. To address the need for further cuts in GHG emissions, more than 10,000 international delegates and interested onlookers representing thirty-eight industrialized countries convened in Japan for ten days in December 1987 to craft the Kyoto Protocol to the United Nations Framework Convention on Climate Change.

During the negotiations, it was already well-known that many politicians in the United States were leery of the intentions of the delegates in Kyoto, as evidenced by the passage of the bipartisan Byrd-Hagel Resolution. U.S. negotiators knew that in order to obtain ratification of the Kyoto Protocol by the Senate, they had to ensure that developing countries were required to make emission cuts using the same time frame assigned to industrialized countries. Unfortunately for negotiators from the United States, other countries were intent on pursuing their national interests. Some developing nations, dubbed "The Group of 77 Plus China," demanded that wealthy industrialized nations, such as Germany, Great Britain, and the United States, reduce their GHG emissions by 35 percent of 1990 levels by 2020. In their eyes, the industrialized countries were responsible for more than a century of GHG emissions and thus should bear the brunt of the GHG emission reductions. Thusly, the Group of 77 Plus China argued that they should be allowed to continue emitting GHGs into the atmosphere as they developed economically. Their negotiating strategy proved successful, as 38 industrialized countries, including the United States, agreed that they were largely responsible for the problem and thus should take the lead in rectifying the sins of the past. In order to assist developing countries, their wealthier partners also promised to encourage economic investments that would help their poorer counterparts cut their levels of GHG emissions.

The Kyoto Protocol ultimately called for industrialized countries to cut their GHG emission levels by five percent of 1990 levels by 2012. Developing countries were exempted from such requirements because the majority agreed that the respective countries needed to use what monies they had available to combat more pressing needs, such as endemic poverty. The disparity between the expectations borne by different countries doomed the agreement's prospects within the United States. Opponents of the Kyoto Protocol charged that the agreement was unfair because it required the United States to hurt itself economically while sparing other countries similar pain. Furthermore, even if the United States made the cuts demanded in the agreement, it would ultimately not curb climate change in any fashion because a majority of the countries around the world would offset the United States cuts by continuing to emit excessive amounts of GHGs into the atmosphere. Due to the political firestorm, President Clinton decided to avoid the political embarrassment of having the agreement rejected by not submitting it to the Senate.

FURTHER READING

Afionis, Stavros. 2016. *The European Union in International Climate Change Negotiations*. New York: Routledge.

Bartsch, Ulrich, and Benito Müller. 2000. *Fossil Fuels in a Changing Climate: Impacts of the Kyoto Protocol and Developing Country Participation*. New York: Oxford University Press.

Chichilnisky, Graciela, and Kristin A. Sheeran. 2009. *Saving Kyoto: An Insider's Guide to How It Works, Why It Matters and What It Means for the Future*. London: New Holland.

Eastwood, Lauren. 2016. *Negotiating the Environment: Civil Society, Globalization, and the UN*. New York: Routledge.

Grover, Velma, ed. 2008. *Global Warming and Climate Change: Ten Years After Kyoto and Still Counting*. Enfield, NH: Science.

Kutney, Gerald. 2014. *Carbon Politics and the Failure of the Kyoto Protocol*. New York: Routledge.

Oberthür, Sebastian, Hermann E. Ott, with Richard T. Tarasofsky. 1999. *The Kyoto Protocol: International Climate Policy for the 21st Century*. New York: Springer.

Peloso, Chris. 2010. "Crafting an International Climate Change Protocol: Applying the Lessons Learned from the Success of the Montreal Protocol and the Ozone Depletion Problem." *Journal of Land Use & Environmental Law* 25: 305–329.

Victor, David G. 2001. *The Collapse of the Kyoto Protocol and the Struggle to Slow Global Warming*. Princeton, NJ: Princeton University Press.

Statement on the Kyoto Protocol on Climate Change

Date: December 10, 1997
Location: Washington, D.C.
Significance: Despite strong resistance to the ratification of the Kyoto Protocol by congressmen from both the Democratic and Republican Parties, the Clinton Administration was touting the agreement as evidence of United States leadership at the international level on environmental issues.
Source: Clinton, William J. 1999. *Public Papers of the Presidents, William J. Clinton: 1997, Book 2—July 1 to December 31, 1997.* Washington, D.C.: Government Printing Office.

I am very pleased that the United States has reached an historic agreement with other nations of the world to take unprecedented action to address global warming. This agreement is environmentally strong and economically sound. It reflects a commitment by our generation to act in the interests of future generations.

No nation is more committed to this effort than the United States. In Kyoto, our mission was to persuade other nations to find common ground so we could make realistic and achievable commitments to reduce greenhouse gas emissions. That mission was accomplished. The United States delegation, at the direction of Vice President Gore and with the skilled leadership of Under Secretary Stuart Eizenstat, showed the way. The momentum generated by Vice President Gore's visit helped move the negotiation to a successful conclusion, and I thank him. I am particularly pleased the agreement strongly reflects the commitment of the United States to use the tools of the free market to tackle this difficult problem.

There are still hard challenges ahead, particularly in the area of involvement by developing nations. It is essential that these nations participate in a meaningful way if we are to truly tackle this global environmental challenge. But the industrialized nations have come together, taken a strong step, and that is real progress.

Finally, let me thank Prime Minister Hashimoto and the people of Japan for their spirit and dedication to the task.

Analysis

As documented by the Byrd-Hagel Resolution, the Senate had strong misgivings about the negotiations in Kyoto, Japan and insisted that the Kyoto Protocol apply to all nations equally. In order to assuage congressional opposition, U.S. negotiators were instructed to make every effort to meet the desires of the Senators in order to ensure that they would be willing to ratify the agreement when finalized. Despite the political needs of the Clinton administration's negotiators, their international counterparts insisted on creating scenarios that exempted "developing" countries from the agreement's dictates. Late in the process, United States Vice President Al Gore went to Kyoto and personally helped negotiate some of the protocol's sticking points. Being an ardent environmentalist, Gore acceded to far more of the international community's demands than the Senate would ever accept. Politically, Gore's efforts tied the Clinton administration to the agreement, which required them to embrace it. The episode proved a huge setback to the administration as, although it signed the agreement, it ultimately opted to avoid certain political defeat by not submitting the Kyoto Protocol to the Senate for ratification.

Further Reading

Bartsch, Ulrich, and Benito Müller. 2000. *Fossil Fuels in a Changing Climate: Impacts of the Kyoto Protocol and Developing Country Participation.* New York: Oxford University Press.

Victor, David G. 2001. *The Collapse of the Kyoto Protocol and the Struggle to Slow Global Warming.* Princeton, NJ: Princeton University Press.

Climate Change Impacts on the United States
The Potential Consequences of Climate Variability and Change (Summary)

Date: June 2000
Location: Washington, D.C.
Significance: The first National Assessment of the Potential Consequences of Climate Variability produced by the United States Global Change Research Program provides a detailed overview of how the federal government perceived the present and future impacts of climate change on both the country as a whole and its individual geographic regions in 2000.
Source: National Assessment Synthesis Team. 2000. *Climate Change Impacts on the United States: The Potential Consequences of Climate Variability and Change.* Washington, D.C.: U.S. Global Change Research Program. http://data.globalchange.gov/assets/9a/aa/ec5b4bb3b895bc8369be2ddac377/nca-2000-report-overview.pdf (accessed October 2, 2015).

Climate Change and Our Nation

Long-term observations confirm that our climate is now changing at a rapid rate. Over the 20th century, the average annual U.S. temperature has risen by almost 1°F (0.6°C) and precipitation has increased nationally by 5 to 10%, mostly due to increases in heavy downpours. These trends are most apparent over the past few decades. The science indicates that the warming in the 21st century will be significantly larger than in the 20th century. Scenarios examined in this Assessment, which assume no major interventions to reduce continued growth of world green-

house gas emissions, indicate that temperatures in the U.S. will rise by about 5–9°F (3–5°C) on average in the next 100 years, which is more than the projected global increase. This rise is very likely to be associated with more extreme precipitation and faster evaporation of water, leading to greater frequency of both very wet and very dry conditions.

This Assessment reveals a number of national-level impacts of climate variability and change including impacts to natural ecosystems and water resources. Natural ecosystems appear to be the most vulnerable to the harmful effects of climate change, as there is often little that can be done to help them adapt to the projected speed and amount of change. Some ecosystems that are already constrained by climate, such as alpine meadows in the Rocky Mountains, are likely to face extreme stress, and disappear entirely in some places. It is likely that other more widespread ecosystems will also be vulnerable to climate change.

One of the climate scenarios used in this Assessment suggests the potential for the forests of the Southeast to break up into a mosaic of forests, savannas, and grasslands. Climate scenarios suggest likely changes in the species composition of the Northeast forests, including the loss of sugar maples. Major alterations to natural ecosystems due to climate change could possibly have negative consequences for our economy, which depends in part on the sustained bounty of our nation's lands, waters, and native plant and animal communities.

A unique contribution of this first U.S. Assessment is that it combines national-scale analysis with an examination of the potential impacts of climate change on different regions of the U.S. For example, sea-level rise will very likely cause further loss of coastal wetlands (ecosystems that provide vital nurseries and habitats for many fish species) and put coastal communities at greater risk of storm surges, especially in the Southeast. Reduction in snowpack will very likely alter the timing and amount of water supplies, potentially exacerbating water shortages and conflicts, particularly throughout the western U.S. The melting of glaciers in the high-elevation West and in Alaska represents the loss or diminishment of unique national treasures of the American landscape. Large increases in the heat index (which combines temperature and humidity) and increases in the frequency of heat waves are very likely. These changes will, at minimum, increase discomfort, particularly in cities. It is very probable that continued thawing of permafrost and melting of sea ice in Alaska will further damage forests, buildings, roads, and coastlines, and harm subsistence livelihoods. In various parts of the nation, cold-weather recreation such as skiing will very likely be reduced, and air conditioning usage will very likely increase.

Highly managed ecosystems appear more robust, and some potential benefits have been identified. Crop and forest productivity is likely to increase in some areas for the next few decades due to increased carbon dioxide in the atmosphere and an extended growing season.

It is possible that some U.S. food exports could increase, depending on impacts in other food-growing regions around the world. It is also possible that a rise in crop production in fertile areas could cause prices to fall, benefiting consumers. Other benefits that are possible include extended seasons for construction and warm weather recreation, reduced heating requirements, and reduced cold-weather mortality.

Climate variability and change will interact with other environmental stresses and socioeconomic changes. Air and water pollution, habitat fragmentation, wetland loss, coastal erosion, and reductions in fisheries are likely to be compounded by climate-related stresses. An aging populace nationally, and rapidly growing populations in cities, coastal areas, and across the South and West are social factors that interact with and alter sensitivity to climate variability and change.

There are also very likely to be unanticipated impacts of climate change during the next century.

Such "surprises" may stem from unforeseen changes in the physical climate system, such as major alterations in ocean circulation, cloud distribution, or storms; and unpredicted biological consequences of these physical climate changes, such as massive dislocations of species or pest outbreaks. In addition, unexpected social or economic change, including major shifts in wealth, technology, or political priorities, could affect our ability to respond to climate change.

Greenhouse gas emissions lower than those assumed in this Assessment would result in reduced impacts. The signatory nations of the Framework Convention on Climate Change are negotiating the path they will ultimately take. Even with such reductions, however, the planet and the nation are certain to experience more than a century of climate change, due to the long lifetimes of greenhouse gases already in the atmosphere and the momentum of the climate system.

Adapting to a changed climate is consequently a necessary component of our response strategy.

Adaptation measures can, in many cases, reduce the magnitude of harmful impacts, or take advantage of beneficial impacts. For example, in agriculture, many farmers will probably be able to alter cropping and management practices. Roads, bridges, buildings, and other long-lived infrastructure can be designed taking projected climate change into account. Adaptations, however, can involve trade-offs, and do involve costs. For example, the benefits of building sea walls to prevent sea-level rise from disrupting human coastal communities will need to be weighed against the economic and ecological costs of seawall construction.

The ecological costs could be high as sea walls prevent the inland shifting of coastal wetlands in response to sea-level rise, resulting in the loss of vital fish and bird habitat and other wetland functions, such as protecting shorelines from damage due to storm surges. Protecting against any increased risk of water-borne and insect-borne diseases will require diligent maintenance of our public health system. Many adaptations, notably those that seek to reduce other environmental stresses such as pollution and habitat fragmentation, will have beneficial effects beyond those related to climate change.

Vulnerability in the U.S. is linked to the fates of other nations, and we cannot evaluate national consequences due to climate variability and change without also considering the consequences of changes elsewhere in the world. The U.S. is linked to other nations in many ways, and both our vulnerabilities and our potential responses will likely depend in part on impacts and responses in other nations. For example, conflicts or mass migrations resulting from resource limits, health, and environmental stresses in more vulnerable nations could possibly pose challenges for global security and U.S. policy. Effects of climate variability and change on U.S. agriculture will depend critically on changes in agricultural productivity elsewhere, which can shift international patterns of food supply and demand. Climate-induced changes in water resources available for power generation, transportation, cities, and agriculture are likely to raise potentially delicate diplomatic issues with both Canada and Mexico.

This Assessment has identified many remaining uncertainties that limit our ability to fully understand the spectrum of potential consequences of climate change for our nation. To address these uncertainties, additional research is needed to improve understanding of ecological and social processes that are sensitive to climate, application of climate scenarios and reconstructions of past climates to impacts studies, and assessment strategies and methods. Results from these research efforts will inform future assessments that will continue the process of building our understanding of humanity's impacts on climate, and climate's impacts on us.

Key Findings

1. Increased warming

Assuming continued growth in world greenhouse gas emissions, the primary climate models used in this Assessment project that temperatures in the US will rise 5–9°F (3–5°C) on average in the next 100 years. A wider range of outcomes is possible.

2. Differing regional impacts

Climate change will vary widely across the US. Temperature increases will vary somewhat from one region to the next. Heavy and extreme precipitation events are likely to become more frequent, yet some regions will get drier. The potential impacts of climate change will also vary widely across the nation.

3. Vulnerable ecosystems

Many ecosystems are highly vulnerable to the projected rate and magnitude of climate change. A few, such as alpine meadows in the Rocky Mountains and some barrier islands, are likely to disappear entirely in some areas. Others, such as forests of the Southeast, are likely to experience major species shifts or break up into a mosaic of grasslands, woodlands, and forests. The goods and services lost through the disappearance or fragmentation of certain ecosystems are likely to be costly or impossible to replace.

4. Widespread water concerns

Water is an issue in every region, but the nature of the vulnerabilities varies. Drought is an important concern in every region. Floods and water quality are concerns in many regions. Snowpack changes are especially important in the West, Pacific Northwest, and Alaska.

5. Secure food supply

At the national level, the agriculture sector is likely to be able to adapt to climate change. Overall, US crop productivity is very likely to increase over the next few decades, but the gains will not be uniform across the nation. Falling prices and competitive pressures are very likely to stress some farmers, while benefiting consumers.

6. Near-term increase in forest growth

Forest productivity is likely to increase over the next several decades in some areas as trees respond to higher carbon dioxide levels. Over the longer term, changes in larger-scale processes such as fire, insects, droughts, and disease will possibly decrease forest productivity. In addition, climate change is likely to cause long-term shifts in forest species, such as sugar maples moving north out of the US.

7. Increased damage in coastal and permafrost areas

Climate change and the resulting rise in sea level are likely to exacerbate threats to buildings, roads, powerlines, and other infrastructure in climatically sensitive places. For example, infrastructure damage is related to permafrost melting in Alaska, and to sea-level rise and storm surge in low-lying coastal areas.

8. Adaptation determines health outcomes

A range of negative health impacts is possible from climate change, but adaptation is likely to help protect much of the US population. Maintaining our nation's public health and community infrastructure, from water treatment systems to emergency shelters, will be important for minimizing the impacts of waterborne diseases, heat stress, air pollution, extreme weather events, and diseases transmitted by insects, ticks, and rodents.

9. Other stresses magnified by climate change

Climate change will very likely magnify the cumulative impacts of other stresses, such as air and water pollution and habitat destruction due to human development patterns. For some systems, such as coral reefs, the combined effects of climate change and other stresses are very likely to exceed a critical threshold, bringing large, possibly irreversible impacts.

10. Uncertainties remain and surprises are expected

Significant uncertainties remain in the science underlying regional climate changes and their impacts. Further research would improve understanding and our ability to project societal and ecosystem impacts, and provide the public with additional useful information about options for adaptation. However, it is likely that some aspects and impacts of climate change will be totally unanticipated as complex systems respond to ongoing climate change in unforeseeable ways.

Analysis

In the Global Change Research Act of 1990, Congress tasked the United States Global Change Research Program (USGCRP) with producing a report for the President of the United States on a four-year cycle that identified the current impacts of climate change on the United States and posited what those changes portended for the next 25 years. The first National Assessment of the Potential Consequences of Climate Variability was produced in 2000 by the United States Global Change Research Program. Much of the research presented therein was produced by scientists employed by the federal government, with oversight provided by the Independent Review Board of the President's Committee of Advisers on Science & Technology. The report proved extremely influential since it not only detailed how global climate change was impacting the nation as a whole, but also addressed its manifestations regionally. This helped the citizenry understand why they needed to be concerned about the long-term threats posed by global warming. If one was living on the Atlantic or Pacific Coast, it was easy to envision how the rise in sea levels caused by the melting of glaciers threatened their way of life. In the West, where the availability of water has been an issue for humans since they first arrived in the region, the potential of increasing drought conditions portended that large areas would be uninhabitable if the increasing pace of global climate change was not stemmed.

Further Reading

National Assessment Synthesis Team. 2000. *Climate Change Impacts on the United States: The Potential Consequences of Climate Variability and Change: Overview.* New York: Cambridge University Press.

Letter to Members of the Senate on the Kyoto Protocol on Climate Change

Date: March 13, 2001
Location: Washington, D.C.
Significance: The letter detailed President George W. Bush's views on both the Kyoto Protocol on Climate Change and the Clean Air Act.
Source: Bush, George W. 2001. *Letter to Members of the Senate on the Kyoto Protocol on Climate Change.* http://www.gpo.gov/fdsys/pkg/WCPD-2001-03-19/html/WCPD-2001-03-19-Pg444-2.htm (accessed August 24, 2015).

March 13, 2001
Dear _____ :
Thank you for your letter of March 6, 2001, asking for the Administration's views on global climate change, in particular the Kyoto Protocol and efforts to regulate carbon dioxide under the Clean Air Act. My Administration takes the issue of global climate change very seriously.

As you know, I oppose the Kyoto Protocol because it exempts 80 percent of the world, including major population centers such as China and India, from compliance, and would cause serious harm to the U.S. economy. The Senate's vote, 95–0, shows that there is a clear consensus that the Kyoto Protocol is an unfair and ineffective means of addressing global climate change concerns.

As you also know, I support a comprehensive and balanced national energy policy that takes into account the importance of improving air quality. Consistent with this balanced approach, I intend to work with the Congress on a multipollutant strategy to require power plants to reduce emissions of sulfur dioxide, nitrogen oxides, and mercury. Any such strategy would include phasing in reductions over a reasonable period of time, providing regulatory certainty, and offering market-based incentives to help industry meet the targets. I do not believe, however, that the government should impose on power plants mandatory emissions reductions for carbon dioxide, which is not a "pollutant" under the Clean Air Act.

A recently released Department of Energy Report, "Analysis of Strategies for Reducing Multiple Emissions from Power Plants," concluded that including caps on carbon dioxide emissions as part of a multiple emissions strategy would lead to an even more dramatic shift from coal to natural gas for electric power generation and significantly higher electricity prices compared to scenarios in which only sulfur dioxide and nitrogen oxides were reduced.

This is important new information that warrants a reevaluation, especially at a time of rising energy prices and a serious energy shortage. Coal generates more than half of America's electricity supply. At a time when California has already experienced energy shortages, and other Western states are worried about price and availability of energy this summer, we must be very careful not to take actions that could harm consumers. This is especially true given the incomplete state of scientific knowledge of the causes of, and solutions to, global climate change and the lack of commercially available technologies for removing and storing carbon dioxide.

Consistent with these concerns, we will continue to fully examine global climate change issues—including the science, technologies, market-based systems, and innovative options for addressing concentrations of greenhouse gases in the atmosphere. I am very optimistic that, with the proper focus and working with our friends and allies, we will be able to develop technologies, market incentives, and other creative ways to address global climate change.

I look forward to working with you and others to address global climate change issues in the context of a national energy policy that protects our environment, consumers, and economy.

Sincerely,

George W. Bush

Note: Identical letters were sent to Senators Jesse Helms, Larry E. Craig, Pat Roberts, and Chuck Hagel.

Analysis

When asked about his administration's views on global climate change, President George W. Bush reiterated the statements made while pursuing the presidency that he was unequivocally opposed to the Kyoto Protocol to the United Nations Framework Convention on Climate Change. Like many critics of the pact, Bush complained that the agreement exempted most of the countries in the world from its dictates, including China and India. He added that if most other countries were not going to bear any responsibility for cutting their emissions of greenhouse gases, it was not fair of them to expect the United States to be bound to the cuts agreed to in Kyoto, Japan.

Further Reading

Energy Information Administration, Office of Integrated Analysis and Forecasting, U.S. Department of Energy. 2000. *Analysis of Strategies for Reducing Multiple Emissions from Power Plants: Sulfur Dioxide, Nitrogen Oxides, and Carbon Dioxide.* Washington, D.C.: Energy Information Administration. http://www.eia.gov/oiaf/servicerpt/powerplants/pdf/sroiaf%282000%2905.pdf.

Kahn, Greg. 2003. "The Fate of the Kyoto Protocol Under the Bush Administration." *Berkeley Journal of International Law* 21: 548–571.

Matsuo, Naoki. 2001. "Rationality in the Statements of President Bush? An Assessment of Whether the Kyoto Protocol is Fatally Flawed." *International Review for Environmental Strategies* 2: 173–179.

Global Climate Change Policy Book
Executive Summary

Date: February 14, 2002
Location: Washington, D.C.
Significance: In lieu of joining the Kyoto Protocol, President George W. Bush's administration opted to craft its own strategy to combat climate change.
Source: Bush, George W. *Global Climate Change Policy Book*. 2002. http://georgewbush-whitehouse.archives.gov/news/releases/2002/02/print/climatechange.html (accessed July 18, 2015).

Executive Summary

"Addressing global climate change will require a sustained effort, over many generations. My approach recognizes that sustained economic growth is the solution, not the problem—because a nation that grows its economy is a nation that can afford investments in efficiency, new technologies, and a cleaner environment."

President George W. Bush

The President announced a new approach to the challenge of global climate change. This approach is designed to harness the power of markets and technological innovation. It holds the promise of a new partnership with the developing world. And it recognizes that climate change is a complex, long-term challenge that will require a sustained effort over many generations. As the President has said, "The policy challenge is to act in a serious and sensible way, given the limits of our knowledge. While scientific uncertainties remain, we can begin now to address the factors that contribute to climate change."

While investments today in science will increase our understanding of this challenge, our investments in advanced energy and sequestration technologies will provide the breakthroughs we need to dramatically reduce our emissions in the longer term. In the near term, we will vigorously pursue emissions reductions even in the absence of complete knowledge. Our approach recognizes that sustained economic growth is an essential part of the solution, not the problem. Economic growth will make possible the needed investment in research, development, and deployment of advanced technologies. This strategy is one that should offer developing countries the incentive and means to join with us in tackling this challenge together. Significantly, the President's plan will:

Reduce the Greenhouse Gas Intensity of the U.S. Economy by 18 Percent in the Next Ten Years. Greenhouse gas intensity measures the ratio of greenhouse gas (GHG) emissions to economic output. This new approach focuses on reducing the growth of GHG emissions, while sustaining the economic growth needed to finance investment in new, clean energy technologies. It sets America on a path to slow the growth of greenhouse gas emissions, and—as the science justifies—to stop and then reverse that growth:
- In efficiency terms, the 183 metric tons of emissions per million dollars GDP that we emit today will be lowered to 151 metric tons per million dollars GDP in 2012.
- Existing trends and efforts in technology improvement will play a significant role. Beyond that, the President's commitment will achieve 100 million metric tons of reduced emissions in 2012 alone, with more than 500 million metric tons in cumulative savings over the entire decade.
- This goal is comparable to the average progress that nations participating in the Kyoto Protocol are required to achieve.

Substantially Improve the Emission Reduction Registry. The President directed the Secretary of Energy, in consultation with the Secretary of Commerce, the Secretary of Agriculture, and the Administrator of the Environmental Protection Agency, to propose improvements to the current voluntary emission reduction registration program under section 1605(b) of the 1992 Energy Policy Act within 120 days. These improvements will enhance measurement accuracy, reliability and verifiability, working with and taking into account emerging domestic and international approaches.

Protect and Provide Transferable Credits for Emissions Reduction. The President directed the Secretary of Energy to recommend reforms to ensure that businesses and individuals that register reductions are not penalized under a future climate policy, and to give transferable credits to companies that can show real emissions reductions.

Review Progress Toward Goal and Take Additional Action if Necessary. If, in 2012, we find that we are not on track toward meeting our goal, and sound science justifies further policy action, the United States will respond with additional measures that may include a broad, market-based program as well as additional incentives and voluntary measures designed to accelerate technology development and deployment.

Increase Funding for America's Commitment to Climate Change. The President's FY '03 budget seeks $4.5 billion in total climate spending—an increase of $700 million. This commitment is unmatched in the world, and is particularly notable given America's focus on international and homeland security and domestic economic issues in the President's FY '03 budget proposal.

Take Action on the Science and Technology Review. The Secretary of Commerce and Secretary of Energy have completed their review of the federal government's science and technology research portfolios and recommended a path forward. As a result of their review, the President has established a new management structure to advance and coordinate climate change science and technology research.
- The President has established a Cabinet-level Committee on Climate Change Science and Technology Integration to oversee this effort. The Secretary of Commerce and Secretary of Energy will lead the effort, in close coordination with the President's Science Advisor. The research effort will continue to be coordinated through the National Science and Technology Council in accordance with the Global Change Research Act of 1990.
- The President's FY '03 budget proposal dedicates $1.7 billion to fund basic scientific

research on climate change and $1.3 billion to fund research on advanced energy and sequestration technologies.
- This includes $80 million in new funding dedicated to implementation of the Climate Change Research Initiative (CCRI) and the National Climate Change Technology Initiative (NCCTI) announced last June. This funding will be used to address major gaps in our current understanding of the natural carbon cycle and the role of black soot emissions in climate change. It will also be used to promote the development of the most promising "breakthrough" technologies for clean energy generation and carbon sequestration.

Implement a Comprehensive Range of New and Expanded Domestic Policies, Including:
- *Tax Incentives for Renewable Energy, Cogeneration, and New Technology.* The President's FY '03 budget seeks $555 million in clean energy tax incentives, as the first part of a $4.6 billion commitment over the next five years ($7.1 billion over the next 10 years). These tax credits will spur investments in renewable energy (solar, wind, and biomass), hybrid and fuel cell vehicles, cogeneration, and landfill gas conversion. Consistent with the National Energy Policy, the President has directed the Secretary of the Treasury to work with Congress to extend and expand the production tax credit for electricity generation from wind and biomass, to develop a new residential solar energy tax credit, and to encourage cogeneration projects through investment tax credits.
- *Business Challenges.* The President has challenged American businesses to make specific commitments to improving the greenhouse gas intensity of their operations and to reduce emissions. Recent agreements with the semi-conductor and aluminum industries and industries that emit methane already have significantly reduced emissions of some of the most potent greenhouse gases. We will build upon these successes with new agreements, producing greater reductions.
- *Transportation Programs.* The Administration is promoting the development of fuel-efficient motor vehicles and trucks, researching options for producing cleaner fuels, and implementing programs to improve energy efficiency. The President is committed to expanding federal research partnerships with industry, providing market-based incentives and updating current regulatory programs that advance our progress in this important area. This commitment includes expanding fuel cell research, in particular through the "FreedomCAR" initiative. The President's FY '03 budget seeks more than $3 billion in tax credits over 11 years for consumers to purchase fuel cell and hybrid vehicles. The Secretary of Transportation has asked the Congressional leadership to work with him on legislation that would authorize the Department of Transportation to reform the Corporate Average Fuel Economy (CAFE) program, fully considering the recent National Academy Sciences report, so that we can safely improve fuel economy for cars and trucks.
- *Carbon Sequestration.* The President's FY '03 budget requests over $3 billion—a $1 billion increase above the baseline—as the first part of a ten year (2002–2011) commitment to implement and improve the conservation title of the Farm Bill, which will significantly enhance the natural storage of carbon. The President also directed the Secretary of Agriculture to provide recommendations for further, targeted incentives aimed at forest and agricultural sequestration of greenhouse gases. The President further directed the Secretary of Agriculture, in consultation

with the Environmental Protection Agency and the Department of Energy, to develop accounting rules and guidelines for crediting sequestration projects, taking into account emerging domestic and international approaches.

Promote New and Expanded International Policies to Complement Our Domestic Program. The President's approach seeks to expand cooperation internationally to meet the challenge of climate change, including:

- *Investing $25 Million in Climate Observation Systems in Developing Countries.* In response to the National Academy of Sciences' recommendation for better observation systems, the President has allocated $25 million and challenged other developed nations to match the U.S. commitment.
- *Tripling Funding for "Debt-for-Nature" Forest Conservation Programs.* Building upon recent Tropical Forest Conservation Act (TFCA) agreements with Belize, El Salvador, and Bangladesh, the President's FY '03 budget request of $40 million to fund "debt for nature" agreements with developing countries nearly triples funding for this successful program. Under TFCA, developing countries agree to protect their tropical forests from logging, avoiding emissions and preserving the substantial carbon sequestration services they provide. The President also announced a new agreement with the Government of Thailand, which will preserve important mangrove forest in Northeastern Thailand in exchange for debt relief worth $11.4 million.
- *Fully Funding the Global Environmental Facility.* The Administration's FY '03 budget request of $178 million for the GEF is more than $77 million above this year's funding and includes a substantial $70 million payment for arrears incurred during the prior administration. The GEF is the primary international institution for transferring energy and sequestration technologies to the developing world under the United Nations Framework Convention on Climate Change (UNFCCC).
- *Dedicating Significant Funds to the United States Agency for International Development (USAID).* The President's FY'03 budget requests $155 million in funding for USAID climate change programs. USAID serves as a critical vehicle for transferring American energy and sequestration technologies to developing countries to promote sustainable development and minimize their GHG emissions growth.
- *Pursue Joint Research with Japan.* The U.S. and Japan continue their High-Level Consultations on climate change issues. Later this month, a team of U.S. experts will meet with their Japanese counterparts to discuss specific projects within the various areas of climate science and technology, to identify the highest priorities for collaborative research.
- *Pursue Joint Research with Italy.* Following up on a pledge of President Bush and Prime Minister Berlusconi to undertake joint research on climate change, the U.S. and Italy convened a Joint Climate Change Research Meeting in January 2002. The delegations for the two countries identified more than 20 joint climate change research activities for immediate implementation, including global and regional modeling.
- *Pursue Joint Research with Central America.* The United States and Central American Heads of Government signed the Central American-United States of America Joint Accord (CONCAUSA) on December 10, 1994. The original agreement covered cooperation under action plans in four major areas: conservation of biodiversity, sound use of energy, environmental legislation, and sustainable economic development. On June 7, 2001, the United States and its Central American partners signed an expanded and renewed CONCAUSA Declaration, adding disaster relief and climate change

as new areas for cooperation. The new CONCAUSA Declaration calls for intensified cooperative efforts to address climate change through scientific research, estimating and monitoring greenhouse gases, investing in forestry conservation, enhancing energy efficiency, and utilizing new environmental technologies.

Analysis

The *Global Climate Change Policy Book* was written to demonstrate how George W. Bush's administration planned to address global climate change. The plan called for an 18% reduction in greenhouse gas intensity, which sounded impressive to individuals unacquainted with how the international community measured greenhouse gases in the atmosphere. Instead of pledging to make a percentage decrease in overall greenhouse gas emissions, as was called for in the Kyoto Protocol, the Bush administration was calling for a decrease in the ratio of greenhouse gasses found in the air, which was a far lower standard.

Included in the proposal was money for other countries to join the United States in addressing climate change without threatening their economies. Through partnerships, such as one with Central American countries to jointly work on climate change research, the Bush administration hoped to demonstrate that it was possible to address climate change responsibly without acceding to what they considered were the draconian demands contained of the Kyoto Protocol. At the same time, the administration was undercutting adoption of the Kyoto Protocol by other countries through the presentation and financing of this alternative.

In practice, the Bush administration did not aggressively implement even this watered-down plan because many of its proposals were in direct conflict with its National Energy Strategy. The administration got around this problem by making the apparent dictates contained within the *Global Climate Change Policy Book* voluntary.

Further Reading

Abraham, Spencer. 2004. "The Bush Administration's Approach to Climate Change." *Science* 305: 616–617.

Kahn, Greg. 2003. "The Fate of the Kyoto Protocol Under the Bush Administration." *Berkeley Journal of International Law* 21: 548–571.

McCright, Aaron M., and Riley E. Dunlap. 2003. "Defeating Kyoto: The Conservative Movement's Impact on U.S. Climate Change Policy." *Social Problems* 50: 348–373.

Climate Stewardship Act of 2003

Date: January 9, 2003
Location: Washington, D.C.
Significance: Since it was obvious that the Senate would never ratify the Kyoto Protocol to the United Nations Framework Convention on Climate Change, the Climate Stewardship Act was offered as an alternative by Senators Joe Lieberman and John McCain. While the legislation required the United States to cut greenhouse gas emissions, the reduction level was much more modest than what was proposed in the Kyoto Protocol.
Source: 108th Congress, 1st Session. 2003. S. *139*. https://www.govtrack.us/congress/bills/108/s139/text (accessed July 30, 2015).

A Bill

To provide for a program of scientific research on abrupt climate change, to accelerate the reduction of greenhouse gas emissions in the United States by establishing a market-driven system of greenhouse gas tradeable allowances that could be used interchangeably with passenger vehicle fuel economy standard credits, to limit greenhouse gas emissions in the United States and reduce dependence upon foreign oil, and ensure benefits to consumers from the trading in such allowances.

Be it enacted by the Senate and House of Representatives of the United States of America in Congress assembled,

Section 1. Short Title.

This Act may be cited as the "Climate Stewardship Act of 2003."

Sec. 2. Table of Contents.

The table of contents for this Act is as follows:
Sec. 1. Short title.
Sec. 2. Table of contents.
Sec. 3. Definitions.

Title I—Federal Climate Change Research and Related Activities

Sec. 101. National Science Foundation scholarships.
Sec. 102. Commerce Department study of technology transfer barriers.
Sec. 103. Report on United States impact of Kyoto protocol.
Sec. 104. Research grants.
Sec. 105. Abrupt climate change research.
Sec. 106. NIST greenhouse gas functions.
Sec. 107. Development of new measurement technologies.
Sec. 108. Enhanced environmental measurements and standards.
Sec. 109. Technology development and diffusion.

Title II—National Greenhouse Gas Database

Sec. 201. National greenhouse gas database and registry established.
Sec. 202. Inventory of greenhouse gas emissions for covered entities.
Sec. 203. Greenhouse gas reduction reporting.
Sec. 204. Measurement and verification.

Title III—Market-Driven Greenhouse Gas Reductions

Subtitle A—Emission Reduction Requirements; Use of Tradeable Allowances
Sec. 311. Covered entities must submit allowances for emissions.
Sec. 312. Compliance.
Sec. 313. Tradeable allowances and fuel economy standard credits.
Sec. 314. Borrowing against future reductions.
Sec. 315. Other uses of tradeable allowances.
Sec. 316. Exemption of source categories.

Subtitle B—Establishment and Allocation of Tradeable Allowances
Sec. 331. Establishment of tradeable allowances.
Sec. 332. Determination of tradeable allowance allocations.
Sec. 333. Allocation of tradeable allowances.
Sec. 334. Initial allocations for early participation and accelerated participation.
Sec. 335. Bonus for accelerated participation.
Sec. 336. Ensuring target adequacy.

Subtitle C—Climate Change Credit Corporation
Sec. 351. Establishment.
Sec. 352. Purposes and functions.

Subtitle D—Sequestration Accounting; Penalties
Sec. 371. Sequestration accounting.
Sec. 372. Penalties.

Sec. 3. Definitions.

In this Act:
(1) ADMINISTRATOR—The term "Administrator" means the Administrator of the Environmental Protection Agency.
(2) BASELINE—The term "baseline" means the historic greenhouse gas emission levels of an entity, as adjusted upward by the Administrator to reflect actual reductions that are verified in accordance with—
   (A) regulations promulgated under section 201(c)(1); and
   (B) relevant standards and methods developed under this title.
(3) COVERED SECTORS—The term "covered sectors" means the electricity, transportation, industry, and commercial sectors, as such terms are used in the Inventory.
(4) COVERED ENTITY—The term "covered entity" means an entity (including a branch, department, agency, or instrumentality of Federal, State, or local government) that—
   (A) owns or controls a source of greenhouse gas emissions in the electric power, industrial, or commercial sectors of the United States economy (as defined in the Inventory), refines or imports petroleum products for use in transportation, or produces or imports hydrofluorocarbons, perfluorocarbons, or sulfur hexafluoride; and
   (B) emits over 10,000 metric tons of greenhouse gas per year, measured in units of carbon dioxide equivalence, or produces or imports—
      (i) petroleum products that, when combusted, will emit,
      (ii) hydrofluorocarbons, perfluorocarbons, or sulfur hexafluoride that, when used, will emit, or
      (iii) other greenhouse gases that, when used, will emit, over 10,000 metric tons of greenhouse gas per year, measured in units of carbon dioxide equivalence.
(5) DATABASE—The term "database" means the National Greenhouse Gas Database established under section 201.
(6) DIRECT EMISSIONS—The term "direct emissions" means greenhouse gas emissions by an entity from a facility that is owned or controlled by that entity.
(7) FACILITY—The term "facility" means a building, structure, or installation located on any 1 or more contiguous or adjacent properties of an entity in the United States.
(8) GREENHOUSE GAS—The term "greenhouse gas" means—
   (A) carbon dioxide;
   (B) methane;
   (C) nitrous oxide;
   (D) hydrofluorocarbons;
   (E) perfluorocarbons; and
   (F) sulfur hexafluoride.

(9) INDIRECT EMISSIONS—The term "indirect emissions" means greenhouse gas emissions that are—
   (A) a result of the activities of an entity; but
   (B) emitted from a facility owned or controlled by another entity; and
   (C) not reported as direct emissions by the entity from which they were emitted.
(10) INVENTORY—The term "Inventory" means the Inventory of U.S. Greenhouse Gas Emissions and Sinks, prepared in compliance with the United Nations Framework Convention on Climate Change Decision 3/CP.5).
(11) PHASE I ALLOTMENT—The term "Phase I allotment" means—
   (A) the amount of emissions emitted by a covered sector, as identified in the Inventory for the calendar year preceding the calendar year in which this Act is enacted (reduced by the amount of allowances allocated to early and accelerated participants under section 334 of this Act); multiplied by—
   (B) the result of—
      (i) the total greenhouse emissions for all covered sectors for the year 2000, as identified in the 2000 Inventory; divided by
      (ii) the total greenhouse emissions for all covered sectors for the calendar year preceding the date of enactment of this Act, as identified in the Inventory.
(12) PHASE II ALLOTMENT—The term "Phase II allotment" means—
   (A) the amount of emissions emitted by a covered sector, as identified in the Inventory for the calendar year preceding the calendar year in which this Act is enacted (reduced by the amount of allowances allocated to early and accelerated participants under section 334 of this Act); multiplied by—
   (B) the result of—
      (i) the total greenhouse emissions for all covered sectors for the year 1990, as identified in the 1990 Inventory; divided by
      (ii) the total greenhouse emissions for all covered sectors for the calendar year preceding the date of enactment of this Act, as identified in the Inventory.
(13) REGISTRY—The term "registry" means the registry of greenhouse gas emission reductions established under section 201(b)(2).
(14) SECRETARY—The term "Secretary" means the Secretary of Commerce.
(15) SEQUESTRATION—
   (A) IN GENERAL—The term "sequestration" means the capture, long-term separation, isolation, or removal of greenhouse gases from the atmosphere.
   (B) INCLUSIONS—The term "sequestration" includes—
      (i) agricultural and conservation practices;
      (ii) reforestation;
      (iii) forest preservation; and
      (iv) any other appropriate method of capture, long-term separation, isolation, or removal of greenhouse gases from the atmosphere, as determined by the Administrator.
   (C) EXCLUSIONS—The term "sequestration" does not include—
      (i) any conversion of, or negative impact on, a native ecosystem; or
      (ii) any introduction of non-native species or genetically modified organisms.

(16) SOURCE CATEGORY—The term "source category" means a process or activity that leads to direct emissions of greenhouse gases, as listed in the Inventory.

Title I—Federal Climate Change Research and Related Activities

Sec. 101. National Science Foundation Scholarships.

The Director of the National Science Foundation shall establish a scholarship program for post-secondary students studying global climate change, including capability in observation, analysis, modeling, paleoclimatology, consequences, and adaptation.

Sec. 102. Commerce Department Study of Technology Transfer Barriers.

(a) STUDY—The Assistant Secretary of Technology Policy at Department of Commerce shall conduct a study of technology transfer barriers, best practices, and outcomes of technology transfer activities at Federal laboratories related to the licensing and commercialization of energy efficient technologies. The study shall be submitted to the Senate Committee on Commerce, Science, and Transportation and the House of Representatives Committee on Science within 6 months after the date of enactment of this Act. The Assistant Secretary shall work with the existing interagency working group to address identified barriers.

(b) AGENCY REPORT TO INCLUDE INFORMATION ON TECHNOLOGY TRANSFER INCOME AND ROYALTIES—Paragraph (2)(B) of section 11(f) of the Stevenson-Wydler Technology Innovation Act of 1980 (15 U.S.C. 3710(f)) is amended—

    (1) by striking "and" after the semicolon in clause (vi);

    (2) by redesignating clause (vii) as clause (ix); and

    (3) by inserting after clause (vi) the following:

        "(vii) the number of fully-executed licenses which received royalty income in the preceding fiscal year for climate-change or energy-efficient technology;

        "(viii) the total earned royalty income for climate-change or energy-efficient technology; and."

(c) INCREASED INCENTIVES FOR DEVELOPMENT OF CLIMATE-CHANGE OR ENERGY-EFFICIENT TECHNOLOGY—Section 14(a) of the Stevenson-Wydler Technology Innovation Act of 1980 (15 U.S.C. 3710c(a)) is amended—

    (1) by striking "15 percent," in paragraph (1)(A) and inserting "15 percent (25 percent for climate change-related technologies),"; and

    (2) by inserting "($250,000 for climate change-related technologies)" after "$150,000" each place it appears in paragraph (3).

Sec. 103. Report On United States Impact of Kyoto Protocol.

Within 6 months after the date of enactment of this Act, the Secretary shall submit a report to the Senate Committee on Commerce, Science, and Transportation and the House of Representatives Committee on Science on the effects that the entry into force of the Kyoto Protocol will have on—

(1) United States industry and its ability to compete globally;

(2) international cooperation on scientific research and development; and

(3) United States participation in international environmental climate change mitigation efforts and technology deployment.

Sec. 104. Research Grants.

Section 105 of the Global Change Research Act of 1990 (15 U.S.C. 2935) is amended—
(1) by redesignating subsection (c) as subsection (d); and
(2) by inserting after subsection (b) the following:
"(c) Research Grants-
"(1) COMMITTEE TO DEVELOP LIST OF PRIORITY RESEARCH AREAS—The Committee shall develop a list of priority areas for research and development on climate change that are not being addressed by Federal agencies.
"(2) DIRECTOR OF OSTP TO TRANSMIT LIST TO NSF—The Director of the Office of Science and Technology Policy shall transmit the list to the National Science Foundation.
"(3) Funding through NSF-
"(A) BUDGET REQUEST—The National Science Foundation shall include, as part of the annual request for appropriations for the Science and Technology Policy Institute, a request for appropriations to fund research in the priority areas on the list developed under paragraph (1).
"(B) AUTHORIZATION—For fiscal year 2004 and each fiscal year thereafter, there are authorized to be appropriated to the National Science Foundation not less than $17,000,000, to be made available through the Science and Technology Policy Institute, for research in those priority areas."

Sec. 105. Abrupt Climate Change Research.

(a) IN GENERAL—The Secretary, through the National Oceanic and Atmospheric Administration, shall carry out a program of scientific research on potential abrupt climate change designed—
(1) to develop a global array of terrestrial and oceanographic indicators of paleoclimate in order sufficiently to identify and describe past instances of abrupt climate change;
(2) to improve understanding of thresholds and nonlinearities in geophysical systems related to the mechanisms of abrupt climate change;
(3) to incorporate these mechanisms into advanced geophysical models of climate change; and
(4) to test the output of these models against an improved global array of records of past abrupt climate changes.
(b) ABRUPT CLIMATE CHANGE DEFINED—In this section, the term "abrupt climate change" means a change in climate that occurs so rapidly or unexpectedly that human or natural systems may have difficulty adapting to it.

Sec. 106. NIST Greenhouse Gas Functions.

Section 2(c) of the National Institute of Standards and Technology Act (15 U.S.C. 272(c)) is amended—

(1) by striking "and" after the semicolon in paragraph (21);
(2) by redesignating paragraph (22) as paragraph (23); and
(3) by inserting after paragraph (21) the following:
> "(22) perform research to develop enhanced measurements, calibrations, standards, and technologies which will enable the reduced production in the United States of greenhouse gases associated with global warming, including carbon dioxide, methane, nitrous oxide, ozone, perfluorocarbons, hydrofluorocarbons, and sulfur hexafluoride; and."

Sec. 107. Development of New Measurement Technologies.

The Secretary shall initiate a program to develop, with technical assistance from appropriate Federal agencies, innovative standards and measurement technologies (including technologies to measure carbon changes due to changes in land use cover) to calculate—
(1) greenhouse gas emissions and reductions from agriculture, forestry, and other land use practices;
(2) noncarbon dioxide greenhouse gas emissions from transportation;
(3) greenhouse gas emissions from facilities or sources using remote sensing technology; and
(4) any other greenhouse gas emission or reductions for which no accurate or reliable measurement technology exists.

Sec. 108. Enhanced Environmental Measurements and Standards.

The National Institute of Standards and Technology Act (15 U.S.C. 271 et seq.) is amended—
(1) by redesignating sections 17 through 32 as sections 18 through 33, respectively; and
(2) by inserting after section 16 the following:

"**Sec. 17. Climate Change Standards and Processes.**

"(a) In General—The Director shall establish within the Institute a program to perform and support research on global climate change standards and processes, with the goal of providing scientific and technical knowledge applicable to the reduction of greenhouse gases (as defined in section 3(8) of the Climate Stewardship Act of 2003).

"(b) Research Program-
> "(1) In general—The Director is authorized to conduct, directly or through contracts or grants, a global climate change standards and processes research program.
> "(2) Research projects—The specific contents and priorities of the research program shall be determined in consultation with appropriate Federal agencies, including the Environmental Protection Agency, the National Oceanic and Atmospheric Administration, and the National Aeronautics and Space Administration. The program generally shall include basic and applied research—
>> "(A) to develop and provide the enhanced measurements, calibrations, data, models, and reference material standards which will enable the monitoring of greenhouse gases;
>> "(B) to assist in establishing a baseline reference point for future trading in greenhouse gases and the measurement of progress in emissions reduction;
>> "(C) that will be exchanged internationally as scientific or technical information which has the stated purpose of developing mutually recognized

measurements, standards, and procedures for reducing greenhouse gases; and

"(D) to assist in developing improved industrial processes designed to reduce or eliminate greenhouse gases.

"(c) National Measurement Laboratories-

"(1) In general—In carrying out this section, the Director shall utilize the collective skills of the National Measurement Laboratories of the National Institute of Standards and Technology to improve the accuracy of measurements that will permit better understanding and control of these industrial chemical processes and result in the reduction or elimination of greenhouse gases.

"(2) Material, process, and building research—The National Measurement Laboratories shall conduct research under this subsection that includes—

"(A) developing material and manufacturing processes which are designed for energy efficiency and reduced greenhouse gas emissions into the environment;

"(B) developing environmentally-friendly, 'green' chemical processes to be used by industry; and

"(C) enhancing building performance with a focus in developing standards or tools which will help incorporate low- or no-emission technologies into building designs.

"(3) Standards and tools—The National Measurement Laboratories shall develop standards and tools under this subsection that include software to assist designers in selecting alternate building materials, performance data on materials, artificial intelligence-aided design procedures for building subsystems and "smart buildings," and improved test methods and rating procedures for evaluating the energy performance of residential and commercial appliances and products.

"(d) National Voluntary Laboratory Accreditation Program—The Director shall utilize the National Voluntary Laboratory Accreditation Program under this section to establish a program to include specific calibration or test standards and related methods and protocols assembled to satisfy the unique needs for accreditation in measuring the production of greenhouse gases. In carrying out this subsection the Director may cooperate with other departments and agencies of the Federal Government, State and local governments, and private organizations."

Sec. 109. Technology Development and Diffusion.

The Director of the National Institute of Standards and Technology, through the Manufacturing Extension Partnership Program, may develop a program to support the implementation of new "green" manufacturing technologies and techniques by the more than 380,000 small manufacturers.

Title II—National Greenhouse Gas Database

Sec. 201. National Greenhouse Gas Database and Registry Established.

(a) ESTABLISHMENT—As soon as practicable after the date of enactment of this Act, the Administrator, in coordination with the Secretary, the Secretary of Energy, the

Secretary of Agriculture, and private sector and nongovernmental organizations, shall establish, operate, and maintain a database, to be known as the 'National Greenhouse Gas Database,' to collect, verify, and analyze information on greenhouse gas emissions by entities.

(b) NATIONAL GREENHOUSE GAS DATABASE COMPONENTS—The database shall consist of—

    (1) an inventory of greenhouse gas emissions; and

    (2) a registry of greenhouse gas emission reductions and increases in greenhouse gas sequestrations.

(c) COMPREHENSIVE SYSTEM-

    (1) IN GENERAL—Not later than 2 years after the date of enactment of this Act, the Administrator shall promulgate regulations to implement a comprehensive system for greenhouse gas emissions reporting, inventorying, and reductions registration.

    (2) REQUIREMENTS—The Administrator shall ensure, to the maximum extent practicable, that—

        (A) the comprehensive system described in paragraph (1) is designed to—

            (i) maximize completeness, transparency, and accuracy of information reported; and

            (ii) minimize costs incurred by entities in measuring and reporting greenhouse gas emissions; and

        (B) the regulations promulgated under paragraph (1) establish procedures and protocols necessary—

            (i) to prevent the reporting of some or all of the same greenhouse gas emissions or emission reductions by more than 1 reporting entity;

            (ii) to provide for corrections to errors in data submitted to the database;

            (iii) to provide for adjustment to data by reporting entities that have had a significant organizational change (including mergers, acquisitions, and divestiture), in

order to maintain comparability among data in the database over time;

            (iv) to provide for adjustments to reflect new technologies or methods for measuring or calculating greenhouse gas emissions;

            (v) to account for changes in registration of ownership of emission reductions resulting from a voluntary private transaction between reporting entities; and

            (vi) to clarify the responsibility for reporting in the case of any facility owned or controlled by more than 1 entity.

    (3) SERIAL NUMBERS—Through regulations promulgated under paragraph (1), the Administrator shall develop and implement a system that provides—

        (A) for the verification of submitted emissions reductions;

        (B) for the provision of unique serial numbers to identify the verified emission reductions made by an entity relative to the baseline of the entity; and

        (C) for the tracking of the reductions associated with the serial numbers.

Sec. 202. Inventory of Greenhouse Gas Emissions for Covered Entities.

(a) IN GENERAL—Not later than July 1st of each calendar year after 2008, a covered

entity shall submit to the Administrator a report that describes, for the preceding calendar year, the entity-wide greenhouse gas emissions (as reported at the facility level), including—
 (1) the total quantity of direct greenhouse gas emissions from stationary sources, expressed in units of carbon dioxide equivalence;
 (2) the amount of petroleum products sold or imported and the amount of greenhouse gases, expressed in carbon dioxide equivalents, that would be produced when these products are used for transportation; and
 (3) such other categories of emissions as the Administrator determines in the regulations promulgated under section 201(c)(1) may be practicable and useful for the purposes of this Act, such as—
  (A) indirect emissions from imported electricity, heat, and steam;
  (B) process and fugitive emissions; and
  (C) production or importation of greenhouse gases.
(b) COLLECTION AND ANALYSIS OF DATA—The Administrator shall collect and analyze information reported under subsection (a) for use under title III.

Sec. 203. Greenhouse Gas Reduction Reporting.

(a) IN GENERAL—Subject to the requirements described in subsection (b)—
 (1) a covered entity may register greenhouse gas emission reductions achieved after 1990 and before 2010 under this section; and
 (2) an entity that is not a covered entity may register greenhouse gas emission reductions achieved at any time since 1990 under this section.
(b) REQUIREMENTS-
 (1) IN GENERAL—The requirements referred to in subsection (a) are that an entity (other than an entity described in paragraph (2)) shall—
  (A) establish a baseline; and
  (B) submit the report described in subsection (c)(1).
 (2) REQUIREMENTS APPLICABLE TO ENTITIES ENTERING INTO CERTAIN AGREEMENTS—An entity that enters into an agreement with a participant in the registry for the purpose of a carbon sequestration project shall not be required to comply with the requirements specified in paragraph (1) unless that entity is required to comply with the requirements by reason of an activity other than the agreement.
(c) REPORTS-
 (1) REQUIRED REPORT—Not later than July 1st of the each calendar year beginning more than 2 years after the date of enactment of this Act, but subject to paragraph (3), an entity described in subsection (a) shall submit to the Administrator a report that describes, for the preceding calendar year, the entity-wide greenhouse gas emissions (as reported at the facility level), including—
  (A) the total quantity of direct greenhouse gas emissions from stationary sources, expressed in units of carbon dioxide equivalence;
  (B) the amount of petroleum products sold or imported and the amount of greenhouse gases, expressed in carbon dioxide equivalents, that would be produced when these products are used by vehicles; and
  (C) such other categories of emissions as the Administrator determines in

the regulations promulgated under section 201(c)(1) may be practicable and useful for the purposes of this Act, such as—
- (i) indirect emissions from imported electricity, heat, and steam;
- (ii) process and fugitive emissions; and
- (iii) production or importation of greenhouse gases.

(2) VOLUNTARY REPORTING—An entity described in subsection (a) may (along with establishing a baseline and reporting emissions under this section)—
- (A) submit a report described in paragraph (1) before the date specified in that paragraph for the purposes of achieving and commoditizing greenhouse gas reductions through use of the registry; and
- (B) submit to the Administrator, for inclusion in the registry, information that has been verified in accordance with regulations promulgated under section 201(c)(1) and that relates to—
  - (i) any entity-wide greenhouse gas emission reductions activities of the entity that were carried out during or after 1990 and before the establishment of the National Greenhouse Gas Database, verified in accordance with regulations promulgated under section 201(c)(1), and submitted to the Administrator before the date that is 4 years after the date of enactment of this Act; and
  - (ii) with respect to the calendar year preceding the calendar year in which the information is submitted, any project or activity that results in an entity-wide reduction of greenhouse gas emissions or an increase in net sequestration of a greenhouse gas that is carried out by the entity.

(3) PROVISION OF VERIFICATION INFORMATION BY REPORTING ENTITIES—Each entity that submits a report under this subsection shall provide information sufficient for the Administrator to verify, in accordance with measurement and verification methods and standards developed under section 203, that the greenhouse gas report of the reporting entity—
- (A) has been accurately reported; and
- (B) in the case of each voluntary report under paragraph (2), represents—
  - (i) actual reductions in direct greenhouse gas emissions—
    - (I) relative to historic emission levels of the entity; and
    - (II) after accounting for any increases in indirect emissions described in paragraph (1)(C)(i); or
  - (ii) actual increases in net sequestration.

(4) FAILURE TO SUBMIT REPORT—An entity that participates or has participated in the registry and that fails to submit a report required under this subsection shall be prohibited from using, or allowing another entity to use, its registered emissions reductions or increases in sequestration to satisfy the requirements of section 311.

(5) INDEPENDENT THIRD-PARTY VERIFICATION—To meet the requirements of this section and section 203, an entity that is required to submit a report under this section may—
- (A) obtain independent third-party verification; and
- (B) present the results of the third-party verification to the Administrator.

(6) AVAILABILITY OF DATA—
   (A) IN GENERAL—The Administrator shall ensure that information in the database is—
      (i) published; and
      (ii) accessible to the public, including in electronic format on the Internet.
   (B) EXCEPTION—Subparagraph (A) shall not apply in any case in which the Administrator determines that publishing or otherwise making available information described in that subparagraph poses a risk to national security.
(7) DATA INFRASTRUCTURE—The Administrator shall ensure, to the maximum extent practicable, that the database uses, and is integrated with, Federal, State, and regional greenhouse gas data collection and reporting systems in effect as of the date of enactment of this Act.
(8) ADDITIONAL ISSUES TO BE CONSIDERED—In promulgating the regulations under section 201(c)(1) and implementing the database, the Administrator shall take into consideration a broad range of issues involved in establishing an effective database, including—
   (A) the appropriate allowances for reporting each greenhouse gas;
   (B) the data and information systems and measures necessary to identify, track, and verify greenhouse gas emissions in a manner that will encourage private sector trading and exchanges;
   (C) the greenhouse gas reduction and sequestration methods and standards applied in other countries, as applicable or relevant;
   (D) the extent to which available fossil fuels, greenhouse gas emissions, and greenhouse gas production and importation data are adequate to implement the database; and
   (E) the differences in, and potential uniqueness of, the facilities, operations, and business and other relevant practices of persons and entities in the private and public sectors that may be expected to participate in the database.
(9) ANNUAL REPORT—The Administrator shall publish an annual report that—
   (1) describes the total greenhouse gas emissions and emission reductions reported to the database during the year covered by the report;
   (2) provides entity-by-entity and sector-by-sector analyses of the emissions and emission reductions reported;
   (3) describes the atmospheric concentrations of greenhouse gases; and
   (4) provides a comparison of current and past atmospheric concentrations of greenhouse gases.

Sec. 204. Measurement and Verification.
(a) STANDARDS-
   (1) IN GENERAL—Not later than 1 year after the date of enactment of this Act, the Secretary shall develop comprehensive measurement and verification methods and standards to ensure a consistent and technically accurate record of greenhouse gas emissions, emission reductions, sequestration, and atmospheric concentrations for use in the registry.

(2) REQUIREMENTS—The development of methods and standards under paragraph (1) shall include—
  (A) a requirement that a covered entity use a continuous emissions monitoring system, or another system of measuring or estimating emissions that is determined by the Secretary to provide information with the same precision, reliability, accessibility, and timeliness as a continuous emissions monitoring system provides;
  (B) establishment of standardized measurement and verification practices for reports made by all entities participating in the registry, taking into account—
    (i) protocols and standards in use by entities desiring to participate in the registry as of the date of development of the methods and standards under paragraph (1);
    (ii) boundary issues, such as leakage and shifted use;
    (iii) avoidance of double counting of greenhouse gas emissions and emission reductions;
    (iv) protocols to prevent a covered entity from avoiding the requirements of this Act by reorganization into multiple entities that are under common control; and
    (v) such other factors as the Secretary, in consultation with the Administrator, determines to be appropriate;
  (C) establishment of measurement and verification standards applicable to actions taken to reduce, avoid, or sequester greenhouse gas emissions;
  (D) in coordination with the Secretary of Agriculture, standards to measure the results of the use of carbon sequestration and carbon recapture technologies, including—
    (i) organic soil carbon sequestration practices; and
    (ii) forest preservation and reforestation activities that adequately address the issues of permanence, leakage, and verification;
  (E) establishment of such other measurement and verification standards as the Secretary, in consultation with the Secretary of Agriculture, the Administrator, and the Secretary of Energy, determines to be appropriate;
  (F) establishment of standards for obtaining the Secretary's approval of the suitability of geological storage sites that include evaluation of both the geology of the site and the entity's capacity to manage the site; and
  (G) establishment of other features that, as determined by the Secretary, will allow entities to adequately establish a fair and reliable measurement and reporting system.
(b) REVIEW AND REVISION—The Secretary shall periodically review, and revise as necessary, the methods and standards developed under subsection (a).
(c) PUBLIC PARTICIPATION—The Secretary shall—
  (1) make available to the public for comment, in draft form and for a period of at least 90 days, the methods and standards developed under subsection (a); and
  (2) after the 90-day period referred to in paragraph (1), in coordination with the Secretary of Energy, the Secretary of Agriculture, and the Administrator,

adopt the methods and standards developed under subsection (a) for use in implementing the database.

(d) EXPERTS AND CONSULTANTS-

(1) IN GENERAL—The Secretary may obtain the services of experts and consultants in the private and nonprofit sectors in accordance with section 3109 of title 5, United States Code, in the areas of greenhouse gas measurement, certification, and emission trading.

(2) AVAILABLE ARRANGEMENTS—In obtaining any service described in paragraph (1), the Secretary may use any available grant, contract, cooperative agreement, or other arrangement authorized by law.

Title III—Market-Driven Greenhouse Gas Reductions: Subtitle A—Emission Reduction Requirements; Use of Tradeable Allowances

Sec. 311. Covered Entities Must Submit Allowances for Emissions.

(a) IN GENERAL—Beginning with calendar year 2010—

(1) each covered entity in the electric generation, industrial, and commercial sectors shall submit to the Administrator one tradeable allowance for every metric ton of greenhouse gases, measured in units of carbon dioxide equivalence, that it emits;

(2) producer or importer of hydrofluorocarbons, perfluorocarbons, or sulfur hexafluoride that is a covered entity shall submit to the Administrator one tradeable allowance for every metric ton of hydrofluorocarbons, perfluorocarbons, or sulfur hexafluoride it produces or imports, measured in units of carbon dioxide equivalence; and

(3) each petroleum refiner or importer that is a covered entity shall submit one tradeable allowance for every unit of petroleum product it sells that will produce one metric ton of greenhouse gases, measured in units of carbon dioxide equivalence, when used for transportation.

(b) DETERMINATION OF TRANSPORTATION SECTOR AMOUNT—For the transportation sector, the Administrator shall determine the amount of greenhouse gases, measured in units of carbon dioxide equivalence, that will be emitted when petroleum products are used for transportation.

(c) EXCEPTION FOR CERTAIN DEPOSITED EMISSIONS—Notwithstanding subsection (a), a covered entity is not required to submit a tradeable allowance for any amount of greenhouse gas that would otherwise have been emitted from a source under the ownership or control of that entity if—

(1) the emission is deposited in a geological storage facility approved by the Administrator under section 204(a)(2)(F); and

(2) the entity agrees to submit tradeable allowances for any portion of the deposited emission that is subsequently emitted from that facility.

Sec. 312. Compliance.

(a) In General-

(1) SOURCE OF TRADEABLE ALLOWANCES USED—A covered entity may

use a tradeable allowance to meet the requirements of this section without regard to whether the tradeable allowance was allocated to it under subtitle B or acquired from another entity or the Climate Change Credit Corporation established under section 351.

(2) VERIFICATION BY ADMINISTRATOR—At various times during each year, the Administrator shall determine whether each covered entity has met the requirements of this section. In making that determination, the Administrator shall—

(A) take into account tradeable allowances allocated to, or acquired by, that covered entity; and

(B) retire the serial number assigned to each such tradeable allowance so used.

(b) Alternative Means of Compliance From 2010 Through 2015—For the years 2010, 2011, 2012, 2013, 2014, and 2015, a covered entity may satisfy 15 percent of its total allowance submission requirement under this section by—

(1) submitting tradeable allowances from another nation's market in greenhouse gas emissions if—

(A) the Secretary certifies that the other nation's system for trading in greenhouse gas emissions is complete, accurate, and transparent and reviews that determination at least once every 5 years;

(B) the other nation has adopted enforceable limits on its greenhouse gas emissions which the tradeable allowances were issued to implement; and

(C) the covered entity certifies that the tradeable allowance has been retired unused in the other nation's market;

(2) submitting a registered net increase in sequestration, as registered in the National Greenhouse Gas Database established under section 201, adjusted, if necessary, to comply with the accounting standards and methods established under section 372;

(3) submitting a greenhouse gas emissions reduction (other than a registered net increase in sequestration) that was registered in the National Greenhouse Gas Database by a person that is not a covered entity; or

(4) submitting credits obtained from the Administrator under section 314

(c) Alternative Means of Compliance After 2015—For years beginning after 2015, a covered entity may meet the requirements of this section by any means described in subsection (b), except that for the purpose of applying subsection (d) after 2015, "10 percent" shall be substituted for "15 percent."

Sec. 313. Tradeable Allowances and Fuel Economy Standard Credits.

(a) IN GENERAL—Section 32903 of title 49, United States Code, is amended by striking the second sentence of subsection (a) and inserting "The credits may be—

"(1) applied to any of the 3 model years immediately following the model year for which the credits are earned; or

"(2) if the average fuel economy of a manufacturer exceeds the fuel efficiency standards by more than 20 percent, sold to the registry established under section 201 of the Climate Stewardship Act of 2003."

(b) CONVERSION RATIO—The Secretary of Transportation, in consultation with the Administrator, shall determine the conversion factor to be used for purposes

of credits purchased from, or sold to, the registry established under section 201 of this Act and fuel economy standard credits under section 32903 of title 49, United States Code.
  (c) REDUCTION OF TRANSPORTATION SECTOR ALLOCATION—If any manufacturer sells credits under section 32903(a)(2) of title 49, United States Code, to the registry established under section 201 of this Act in any calendar year, the amount of tradeable allowances allocated to the transportation sector under section 311(b) for the next calendar year, and the total allocation of tradeable allowance available for allocation in the next calendar year, shall be reduced by an amount equivalent to the sum of the credits, measured in units of carbon dioxide equivalents, sold to the registry by such manufacturers during the preceding calendar year.

Sec. 314. Borrowing Against Future Reductions.

  (a) IN GENERAL—The Administrator shall establish a program under which a covered entity may—
    (1) receive a credit in the current calendar year for anticipated reductions in emissions in a future calendar year; and
    (2) use the credit in lieu of a tradeable allowance to meet the requirements of this Act for the current calendar year, subject to the limitation imposed by section 312(b).
  (b) DETERMINATION OF TRADEABLE ALLOWANCE CREDITS—The Administrator may make credits available under subsection (a) only for anticipated reductions in emissions that—
    (1) are attributable to the realization of capital investments in equipment, the construction, reconstruction, or acquisition of facilities, or the deployment of new technologies—
      (A) for which the covered entity has executed a binding contract and secured, or applied for, all necessary permits and operating or implementation authority;
      (B) that will not become operational within the current calendar year; and
      (C) that will become operational and begin to reduce emissions from the covered source within 5 years after the year in which the credit is used; and
    (2) will be realized within 5 years after the year in which the credit is used.
  (c) CARRYING COST—If a covered entity uses a credit under this section to meet the requirements of this Act for a calendar year (referred to as the use year), the tradeable allowance requirement for the year from which the credit was taken (referred to as the source year) shall be increased by an amount equal to—
    (1) 10 percent for each credit borrowed from the source year; multiplied by
    (2) the number of years beginning after the use year and before the source year.
  (d) MAXIMUM BORROWING PERIOD—A credit from a year beginning more than 5 years after the current year may not be used to meet the requirements of this Act for the current year.
  (e) FAILURE TO ACHIEVE REDUCTIONS GENERATING CREDIT—If a covered

entity that uses a credit under this section fails to achieve the anticipated reduction for which the credit was granted for the year from which the credit was taken, then—
- (1) the covered entity's requirements under this Act for that year shall be increased by the amount of the credit, plus the amount determined under subsection (c);
- (2) any tradeable allowances submitted by the covered entity for that year shall be counted first against the increase in those requirements; and
- (3) the covered entity may not use credits under this section to meet the increased requirements.

Sec. 315. Other Uses of Tradeable Allowances.
- (a) IN GENERAL—Tradeable allowances may be sold, exchanged, purchased, retired, or used as provided in this section.
- (b) INTERSECTOR TRADING—Covered entities may purchase or otherwise acquire tradeable allowances from other covered sectors to satisfy the requirements of section 311.
- (c) CLIMATE CHANGE CREDIT ORGANIZATION—The Climate Change Credit Corporation established under section 351 may sell tradeable allowances allocated to it under section 332(a)(2) to any covered entity or to any investor, broker, or dealer in such tradeable allowances. The Climate Change Credit Corporation shall use all proceeds from such sales in accordance with the provisions of section 352.
- (d) BANKING OF TRADEABLE ALLOWANCES—Notwithstanding the requirements of section 311, a covered entity that has more than a sufficient amount of tradeable allowances to satisfy the requirements of section 311, may refrain from submitting a tradeable allowance to satisfy the requirements in order to sell, exchange, or use the tradeable allowance in the future.

Sec. 316. Exemption of Source Categories.
- (a) IN GENERAL—The Administrator may grant an exemption from the requirements of this Act to a source category if the Administrator determines, after public notice and comment, that it is not feasible to measure or estimate emissions from that source category.
- (b) REDUCTION OF LIMITATIONS—If the Administrator exempts a source category under subsection (a), the Administrator shall also reduce the total tradeable allowances under section 321(a) as follows:
    - (1) 2010 limitation—For the tradeable allowances under section 311(a)(1), the Administrator shall reduce the total by the amount of greenhouse gas emissions that the exempted source category emitted in calendar year 2000, as identified in the 2000 Inventory.
    - (2) 2016 limitation—For the tradeable allowances under subsection 311(a)(2), the Administrator shall reduce the total by the amount of greenhouse gas emissions that the exempted source category emitted in calendar year 1990, as identified in the 1990 Inventory.
- (c) LIMITATION ON EXEMPTION—The Administrator may not grant an exemption under subsection (a) to carbon dioxide produced from fossil fuel.

Subtitle B—Establishment and Allocation of Tradeable Allowances

Sec. 331. Establishment of Tradeable Allowances.
 (a) IN GENERAL—The Administrator shall promulgate regulations to establish tradeable allowances, denominated in units of carbon dioxide equivalence—
  (1) for calendar years beginning after 2009 and before 2016, equal to—
   (A) 5896 million metric tons, measured in units of carbon dioxide equivalence, reduced by
   (B) the amount of emissions of greenhouse gases in calendar year 2000 from non-covered entities; and
  (2) for calendar years beginning after 2015, equal to—
   (A) 5123 million metric tons, measured in units of carbon dioxide equivalence, reduced by
   (B) the amount of emissions of greenhouse gases in calendar year 1990 from non-covered entities.
 (b) SERIAL NUMBERS—The Administrator shall assign a unique serial number to each tradeable allowance established under subsection (a), and shall take such action as may be necessary to prevent counterfeiting of tradeable allowances.
 (c) NATURE OF TRADEABLE ALLOWANCES—A tradeable allowance is not a property right, and nothing in this title or any other provision of law limits the authority of the United States to terminate or limit a tradeable allowance.
 (d) NON-COVERED ENTITY—In this section:
  (1) IN GENERAL—The term 'non-covered entity' means an entity that—
   (A) owns or controls a source of greenhouse gas emissions in the electric power, industrial, or commercial sectors of the United States economy (as defined in the Inventory), refines or imports petroleum products for use in transportation, or produces or imports hydrofluorocarbons, perfluorocarbons, or sulfur hexafluoride; and
   (B) is not a covered entity, determined by applying the definition in section 3(4) for the year 2000 (for the purpose of subsection (a)(1)(B)) or the year 1990 (for the purpose of subsection (a)(2)(B)).
  (2) EXCEPTION—Notwithstanding paragraph (1), an entity that is a covered entity for any calendar year beginning after 2009 shall not be considered to be a non-covered entity for the purpose of either subsection (a)(1)(B) or subsection (a)(2)(B) only because it emitted, or its products would have emitted, 10,000 metric tons or less of greenhouse gas, measured in units of carbon dioxide equivalence, in the year 2000 or 1990, respectively.

Sec. 332. Determination of Tradeable Allowance Allocations.
 (a) IN GENERAL—The Secretary shall determine—
  (1) the amount of tradeable allowances to be allocated to each covered sector of that sector's Phase I and Phase II allotments; and
  (2) the amount of tradeable allowances to be allocated to the Climate Change Credit Corporation established under section 351.
 (b) ALLOCATION FACTORS—In making the determination required by subsection (a), the Secretary shall consider—

(1) the distributive effect of the allocations on household income and net worth of individuals;
(2) the impact of the allocations on corporate income, taxes, and asset value;
(3) the impact of the allocations on income levels of consumers and on their energy consumption;
(4) the effects of the allocations in terms of economic efficiency;
(5) the ability of covered entities to pass through compliance costs to their customers; and
(6) the degree to which the amount of allocations to the covered sectors should decrease over time.

(c) ALLOCATUION RECOMMENDATIONS AND IMPLEMENTATION—Before allocating or providing tradeable allowances under subsection (a) and within 24 months after the date of enactment of this Act, the Secretary shall submit the determinations under subsection (a) to the Senate Committee on Commerce, Science, and Transportation, the Senate Committee on Environment and Public Works, the House of Representatives Committee on Science, and the House of Representatives Committee on Energy and Commerce. The Secretary's determinations under paragraph (1), including the allocations and provision of tradeable allowances pursuant to that determination, are deemed to be a major rule (as defined in section 804(2) of title 5, United States Code), and subject to the provisions of chapter 8 of that title.

Sec. 333. Allocation of Tradeable Allowances.

(a) IN GENERAL—Beginning with calendar year 2010 and after taking into account any initial allocations under section 334, the Administrator shall—
(1) allocate to each covered sector that sector's Phase I and Phase II allotments determined by the Administrator under section 332 (adjusted for any such initial allocations and the allocation to the Climate Change Credit Corporation established under section 351); and
(2) allocate to the Climate Change Credit Corporation established under section 351 the tradeable allowances allocable to that Corporation.

(b) INTRASECTORIAL ALLOTMENTS—The Administrator shall, by regulation, establish a process for the allocation of tradeable allowances under this section, without cost to facilities within each sector, that will—
(1) encourage investments that increase the efficiency of the processes that produce greenhouse gas emissions;
(2) minimize the costs to the government of allocating the tradeable allowances;
(3) not penalize a covered entity for registered emissions reductions made before 2010; and
(4) provide sufficient allocation for new entrants into the sector.

(c) POINT SOURCE ALLOCATION—The Administrator shall allocate the tradeable allowances for the electricity generation, industrial, and commercial sectors to the entities owning or controlling the point sources of greenhouse gas emissions within that sector.

(d) HYDROFLUOROCARBONS, PERFLUOROCARBONS, AND SULFUR HEXAFLUORIDE—The Administrator shall allocate the tradeable allowances for producers or importers of hydrofluorocarbons, perfluorocarbons, or sulfur hexafluoride one tradeable allowance for every metric ton of hydrofluorocarbons, perfluorocarbons, or sulfur hexafluoride produced or imported, measured in units of carbon dioxide equivalence.

(e) SPECIAL RULE FOR ALLOCATION WITHIN THE TRANSPORTATION SECTOR—
The Administrator shall allocate the tradeable allowances for the transportation sector to petroleum refiners or importers that produce or import petroleum products that will be used as fuel for transportation.

Sec. 334. Initial Allocations for Early Participation and Accelerated Participation.

Before making any allocations under section 333, the Administrator shall allocate—
(1) to any covered entity an amount of tradeable allowances equivalent to the amount of greenhouse gas emissions reductions registered by that covered entity in the national greenhouse gas database if—
   (A) the covered entity has requested to use the registered reduction in the year of allocation;
   (B) the reduction was registered prior to 2010; and
   (C) the Administrator retires the unique serial number assigned to the reduction under section 201(c)(3); and
(2) to any covered entity that has entered into an accelerated participation agreement under section 335, such tradeable allowances as the Administrator has determined to be appropriate under that section.

Sec. 335. Bonus for Accelerated Participation.

(a) IN GENERAL—If a covered entity executes an agreement with the Administrator under which it agrees to reduce its level of greenhouse gas emissions to a level no greater than the level of its greenhouse gas emissions for calendar year 1990 by the year 2010, then, for the 6-year period beginning with calendar year 2010, the Administrator shall—
   (1) provide additional tradeable allowances to that entity when allocating allowances under section 334 in order to recognize the additional emissions reductions that will be required of the covered entity;
   (2) allow that entity to satisfy 20 percent of its requirements under section 311 by—
      (A) submitting tradeable allowances from another nation's market in greenhouse gas emissions under the conditions described in section 312(b)(1);
      (B) submitting a registered net increase in sequestration, as registered in the National Greenhouse Gas Database established under section 201, and as adjusted by the appropriate sequestration discount rate established under section 372; or
      (C) submitting a greenhouse gas emission reduction (other than a registered net increase in sequestration) that was registered in the National Greenhouse Gas Database by a person that is not a covered entity.
(b) TERMINATION—An entity that executes an agreement described in subsection (a) may terminate the agreement at any time.
(c) FAILURE TO MEET COMMITMENT—If an entity that executes an agreement described in subsection (a) fails to achieve the level of emissions to which it committed by calendar year 2010—

(1) its requirements under section 311 shall be increased by the amount of any tradeable allowances provided to it under subsection (a)(1); and

(2) any tradeable allowances submitted thereafter shall be counted first against the increase in those requirements.

Sec. 336. Ensuring Target Adequacy.

(a) IN GENERAL—Beginning 2 years after the date of enactment of this Act, the Under Secretary of Commerce for Oceans and Atmosphere shall review the allowances established by subsection (a) no less frequently than biennially—

(1) to re-evaluate the levels established by that subsection, after taking into account the best available science and the most currently available data, and

(2) to re-evaluate the environmental and public health impacts of specific concentration levels of greenhouse gases, to determine whether the allowances established by subsection (a) continue to be consistent with the objective of the United Nations' Framework Convention on Climate Change of stabilizing levels of greenhouse gas emissions at a level that will prevent dangerous anthropogenic interference with the climate system.

(b) Review of 2010 and 2016 Levels—The Under Secretary shall specifically review in 2008 the level established under section 311(a)(1) and, in 2012, the level established under section 311(a)(2), and transmit a report on his reviews, together with any recommendations, including legislative recommendations, for modification of the levels, to the Senate Committee on Commerce, Science, and Transportation, the Senate Committee on Environment and Public Works, the House of Representatives Committee on Science, and the House of Representatives Committee on Energy and Commerce.

Subtitle C—Climate Change Credit Corporation

Sec. 351. Establishment.

(a) IN GENERAL—The Climate Change Credit Corporation is established as a nonprofit corporation without stock. The Corporation shall not be considered to be an agency or establishment of the United States Government.

(b) APPLICABLE LAWS—The Corporation shall be subject to the provisions of this title and, to the extent consistent with this title, to the District of Columbia Business Corporation Act.

(c) BOARD OF DIRECTORS—The Corporation shall have a board of directors of 5 individuals who are citizens of the United States, of whom 1 shall be elected annually by the board to serve as chairman. No more than 3 members of the board serving at any time may be affiliated with the same political party. The members of the board shall be appointed by the President of the United States, by and with the advice and consent of the Senate and shall serve for terms of 5 years.

Sec. 352. Purposes and Functions.

(a) TRADING—The Corporation—

(1) shall receive and manage tradeable allowances allocated to it under section 333(a)(2); and

(2) shall buy and sell tradeable allowances, whether allocated to it under that section or obtained by purchase, trade, or donation from other entities; but

(3) may not retire tradeable allowances unused.

(b) Use of Tradeable Allowances and Proceeds-

(1) IN GENERAL—The Corporation shall use the tradeable allowances, and proceeds derived from its trading activities in tradeable allowances, to reduce costs borne by consumers as a result of the greenhouse gas reduction requirements of this Act. The reductions—

(A) may be obtained by buy-down, subsidy, negotiation of discounts, consumer rebates, or otherwise;

(B) shall be, as nearly as possible, equitably distributed across all regions of the United States; and

(C) may include arrangements for preferential treatment to consumers who can least afford any such increased costs.

(2) TRANSITION ASSISTANCE TO DISLOCATED WORKERS AND COMMUNITIES—The Corporation shall allocate a percentage of the proceeds derived from its trading activities in tradeable allowances to provide transition assistance to dislocated workers and communities. Transition assistance may take the form of—

(A) grants to employers, employer associations, and representatives of employees—

(i) to provide training, adjustment assistance, and employment services to dislocated workers; and

(ii) to make income-maintenance and needs-related payments to dislocated workers; and

(B) grants to State and local governments to assist communities in attracting new employers or providing essential local government services.

(3) PHASE-OUT OF TRANSITION ASSISTANCE—The percentage allocated by the Corporation under paragraph (2)—

(A) shall be 20 percent for 2010;

(B) shall be reduced by 2 percentage points each year thereafter; and

(C) may not be reduced below zero.

(c) ANNUAL REPORT—The Corporation shall issue an annual report setting forth the results of its operations for the year.

Subtitle D—Sequestration Accounting; Penalties

Sec. 371. Sequestration Accounting.

(a) SEQUESTRATION ACCOUNTING—If a covered entity uses a registered net increase in sequestration to satisfy the requirements of section 311 for any year, that covered entity shall submit information to the Administrator every 5 years thereafter sufficient to allow the Administrator to determine, using the methods and standards created under section 204, whether that net increase in sequestration still exists. Unless the Administrator determines that the net increase in sequestration continues to exist, the covered entity shall offset any loss of sequestration by submitting additional

tradeable allowances of equivalent amount in the calendar year following that determination.
(b) REGULATIONS REQUIRED—The Secretary, acting through the Under Secretary of Commerce for Science and Technology, in coordination with the Secretary of Agriculture, the Secretary of Energy, and the Administrator, shall issue regulations establishing the sequestration accounting rules for all classes of sequestration projects.
(c) CRITERIA FOR REGULATIONS—In issuing regulations under this section, the Secretary shall use the following criteria:
    (1) If the range of possible amounts of net increase in sequestration for a particular class of sequestration project is not more than 10 percent of the median of that range, the amount of sequestration awarded shall be equal to the median value of that range.
    (2) If the range of possible amounts of net increase in sequestration for a particular class of sequestration project is more than 10 percent of the median of that range, the amount of sequestration awarded shall be equal to the fifth percentile of that range.
    (3) The regulations shall include procedures for accounting for potential leakage from sequestration projects and for ensuring that any registered increase in sequestration is in addition that which would have occurred if this Act had not been enacted.
(d) UPDATES—The Secretary shall update the sequestration accounting rules for every class of sequestration project at least once every 5 years.

Sec. 372. Penalties.

Any covered entity that fails to meet the requirements of section 311 for a year shall be liable for a civil penalty, payable to the Administrator, equal to thrice the market value (determined as of the last day of the year at issue) of the tradeable allowances that would be necessary for that covered entity to meet those requirements on the date of the emission that resulted in the violation.

Analysis

Dubbed by critics as "Kyoto Lite," the Climate Stewardship Act was a bipartisan bill drafted by Senators Joe Lieberman and John McCain to require the United States to make minimal cuts to greenhouse gas emissions. While the cuts would be much more modest than what President William Clinton's administration agreed to during the negotiations on the Kyoto Protocol to the United Nations Framework Convention on Climate Change, it would at least mark progress on the road to reducing the impacts of global warming.

The Climate Stewardship Act was modeled on the Clean Air Act. The similarities proved a major problem as some critics noted that greenhouse gases were not viewed as a threat in any fashion in the Clean Air Act of 1990, yet suddenly, at the behest of scientists who had not yet proven definitively the existence of global warming, it was a global calamity. Other opponents of the legislation claimed that following the lead of the United Nations, even at a reduced level, was not worth the resulting harm to the nation's economy and workforce. The senators and lobbyists that worked to scuttle the legislation won the day by a vote of 55 to 43.

FURTHER READING

Choi, Inho. 2005. "Global Climate Change and the Use of Economic Approaches: The Ideal Design Features of Domestic Greenhouse Gas Emissions Trading with an Analysis of the European Union's CO2 Emissions Directive and the Climate Stewardship Act." *Natural Resources Journal* 45: 865–952.

Fisher, Dana R. 2006. "Bringing the Material Back In: Understanding the U.S. Position on Climate Change." *Sociological Forum* 21: 467–494.

Hot & Cold Media Spin Cycle
A Challenge to Journalists Who Cover Global Warming

Date: September 25, 2006
Location: Washington, D.C.
Significance: Senator James Inhofe, Republican from Oklahoma, charged that the press was promoting the existence of global warming by constantly presenting stories filled with dire predictions while ignoring evidence that did not support their political positions.
Source: Inhofe, James. 2006. *Hot & Cold Media Spin Cycle: A Challenge To Journalists Who Cover Global Warming.* http://www.inhofe.senate.gov/download/?id=3a1d408b-a0c2-4d70-9135-0841029472cd& download=1 (accessed July 22, 2015).

I am going to speak today about the most media-hyped environmental issue of all time, global warming. I have spoken more about global warming than any other politician in Washington today. My speech will be a bit different from the previous seven floor speeches, as I focus not only on the science, but on the media's coverage of climate change.

Global Warming—just that term evokes many members in this chamber, the media, Hollywood elites and our pop culture to nod their heads and fret about an impending climate disaster. As the senator who has spent more time speaking about the facts regarding global warming, I want to address some of the recent media coverage of global warming and Hollywood's involvement in the issue. And of course I will also discuss former Vice President Al Gore's movie "An Inconvenient Truth."

Since 1895, the media has alternated between global cooling and warming scares during four separate and sometimes overlapping time periods. From 1895 until the 1930's the media peddled a coming ice age.

From the late 1920's until the 1960's they warned of global warming. From the 1950's until the 1970's they warned us again of a coming ice age. This makes modern global warming the fourth estate's fourth attempt to promote opposing climate change fears during the last 100 years.

Recently, advocates of alarmism have grown increasingly desperate to try to convince the public that global warming is the greatest moral issue of our generation. Last year, the vice president of London's Royal Society sent a chilling letter to the media encouraging them to stifle the voices of scientists skeptical of climate alarmism.

During the past year, the American people have been served up an unprecedented parade of environmental alarmism by the media and entertainment industry, which link every possible weather event to global warming. The year 2006 saw many major organs of the media dismiss any pretense of balance and objectivity on climate change coverage and instead crossed squarely into global warming advocacy.

Summary of Latest Developments of Manmade Global Warming Hockey Stick

First, I would like to summarize some of the recent developments in the controversy over whether or not humans have created a climate catastrophe. One of the key aspects that the United Nations, environmental groups and the media have promoted as the "smoking gun" of proof of catastrophic global warming is the so-called 'hockey stick' temperature graph by climate scientist Michael Mann and his colleagues.

This graph purported to show that temperatures in the Northern Hemisphere remained relatively stable over 900 years, then spiked upward in the 20th century presumably due to human activity. Mann, who also co-publishes a global warming propaganda blog reportedly set up with the help of an environmental group, had his "Hockey Stick" come under severe scrutiny.

The "hockey stick" was completely and thoroughly broken once and for all in 2006. Several years ago, two Canadian researchers tore apart the statistical foundation for the hockey stick. In 2006, both the National Academy of Sciences and an independent researcher further refuted the foundation of the "hockey stick."

The National Academy of Sciences report reaffirmed the existence of the Medieval Warm Period from about 900 AD to 1300 AD and the Little Ice Age from about 1500 to 1850. Both of these periods occurred long before the invention of the SUV or human industrial activity could have possibly impacted the Earth's climate. In fact, scientists believe the Earth was warmer than today during the Medieval Warm Period, when the Vikings grew crops in Greenland.

Climate alarmists have been attempting to erase the inconvenient Medieval Warm Period from the Earth's climate history for at least a decade. David Deming, an assistant professor at the University of Oklahoma's College of Geosciences, can testify firsthand about this effort. Dr. Deming was welcomed into the close-knit group of global warming believers after he published a paper in 1995 that noted some warming in the 20th century. Deming says he was subsequently contacted by a prominent global warming alarmist and told point blank "We have to get rid of the Medieval Warm Period." When the "Hockey Stick" first appeared in 1998, it did just that. END OF LITTLE ICE AGE MEANS WARMING

The media have missed the big pieces of the puzzle when it comes to the Earth's temperatures and mankind's carbon dioxide (CO_2) emissions. It is very simplistic to feign horror and say the one degree Fahrenheit temperature increase during the 20th century means we are all doomed. First of all, the one degree Fahrenheit rise coincided with the greatest advancement of living standards, life expectancy, food production and human health in the history of our planet. So it is hard to argue that the global warming we experienced in the 20th century was somehow negative or part of a catastrophic trend.

Second, what the climate alarmists and their advocates in the media have continued to ignore is the fact that the Little Ice Age, which resulted in harsh winters which froze New York Harbor and caused untold deaths, ended about 1850. So trying to prove man-made global warming by comparing the well-known fact that today's temperatures are warmer than during the Little Ice Age is akin to comparing summer to winter to show a catastrophic temperature trend.

In addition, something that the media almost never addresses are the holes in the theory that C02 has been the driving force in global warming. Alarmists fail to adequately explain why temperatures began warming at the end of the Little Ice Age in about 1850, long before man-made CO2 emissions could have impacted the climate. Then about 1940, just as man-made CO2 emissions rose sharply, the temperatures began a decline that lasted until the 1970's, prompting the media and many scientists to fear a coming ice age.

Let me repeat, temperatures got colder after C02 emissions exploded. If C02 is the driving force of global climate change, why do so many in the media ignore the many skeptical scientists who cite these rather obvious inconvenient truths?

Sixty Scientists

My skeptical views on man-made catastrophic global warming have only strengthened as new science comes in. There have been recent findings in peer-reviewed literature over the last few years showing that the Antarctic is getting colder and the ice is growing and a new study in Geophysical Research Letters found that the sun was responsible for 50% of 20th century warming.

Recently, many scientists, including a leading member of the Russian Academy of Sciences, predicted longterm global cooling may be on the horizon due to a projected decrease in the sun's output.

A letter sent to the Canadian Prime Minister on April 6 of this year by 60 prominent scientists who question the basis for climate alarmism, clearly explains the current state of scientific knowledge on global warming.

The 60 scientists wrote: "If, back in the mid-1990s, we knew what we know today about climate, Kyoto would almost certainly not exist, because we would have concluded it was not necessary." The letter also noted: "'Climate change is real' is a meaningless phrase used repeatedly by activists to convince the public that a climate catastrophe is looming and humanity is the cause. Neither of these fears is justified. Global climate changes occur all the time due to natural causes and the human impact still remains impossible to distinguish from this natural 'noise.'"

Computer Models Threaten Earth

One of the ways alarmists have pounded this mantra of "consensus" on global warming into our pop culture is through the use of computer models which project future calamity. But the science is simply not there to place so much faith in scary computer model scenarios which extrapolate the current and projected buildup of greenhouse gases in the atmosphere and conclude that the planet faces certain doom.

Dr. Vincent Gray, a research scientist and a 2001 reviewer with the UN's Intergovernmental Panel on Climate Change (IPCC) has noted, "The effects of aerosols, and their uncertainties, are such as to nullify completely the reliability of any of the climate models."

Earlier this year, the director of the International Arctic Research Center in Fairbanks Alaska, testified to Congress that highly publicized climate models showing a disappearing Arctic were nothing more than "science fiction."

In fact, after years of hearing about the computer generated scary scenarios about the future of our planet, I now believe that the greatest climate threat we face may be coming from alarmist computer models.

This threat is originating from the software installed on the hard drives of the publicity and grant seeking climate modelers. It is long past the time for us to separate climate change fact from hysteria.

Kyoto: Economic Pain for No Climate Gain

One final point on the science of climate change: I am approached by many in the media and others who ask, "What if you are wrong to doubt the dire global warming predictions? Will you be able to live with yourself for opposing the Kyoto Protocol?"

My answer is blunt. The history of the modern environmental movement is chock full of predictions of doom that never came true. We have all heard the dire predictions about the threat of overpopulation, resource scarcity, mass starvation, and the projected death of our oceans. None of these predictions came true, yet it never stopped the doomsayers from continuing to predict a dire environmental future.

The more the eco-doomsayers' predictions fail, the more the eco-doomsayers predict. These failed predictions are just one reason I respect the serious scientists out there today debunking the latest scaremongering on climate change. Scientists like MIT's Richard Lindzen, former Colorado State climatologist Roger Pielke, Sr., the University of Alabama's Roy Spencer and John Christy, Virginia State Climatologist Patrick Michaels, Colorado State University's William Gray, atmospheric physicist S. Fred Singer, Willie Soon of the Harvard Smithsonian Center for Astrophysics, Oregon State climatologist George Taylor and astrophysicist Sallie Baliunas, to name a few.

But more importantly, it is the global warming alarmists who should be asked the question—"What if they are correct about man-made catastrophic global warming?"— because they have come up with no meaningful solution to their supposed climate crisis in the two decades that they have been hyping this issue.

If the alarmists truly believe that man-made greenhouse gas emissions are dooming the planet, then they must face up to the fact that symbolism does not solve a supposed climate crisis. The alarmists freely concede that the Kyoto Protocol, even if fully ratified and complied with, would not have any meaningful impact on global temperatures. And keep in mind that Kyoto is not even close to being complied with by many of the nations that ratified it, including 13 of the EU-15 nations that are not going to meet their emission reduction promises.

Many of the nations that ratified Kyoto are now realizing what I have been saying all along: The Kyoto Protocol is a lot of economic pain for no climate gain.

Legislation that has been proposed in this chamber would have even less of a temperature effect than Kyoto's undetectable impact. And more recently, global warming alarmists and the media have been praising California for taking action to limit C02. But here again: This costly feel-good California measure, which is actually far less severe than Kyoto, will have no impact on the climate—only the economy.

Symbolism does not solve a climate crisis.

But this symbolism may be hiding a dark side. While greenhouse gas limiting proposals may cost the industrialized West trillions of dollars, it is the effect on the developing world's poor that is being lost in this debate.

The Kyoto Protocol's post 2012 agenda which mandates that the developing world be subjected to restrictions on greenhouse gases could have the potential to severely restrict development in regions of the world like Africa, Asia and South America—where some of the Earth's most energy-deprived people currently reside.

Expanding basic necessities like running water and electricity in the developing world are seen by many in the green movement as a threat to the planet's health that must be avoided.

Energy poverty equals a life of back-breaking poverty and premature death.

If we allow scientifically unfounded fears of global warming to influence policy makers to restrict future energy production and the creation of basic infrastructure in the developing world—billions of people will continue to suffer.

Last week my committee heard testimony from Danish statistician Bjorn Lomborg, who was once a committed left-wing environmentalist until he realized that so much of what that movement preached was based on bad science. Lomborg wrote a book called "The Skeptical Environmentalist" and has organized some of the world's top Nobel Laureates to form the 2004 "Copenhagen Consensus" which ranked the world's most pressing problems.

And guess what?

They placed global warming at the bottom of the list in terms of our planet's priorities. The "Copenhagen Consensus" found that the most important priorities of our planet included: combating disease, stopping malaria, securing clean water, and building infrastructure to help lift the developing nations out of poverty. I have made many trips to Africa, and once you see the devastating poverty that has a grip on that continent, you quickly realize that fears about global warming are severely misguided.

I firmly believe that when the history of our era is written, future generations will look back with puzzlement and wonder why we spent so much time and effort on global warming fears and pointless solutions like the Kyoto Protocol.

French President Jacques Chirac provided the key clue as to why so many in the international community still revere the Kyoto Protocol, who in 2000 said Kyoto represents "the first component of an authentic global governance."

Furthermore, if your goal is to limit C02 emissions, the only effective way to go about it is the use of cleaner, more efficient technologies that will meet the energy demands of this century and beyond.

The Bush administration and my Environment and Public Works Committee have been engaged in these efforts as we work to expand nuclear power and promote the Asia-Pacific Partnership. This partnership stresses the sharing of new technology among member nations including three of the world's top 10 emitters—China, India and South Korea—all of whom are exempt from Kyoto.

Media Coverage of Climate Change

Many in the media, as I noted earlier, have taken it upon themselves to drop all pretense of balance on global warming and instead become committed advocates for the issue.

Here is a quote from Newsweek magazine:

"There are ominous signs that the Earth's weather patterns have begun to change dramatically and that these changes may portend a drastic decline in food production—with serious political implications for just about every nation on Earth."

A headline in the New York Times reads: "Climate Changes Endanger World's Food Output." Here is a quote from Time Magazine:

"As they review the bizarre and unpredictable weather pattern of the past several years, a growing number of scientists are beginning to suspect that many seemingly contradictory meteorological fluctuations are actually part of a global climatic upheaval."

All of this sounds very ominous. That is, until you realize that the three quotes I just read were from articles in 1975 editions of Newsweek Magazine and The New York Times, and Time Magazine in 1974.

They weren't referring to global warming; they were warning of a coming ice age.

Let me repeat, all three of those quotes were published in the 1970's and warned of a coming ice age.

In addition to global cooling fears, Time Magazine has also reported on global warming. Here is an example:

"[Those] who claim that winters were harder when they were boys are quite right... weathermen have no doubt that the world at least for the time being is growing warmer."

Before you think that this is just another example of the media promoting Vice President Gore's movie, you need to know that the quote I just read you from Time Magazine was not a recent quote; it was from January 2, 1939.

Yes, in 1939. Nine years before Vice President Gore was born and over three decades before Time Magazine began hyping a coming ice age and almost five decades before they returned to hyping global warming.

Time Magazine in 1951 pointed to receding permafrost in Russia as proof that the planet was warming.

In 1952, the New York Times noted that the "trump card" of global warming "has been the melting glaciers."

But Media Could Not Decide Between Warming or Cooling Scares

There are many more examples of the media and scientists flip-flopping between warming and cooling scares.

Here is a quote from the New York Times reporting on fears of an approaching ice age.

"Geologists Think the World May be Frozen Up Again."

That sentence appeared over 100 years ago in the February 24, 1895, edition of the New York Times.

Let me repeat. 1895, not 1995.

A front page article in the October 7, 1912, New York Times, just a few months after the Titanic struck an iceberg and sank, declared that a prominent professor "Warns Us of an Encroaching Ice Age."

The very same day in 1912, the Los Angeles Times ran an article warning that the "Human race will have to fight for its existence against cold." An August 10, 1923, Washington Post article declared: "Ice Age Coming Here."

By the 1930's, the media took a break from reporting on the coming ice age and instead switched gears to promoting global warming:

"America in Longest Warm Spell Since 1776; Temperature Line Records a 25-year Rise" stated an article in the New York Times on March 27, 1933. The media of yesteryear was also not above injecting large amounts of fear and alarmism into their climate articles.

An August 9, 1923, front page article in the Chicago Tribune declared:

"Scientist Says Arctic Ice Will Wipe Out Canada." The article quoted a Yale University professor who predicted that large parts of Europe and Asia would be "wiped out" and Switzerland would be "entirely obliterated."

A December 29, 1974, New York Times article on global cooling reported that climatologists believed "the facts of the present climate change are such that the most optimistic experts would assign near certainty to major crop failure in a decade."

The article also warned that unless government officials reacted to the coming catastrophe, "mass deaths by starvation and probably in anarchy and violence" would result. In 1975, the New York Times reported that "A major cooling [was] widely considered to be inevitable." These past predictions of doom have a familiar ring, don't they? They sound strikingly similar to our modern media promotion of former Vice president's brand of climate alarmism.

After more than a century of alternating between global cooling and warming, one would think that this media history would serve a cautionary tale for today's voices in the media and scientific community who are promoting yet another round of eco-doom.

Much of the 100-year media history on climate change that I have documented here today can be found in a publication titled "Fire and Ice" from the Business and Media Institute.

Media Coverage in 2006

Which raises the question: Has this embarrassing 100-year documented legacy of coverage on what turned out to be trendy climate science theories made the media more skeptical of today's sensational promoters of global warming?

You be the judge.

On February 19th of this year, CBS News's "60 Minutes" produced a segment on the North Pole. The segment was a completely one-sided report, alleging rapid and unprecedented melting at the polar cap.

It even featured correspondent Scott Pelley claiming that the ice in Greenland was melting so fast, that he barely got off an ice-berg before it collapsed into the water.

"60 Minutes" failed to inform its viewers that a 2005 study by a scientist named Ola Johannessen and his colleagues showing that the interior of Greenland is gaining ice and mass and that according to scientists, the Arctic was warmer in the 1930's than today.

On March 19th of this year "60 Minutes" profiled NASA scientist and alarmist James Hansen, who was once again making allegations of being censored by the Bush administration.

In this segment, objectivity and balance were again tossed aside in favor of a one-sided glowing profile of Hansen.

The "60 Minutes" segment made no mention of Hansen's partisan ties to former Democrat Vice President Al Gore or Hansen's receiving of a grant of a quarter of a million dollars from the left-wing Heinz Foundation run by Teresa Heinz Kerry. There was also no mention of Hansen's subsequent endorsement of her husband John Kerry for President in 2004.

Many in the media dwell on any industry support given to so-called climate skeptics, but the same media completely fail to note Hansen's huge grant from the left-wing Heinz Foundation.

The foundation's money originated from the Heinz family ketchup fortune. So it appears that the media makes a distinction between oil money and ketchup money.

"60 Minutes" also did not inform viewers that Hansen appeared to concede in a 2003 issue of Natural Science that the use of "extreme scenarios" to dramatize climate change "may have been appropriate at one time" to drive the public's attention to the issue.

Why would "60 Minutes" ignore the basic tenets of journalism, which call for objectivity and balance in sourcing, and do such one-sided segments? The answer was provided by correspondent Scott Pelley. Pelley told the CBS News website that he justified excluding scientists skeptical of global warming alarmism from his segments because he considers skeptics to be the equivalent of "Holocaust deniers."

This year also saw a New York Times reporter write a children's book entitled" The North Pole Was Here." The author of the book, New York Times reporter Andrew Revkin, wrote that it may someday be "easier to sail to than stand on" the North Pole in summer. So here we have a very prominent environmental reporter for the New York Times who is promoting aspects of global warming alarmism in a book aimed at children.

Time Magazine Hypes Alarmism

In April of this year, Time Magazine devoted an issue to global warming alarmism titled "Be Worried, Be Very Worried." This is the same Time Magazine which first warned of a coming ice age in 1920's before switching to warning about global warming in the 1930's before switching yet again to promoting the 1970's coming ice age scare.

The April 3, 2006, global warming special report of Time Magazine was a prime example of the media's shortcomings, as the magazine cited partisan left-wing environmental groups with a vested financial interest in hyping alarmism.

Headlines blared: "More and More Land is Being Devastated by Drought"

"Earth at the Tipping Point"

"The Climate is Crashing,"

Time Magazine did not make the slightest attempt to balance its reporting with any views with scientists skeptical of this alleged climate apocalypse.

I don't have journalism training, but I dare say calling a bunch of environmental groups with an obvious fund-raising agenda and asking them to make wild speculations on how bad global warming might become, is nothing more than advocacy for their left-wing causes. It is a violation of basic journalistic standards.

To his credit, New York Times reporter Revkin saw fit to criticize Time Magazine for its embarrassing coverage of climate science.

So in the end, Time's cover story title of "Be Worried, Be Very Worried," appears to have been apt. The American people should be worried—very worried—of such shoddy journalism.

Al Gore Inconvenient Truth

In May, our nation was exposed to perhaps one of the slickest science propaganda films of all time: former Vice President Gore's "An Inconvenient Truth." In addition to having the backing of Paramount Pictures to market this film, Gore had the full backing of the media, and leading the cheerleading charge was none other than the Associated Press.

On June 27, the Associated Press ran an article by Seth Borenstein that boldly declared "Scientists give two thumbs up to Gore's movie." The article quoted only five scientists praising Gore's science, despite AP's having contacted over 100 scientists.

The fact that over 80% of the scientists contacted by the AP had not even seen the movie or that many scientists have harshly criticized the science presented by Gore did not dissuade the news outlet one bit from its mission to promote Gore's brand of climate alarmism.

I am almost at a loss as to how to begin to address the series of errors, misleading science and unfounded speculation that appear in the former Vice President's film Here is what Richard Lindzen, a meteorologist from MIT has written about "An Inconvenient Truth." "A general characteristic of Mr. Gore's approach is to assiduously ignore the fact that the earth and its climate are dynamic; they are always changing even without any external forcing. To treat all change as something to fear is bad enough; to do so in order to exploit that fear is much worse."

What follows is a very brief summary of the science that the former Vice President promotes in either a wrong or misleading way:

- He promoted the now debunked "hockey stick" temperature chart in an attempt to prove man's overwhelming impact on the climate
- He attempted to minimize the significance of Medieval Warm period and the Little Ice Age
- He insisted on a link between increased hurricane activity and global warming that most sciences believe does not exist.
- He asserted that today's Arctic is experiencing unprecedented warmth while ignoring that temperatures in the 1930's were as warm or warmer
- He claimed the Antarctic was warming and losing ice but failed to note, that is only true of a small region and the vast bulk has been cooling and gaining ice.
- He hyped unfounded fears that Greenland's ice is in danger of disappearing
- He erroneously claimed that ice cap on Mt. Kilimanjaro is disappearing due to global warming, even while the region cools and researchers blame the ice loss on local land-use practices
- He made assertions of massive future sea level rise that is way out side of any supposed scientific "consensus" and is not supported in even the most alarmist literature.
- He incorrectly implied that a Peruvian glacier's retreat is due to global warming, while ignoring the fact that the region has been cooling since the 1930s and other glaciers in South America are advancing
- He blamed global warming for water loss in Africa's Lake Chad, despite NASA scientists concluding that local population and grazing factors are the more likely culprits

- He inaccurately claimed polar bears are drowning in significant numbers due to melting ice when in fact they are thriving
- He completely failed to inform viewers that the 48 scientists who accused President Bush of distorting science were part of a political advocacy group set up to support Democrat Presidential candidate John Kerry in 2004

Now that was just a brief sampling of some of the errors presented in "An Inconvenient Truth." Imagine how long the list would have been if I had actually seen the movie—there would not be enough time to deliver this speech today.

Tom Brokaw

Following the promotion of "An Inconvenient Truth," the press did not miss a beat in their role as advocates for global warming fears.

ABC News put forth its best effort to secure its standing as an advocate for climate alarmism when the network put out a call for people to submit their anecdotal global warming horror stories in June for use in a future news segment.

In July, the Discovery Channel presented a documentary on global warming narrated by former NBC anchor Tom Brokaw. The program presented only those views of scientists promoting the idea that humans are destroying the Earth's climate.

You don't have to take my word for the program's overwhelming bias; a Bloomberg News TV review noted "You'll find more dissent at a North Korean political rally than in this program" because of its lack of scientific objectivity.

Brokaw also presented climate alarmist James Hansen to viewers as unbiased, failing to note his quarter million dollar grant from the partisan Heinz Foundation or his endorsement of Democrat Presidential nominee John Kerry in 2004 and his role promoting former Vice President Gore's Hollywood movie.

Brokaw, however, did find time to impugn the motives of scientists skeptical of climate alarmism when he featured paid environmental partisan Michael Oppenhimer of the group Environmental Defense accusing skeptics of being bought out by the fossil fuel interests.

The fact remains that political campaign funding by environmental groups to promote climate and environmental alarmism dwarfs spending by the fossil fuel industry by a three-to-one ratio. Environmental special interests, through their 527s, spent over $19 million compared to the $7 million that Oil and Gas spent through PACs in the 2004 election cycle.

I am reminded of a question the media often asks me about how much I have received in campaign contributions from the fossil fuel industry. My unapologetic answer is 'Not Enough,'—especially when you consider the millions partisan environmental groups pour into political campaigns.

Engineered "Consensus"

Continuing with our media analysis: On July 24, 2006, The Los Angeles Times featured an op-ed by Naomi Oreskes, a social scientist at the University of California San

Diego and the author of a 2004 Science Magazine study. Oreskes insisted that a review of 928 scientific papers showed there was 100% consensus that global warming was not caused by natural climate variations. This study was also featured in former Vice President Gore's "An Inconvenient Truth."

However, the analysis in Science Magazine excluded nearly 11,000 studies or more than 90 percent of the papers dealing with global warming, according to a critique by British social scientist Benny Peiser. Peiser also pointed out that less than two percent of the climate studies in the survey actually endorsed the so-called "consensus view" that human activity is driving global warming and some of the studies actually opposed that view.

But despite this manufactured "consensus," the media continued to ignore any attempt to question the orthodoxy of climate alarmism.

As the dog days of August rolled in, the American people were once again hit with more hot hype regarding global warming, this time from The New York Times op-ed pages. A columnist penned an August 3rd column filled with so many inaccuracies it is a wonder the editor of the Times saw fit to publish it.

For instance, Bob Herbert's column made dubious claims about polar bears, the snows of Kilimanjaro and he attempted to link this past summer's heat wave in the U.S. to global warming—something even alarmist James Hansen does not support.

Polar Bears Look Tired?

Finally, a September 15, 2006, Reuters News article claimed that polar bears in the Arctic are threatened with extinction by global warming. The article by correspondent Alister Doyle, quoted a visitor to the Arctic who claims he saw two distressed polar bears. According to the Reuters article, the man noted that "one of [the polar bears] looked to be dead and the other one looked to be exhausted." The article did not state the bears were actually dead or exhausted, rather that they "looked" that way.

Have we really arrived at the point where major news outlets in the U.S. are reduced to analyzing whether or not polar bears in the Arctic appear restful? How does reporting like this get approved for publication by the editors at Reuters? What happened to covering the hard science of this issue?

What was missing from this Reuters news article was the fact that according to biologists who study the animals, polar bears are doing quite well. Biologist Dr. Mitchell Taylor from the Arctic government of Nunavut, a territory of Canada, refuted these claims in May when he noted that

"Of the 13 populations of polar bears in Canada, 11 are stable or increasing in number. They are not going extinct, or even appear to be affected at present."

Sadly, it appears that reporting anecdotes and hearsay as fact, has now replaced the basic tenets of journalism for many media outlets.

Alarmism Has Led to Skepticism

It is an inconvenient truth that so far, 2006 has been a year in which major segments of the media have given up on any quest for journalistic balance, fairness and objectivity when it comes to climate change. The global warming alarmists and their friends in the media have attempted to smear scientists who dare question the premise of man-made catastrophic global warming, and as a result some scientists have seen their reputations and research funding dry up.

The media has so relentlessly promoted global warming fears that a British group called the Institute for Public Policy Research—and this from a left leaning group—issued a report in 2006 accusing media 16 outlets of engaging in what they termed "climate porn" in order to attract the public's attention.

Bob Carter, a Paleoclimate geologist from James Cook University in Australia has described how the media promotes climate fear:

"Each such alarmist article is larded with words such as 'if,' 'might,' 'could,' 'probably,' 'perhaps,' 'expected,' 'projected' or 'modeled'—and many involve such deep dreaming, or ignorance of scientific facts and principles, that they are akin to nonsense," professor Carter concluded in an op-ed in April of this year.

Another example of this relentless hype is the reporting on the seemingly endless number of global warming impact studies which do not even address whether global warming is going to happen. They merely project the impact of potential temperature increases.

The media endlessly hypes studies that purportedly show that global warming could increase mosquito populations, malaria, West Nile Virus, heat waves and hurricanes, threaten the oceans, damage coral reefs, boost poison ivy growth, damage vineyards, and global food crops, to name just a few of the global warming linked calamities. Oddly, according to the media reports, warmer temperatures almost never seem to have any positive effects on plant or animal life or food production.

Fortunately, the media's addiction to so-called 'climate porn' has failed to seduce many Americans.

According to a July Pew Research Center Poll, the American public is split about evenly between those who say global warming is due to human activity versus those who believe it's from natural factors or not happening at all.

In addition, an August Los Angeles Times/Bloomberg poll found that most Americans do not attribute the cause of recent severe weather events to global warming, and the portion of Americans who believe global warming is naturally occurring is on the rise.

Yes—it appears that alarmism has led to skepticism.

The American people know when their intelligence is being insulted. They know when they are being used and when they are being duped by the hysterical left.

The American people deserve better—much better—from our fourth estate. We have a right to expect accuracy and objectivity on climate change coverage. We have a right to expect balance in sourcing and fair analysis from reporters who cover the issue.

Above all, the media must roll back this mantra that there is scientific "consensus" of impending climatic doom as an excuse to ignore recent science. After all, there was a so-called scientific "consensus" that there were nine planets in our solar system until Pluto was recently demoted.

Breaking the cycles of media hysteria will not be easy since hysteria sells—it's very profitable. But I want to challenge the news media to reverse course and report on the objective science of climate change, to stop ignoring legitimate voices this scientific debate and to stop acting as a vehicle for unsubstantiated hype.

Analysis

Senator James Inhofe of Oklahoma (Republican) became the chairman of the Senate's Environment and Public Works Committee in 2003. Already viewed as the most ardent denier of climate change in Congress, Inhofe used his post to invite climate change skeptics to testify as experts on the topic before his committee. Angered that the press gave far more credence and coverage to the testimony of environmental activists asking for reductions in the emissions of fossil fuels and greenhouse gases, namely the experts favored by the Democratic Party members of the committee, Inhofe took to the floor of the United Senate and delivered a speech that charged that the supposedly impartial press was pursuing a political agenda by hyping the existence and dire consequences of global warming. He claimed that much of the evidence cited by journalists came not from the environment, but from fundamentally flawed climate models. While hyping "evidence" that supported their political agenda, the press ignored data that would scientifically challenge what the press considered a "consensus" that global warming was already severely altering the climate.

Inhofe was particularly critical of the press' fawning support of former Vice President Al Gore's film *An Inconvenient Truth*. Although, according to Inhofe, the film was rife with falsehoods and mischaracterizations, it was widely hailed and promoted by both the press and Hollywood elites. The film eventually won an Oscar and helped Al Gore earn the Nobel Peace Prize.

Further Reading
Inhofe, James. 2012. *The Greatest Hoax: How the Global Warming Conspiracy Threatens Your Future*. Washington, D.C.: WND.

Atmosphere of Pressure
Political Interference in Federal Climate Science (2007) (Executive Summary)

Date: February 2007
Location: Cambridge, Massachusetts
Significance: In response to allegations from scientists employed by federal agencies that their research was being interfered with by political operatives from the Executive Branch, both the Union of Concerned Scientists and Government Accountability Project launched investigations to determine the merit of the complaints. The results of their separate investigations are presented within this document.
Source: Union of Concerned Scientists. 2007. *Atmosphere of Pressure: Political Interference in Federal Climate Science (2007).* http://www.ucsusa.org/our-work/center-science-and-democracy/promoting-scientific-integrity/atmosphere-of-pressure.html#.Vh1n9CszRQk (accessed October 13, 2015).

Federal climate science research is at the forefront of assessing fundamental causes of global warming and the future dangers it could pose to our nation and the world. Such research is of tremendous value to many Americans planning for these risks, including coastal communities designing infrastructure for protecting against storm surges; civil authorities planning for heat waves; power companies preparing for higher peak energy demands; forest managers planning wildfire management programs; ski resort owners investing in snow-making equipment; and policy makers evaluating energy legislation. Therefore, it is crucial that the best available science on climate change be disseminated to the public, through government websites, reports, and press releases. In recent years, however, this science has been increasingly tailored to reflect political goals rather than scientific fact.

Out of concern that inappropriate political interference and media favoritism are compromising federal climate science, the Union of Concerned Scientists (UCS) and the Government Accountability Project (GAP) undertook independent investigations of federal climate science. UCS mailed a questionnaire to more than 1,600 climate scientists at seven federal agencies to gauge the extent to which politics was playing a role in scientists' research. Surveys were also sent to scientists at the independent (non-federal) National Center for Atmospheric Research (NCAR) to serve as a comparison with the experience of federal scientists. About 19 percent of all scientists responded (279 from federal agencies and 29 from NCAR). At the same time, GAP conducted 40 in-depth interviews with federal climate scientists and other officials and analyzed thousands of pages of government documents, obtained through the Freedom of Information Act (FOIA) and inside sources, regarding agency media policies and congressional communications.

These two complementary investigations arrived at similar conclusions regarding the state of federal climate research and the need for strong policies to protect the integrity of science and the free flow of scientific information.

Political Interference with Climate Science

The federal government needs accurate scientific information to craft effective policies. Political interference with the work of federal scientists threatens the quality and integrity of these policies. As such, no scientist should ever encounter any of the various types of political interference described in our survey questions. Yet unacceptably large numbers of federal climate scientists personally experienced instances of interference over the past five years:

- Nearly half of all respondents (46 percent of all respondents to the question) perceived or personally experienced pressure to eliminate the words "climate change," "global warming," or other similar terms from a variety of communications.
- Two in five (43 percent) perceived or personally experienced changes or edits during review that changed the meaning of scientific findings.
- Nearly half (46 percent) perceived or personally experienced new or unusual administrative requirements that impair climate-related work.
- One-quarter (25 percent) perceived or personally experienced situations in which scientists have actively objected to, resigned from, or removed themselves from a project because of pressure to change scientific findings.
- Asked to quantify the number of incidents of interference of all types, 150 scientists (58 percent) said they had personally experienced one or more such incidents within the past five years, for a total of at least 435 incidents of political interference.

The more frequently a climate scientist's work touches on sensitive or controversial issues, the more interference he or she reported. More than three-quarters (78 percent) of those survey respondents who self-reported that their research "always" or "frequently" touches on issues that could be considered sensitive or controversial also reported they had personally experienced at least one incident of inappropriate interference. More than one-quarter (27 percent) of this same group had experienced six or more such incidents in the past five years.

In contrast to this evidence of widespread interference in climate science at federal agencies, scientists at the independent National Center for Atmospheric Research (NCAR), who are not federal employees, reported far fewer instances of interference. Only 22 percent of all NCAR respondents had personally experienced such incidents over the past five years.

Barriers to Communication

Federal scientists have a constitutional right to speak about their scientific research, and the American public has a right to be informed of the findings of taxpayer-supported research.

Restrictions on scientists who report findings contrary to an administration's preferred policies undermine these basic rights. These practices also contribute to a general misunderstanding of the findings of climate science and degrade our government's ability to make effective policies on topics ranging from public health to agriculture to disaster preparation.

The investigation uncovered numerous examples of public affairs officers at federal agencies taking a highly active role in regulating communications between agency scientists and the media—in effect serving as gatekeepers for scientific information.

Among the examples taken from interviews and FOIA documents:

- One agency scientist, whose research illustrates a possible connection between hurricanes and global warming, was repeatedly barred from speaking to the media. Press inquiries on the subject were routed to another scientist whose views more closely matched official administration policy.
- Government scientists routinely encounter difficulty in obtaining approval for official press releases that highlight research into the causes and consequences of global warming.
- Scientists report that public affairs officers are sometimes present at or listen in on interviews between certain scientists and the media.
- Both scientists and journalists report that restrictive media policies and practices have had the effect of slowing down the process by which interview requests are approved. As a result, the number of contacts between government scientists and the news media has been greatly reduced.

Highly publicized incidents of interference have led at least one agency to implement reforms; in February 2006, NASA adopted a scientific openness policy that affirms the right of open scientific communication. Perhaps as a result, 61 percent of NASA survey respondents said recent policies affirming scientific openness at their agency have improved the environment for climate research. While imperfect, the new NASA media policy stands as a model for the type of action other federal agencies should take in reforming their media policies.

The investigation also highlighted problems with the process by which scientific findings are communicated to policy makers in Congress. One example, taken from internal documents provided to GAP by agency staff, shows edits to official questions for the record by political appointees, which change the meaning of the scientific findings being presented.

Inadequate Funding

When adjusted for inflation, funding for federal climate science research has declined since the mid-1990s. A majority of survey respondents disagreed that the government has done a good job funding climate science, and a large number of scientists warned that inadequate levels of funding are harming the capacity of researchers to make progress in understanding the causes and effects of climate change. Budget cuts that have forced the cancellation of crucial Earth observation satellite programs were of particular concern to respondents.

Poor Morale

Morale among federal climate scientists is generally poor. The UCS survey results suggest a correlation between the deterioration in morale and the politicized environment surrounding federal climate science in the present administration. One primary danger of low morale and decreased funding is that federal agencies may have more difficulty attracting and keeping the best scientists.

A large number of respondents reported decreasing job satisfaction and a worsening environment for climate science in federal agencies:

- Two-thirds of respondents said that today's environment for federal government climate research is worse compared with 5 years ago (67 percent) and 10 years ago (64 percent). Among scientists at NASA, these numbers were higher (79 percent and 77 percent, respectively).
- 45 percent said that their personal job satisfaction has decreased over the past few years. At NASA, three in five (61 percent) reported decreased job satisfaction.
- 36 percent of respondents from NASA, and 22 percent of all respondents, reported that morale in their office was "poor" or "extremely poor." Among NCAR respondents, only seven percent reported such low levels of morale.

Recommendations

This report has brought to light numerous ways in which U.S. federal climate science has been filtered, suppressed, and manipulated in the last five years. Until this political interference ends, the United States will not be able to fully protect Americans and the world from the dangers of a warming planet. Creating systems to ensure long-term independent and accessible science will require the energies of the entire federal government.

UCS and GAP recommend the following reforms and actions:

- The federal government must respect the constitutional right of scientists to speak about any subject, including policy-related matters and those outside their area of expertise, so long as the scientists make it clear that they do so in their private capacity, and such communications do not take from agency time and resources. Scientists should also be made aware of these rights and ensure they are exercised at their agencies.
- Ultimate decisions about the communication of federal scientific information should lie with scientists themselves. While non-scientists may be helpful with various aspects of writing and communication, scientists must have a "right of last review" on agency communications related to their scientific research to ensure scientific accuracy has been maintained.
- Pre-approval and monitoring of media interviews with federal scientists by public affairs officials should be eliminated. Scientists should not be subject to restrictions on media contacts beyond a policy of informing public affairs officials in advance of an interview and summarizing the interaction for them afterward.
- Federal agencies should clearly support the free exchange of scientific information in all venues. They should investigate and correct inappropriate policies, practices,

and incidents that threaten scientific integrity, determine how and why problems have occurred, and make the necessary reforms to prevent further incidents.
- Congress should immediately exert pressure on the Executive branch to comply with its statutory duty under federal law and undertake periodic scientific assessments of climate change that address the consequences for the United States. (The last national assessment was conducted in 2000.)
- Funding decisions regarding climate change programs should be guided by scientific criteria, and must take into account the importance of long-term, continual climate observation programs and models.

The reality of global warming, including the role of heat-trapping gases from human activities in driving climate change, has been repeatedly affirmed by scientific experts. Every day that the government chooses to ignore climate science is a day it fails to protect future generations from the consequences of global warming. Our government must commit to ensuring basic scientific freedoms and support scientists in their endeavors to bring scientific results to the policy arena, scientific fora, and a wide array of other audiences. Addressing climate change is a matter of national preparedness.

Analysis

As a presidential candidate, George W. Bush was unabashedly skeptical of claims that global climate change was underway due to the actions of mankind. He was particularly critical of the Kyoto Protocol, which he perceived as a significant threat to the United States' economy. Upon assuming the presidency, one of his early actions was to formally remove the nation from the Kyoto Protocol's dictates. With a president that was openly hostile to the major climate change agreement of the day, it is not surprising that his administration followed his lead. It was alleged that Bush's subordinates closely monitored the work of the scientists conducting research on climate change within the Executive Branch agencies that they supervised and ensured that their conclusions did not threaten the narrative that the President espoused. This was supposedly accomplished using numerous strategies, including the editing of reports or the outright suppression of documents.

In fairness to the George W. Bush administration, it is important to note that political disagreements over global warming shaped the debate on the censorship of federal scientific research. Since the Republicans were dubious as to whether global warming existed, it is easy to understand why they shifted resources from the study of climate change's causes and impacts. In their eyes, it was "junk science." It was, from their point-of-view, a poor use of limited financial resources. On the other hand, if one thought that the climate science of the day was valid, like the federal scientists and the leaders of the Democratic Party, then the Republican administration was shirking its responsibility to protect the nation's future. The administration certainly contributed to the idea that it was hostile to science through known actions like the quashing of a report produced by the Environmental Protection Agency that supported the science undergirding the Kyoto Protocol.

Simmering tensions between the administration and federal scientists came to the fore in 2006 when Dr. James E. Hansen, a noted climate scientist and head of NASA's Goddard Institute for Space Studies, complained that the George W. Bush administration was actively preventing him from presenting research showing the ongoing impacts of global climate change. Since his reputation was unassailable, he was able to provide a voice to other federal scientists who were much more vulnerable to retribution by politicians that they might cross. Although Hansen was able to spark investigations and subsequent reports such as the one reprinted above, he subsequently alleged that NASA was punished for his actions through the loss of critical funding in the federal budget.

Further Reading

Bowen, Mark. 2008. *Censoring Science: Inside the Political Attack on Dr. James Hansen and the Truth of Global Warming*. New York: Dutton.

Mooney, Chris. 2005. *The Republican War on Science*. New York: Basic.

Mooney, Chris. 2007. *Storm World: Hurricanes, Politics, and the Battle Over Global Warming*. New York: Houghton Mifflin Harcourt.

Massachusetts et al. v. Environmental Protection Agency et al. (Syllabus)

Date: April 2, 2007
Location: Washington, D.C.
Significance: The ruling in this case provided the Environmental Protection Agency the authority to regulate greenhouse gases under the Clean Air Act, as amended in 1990.
Source: Supreme Court of the United States. 2007. *Massachusetts et al. v. Environmental Protection Agency et al.* http://www.supremecourt.gov/opinions/06pdf/05-1120.pdf (accessed April 30, 2016).

Massachusetts et al. v. Environmental Protection Agency et al. Certiorari to the United States Court of Appeals for the District of Columbia Circuit

No. 05-1120. Argued November 29, 2006—Decided April 2, 2007

Based on respected scientific opinion that a well-documented rise in global temperatures and attendant climatological and environmental changes have resulted from a significant increase in the atmospheric concentration of "greenhouse gases," a group of private organizations petitioned the Environmental Protection Agency (EPA) to begin regulating the emissions of four such gases, including carbon dioxide, under §202(a)(1) of the Clean Air Act, which requires that the EPA "shall by regulation prescribe ... standards applicable to the emission of any air pollutant from any class ... of new motor vehicles ... which in [the EPA Administrator's] judgment cause[s], or contribute[s] to, air pollution ... reasonably ... anticipated to endanger public health or welfare," 42 U.S. C. §7521(a)(1). The Act defines "air pollutant" to include "any air pollution agent ... , including any physical, chemical ... substance ... emitted into ... the ambient air." §7602(g). EPA ultimately denied the petition, reasoning that (1) the Act does not authorize it to issue mandatory regulations to address global climate change, and (2) even if it had the authority to set greenhouse gas emission standards, it would have been unwise to do so at that time because a causal link between greenhouse gases and the increase in global surface air temperatures was not unequivocally established. The agency further characterized any EPA regulation of motor-vehicle emissions as a piecemeal approach to climate change that would conflict with the President's comprehensive approach involving additional support for technological innovation, the creation of nonregulatory programs to encourage voluntary private-sector reductions in greenhouse gas emissions, and further research on climate change, and might hamper the President's ability to persuade key developing nations to reduce emissions.

Petitioners, now joined by intervenor Massachusetts and other state and local governments, sought review in the D. C. Circuit. Although each of the three judges on the panel wrote separately, two of them agreed that the EPA Administrator properly exercised his discretion in denying the rulemaking petition. One judge concluded that the Administrator's exercise of "judgment" as to whether a pollutant could "reasonably be anticipated to endanger public health or welfare," §7521(a)(1), could be based on scientific uncertainty as well as other factors, including the concern that unilateral U.S. regulation of motor-vehicle emissions could weaken efforts to reduce other countries' greenhouse gas emissions. The second judge opined that petitioners had failed to demonstrate the particularized injury to them that is necessary to establish standing under Article III, but accepted the contrary view as the law of the case and joined the judgment on the merits as the closest to that which he preferred. The court therefore denied review.

Held:
1. Petitioners have standing to challenge the EPA's denial of their rulemaking petition. Pp. 12–23.

    (a) This case suffers from none of the defects that would preclude it from being a justiciable Article III "Controvers[y]." See, e.g., Luther v. Borden, 7 How. 1. Moreover, the proper construction of a congressional statute is an eminently suitable question for federal-court resolution, and Congress has authorized precisely this type of challenge to EPA action, see 42 U.S. C. §7607(b)(1). Contrary to EPA's argument, standing doctrine presents no insuperable jurisdictional obstacle here. To demonstrate standing, a litigant must show that it has suffered a concrete and particularized injury that is either actual or imminent, that the injury is fairly traceable to the defendant, and that a favorable decision will likely redress that injury. See Lujan v. Defenders of Wildlife, 504 U.S. 555, 560–561. However, a litigant to whom Congress has "accorded a procedural right to protect his concrete interests," id., at 573, n. 7—here, the right to challenge agency action unlawfully withheld, §7607(b)(1)—"can assert that right without meeting all the normal standards for redressability and immediacy," ibid. Only one petitioner needs to have standing to authorize review. See Rumsfeld v. Forum for Academic and Institutional Rights, Inc., 547 U.S. 47, 52, n. 2. Massachusetts has a special position and interest here. It is a sovereign State and not, as in Lujan, a private individual, and it actually owns a great deal of the territory alleged to be affected. The sovereign prerogatives to force reductions in greenhouse gas emissions, to negotiate emissions treaties with developing countries, and (in some circumstances) to exercise the police power to reduce motor-vehicle emissions are now lodged in the Federal Government. Because congress has ordered EPA to protect Massachusetts (among others) by prescribing applicable standards, §7521(a)(1), and has given Massachusetts a concomitant procedural right to challenge the rejection of its rulemaking petition as arbitrary and capricious, §7607(b)(1), petitioners' submissions as they pertain to Massachusetts have satisfied the most demanding standards of the adversarial process. EPA's steadfast refusal to regulate greenhouse gas emissions presents a risk of harm to Massachusetts that is both "actual" and "imminent," Lujan, 504 U.S., at 560, and there is a "substantial likelihood that the judicial relief requested" will prompt EPA to take steps to reduce that risk, Duke Power Co. v. Carolina Environmental Study Group, Inc., 438 U.S. 59, 79. Pp. 12–17.

(b) The harms associated with climate change are serious and well recognized. The Government's own objective assessment of the relevant science and a strong consensus among qualified experts indicate that global warming threatens, inter alia, a precipitate rise in sea levels, severe and irreversible changes to natural ecosystems, a significant reduction in winter snowpack with direct and important economic consequences, and increases in the spread of disease and the ferocity of weather events. That these changes are widely shared does not minimize Massachusetts' interest in the outcome of this litigation. See Federal Election Comm'n v. Akins, 524 U.S. 11, 24. According to petitioners' uncontested affidavits, global sea levels rose between 10 and 20 centimeters over the 20th century as a result of global warming and have already begun to swallow Massachusetts' coastal land. Remediation costs alone, moreover, could reach hundreds of millions of dollars. Pp. 17–19.

(c) Given EPA's failure to dispute the existence of a causal connection between man-made greenhouse gas emissions and global warming, its refusal to regulate such emissions, at a minimum, "contributes" to Massachusetts' injuries. EPA overstates its case in arguing that its decision not to regulate contributes so insignificantly to petitioners' injuries that it cannot be haled into federal court, and that there is no realistic possibility that the relief sought would mitigate global climate change and remedy petitioners' injuries, especially since predicted increases in emissions from China, India, and other developing nations will likely offset any marginal domestic decrease EPA regulation could bring about. Agencies, like legislatures, do not generally resolve massive problems in one fell swoop, see Williamson v. Lee Optical of Okla., Inc., 348 U.S. 483, 489, but instead whittle away over time, refining their approach as circumstances change and they develop a more nuanced understanding of how best to proceed, cf. SEC v. Chenery Corp., 332 U.S. 194, 202–203. That a first step might be tentative does not by itself negate federal-court jurisdiction. And reducing domestic automobile emissions is hardly tentative. Leaving aside the other greenhouse gases, the record indicates that the U.S. transportation sector emits an enormous quantity of carbon dioxide into the atmosphere. Pp. 20–21.

(d) While regulating motor-vehicle emissions may not by itself reverse global warming, it does not follow that the Court lacks jurisdiction to decide whether EPA has a duty to take steps to slow or reduce it. See Larson v. Valente, 456 U.S. 228, 243, n. 15. Because of the enormous potential consequences, the fact that a remedy's effectiveness might be delayed during the (relatively short) time it takes for a new motor-vehicle fleet to replace an older one is essentially irrelevant. Nor is it dispositive that developing countries are poised to substantially increase greenhouse gas emissions: A reduction in domestic emissions would slow the pace of global emissions increases, no matter what happens elsewhere. The Court attaches considerable significance to EPA's espoused belief that global climate change must be addressed. Pp. 21–23.

2. The scope of the Court's review of the merits of the statutory issues is narrow. Although an agency's refusal to initiate enforcement proceedings is not ordinarily subject to judicial review, Heckler v. Chaney, 470 U.S. 821, there are key differences between nonenforcement and denials of rulemaking petitions that are, as in the present circumstances, expressly authorized. EPA concluded alternatively in its petition denial that it lacked authority under §7521(a)(1) to regulate new vehicle emissions because carbon dioxide is not an "air pollutant" under §7602, and that, even if it possessed authority, it would decline to exercise it because regulation would conflict with other administration priorities. Because the Act expressly permits review of such an action, §7607(b)(1), this Court "may reverse [it if it finds it to be] arbitrary, capricious, an abuse of discretion, or otherwise not in accordance with law," §7607(d)(9). Pp. 24–25.

3. Because greenhouse gases fit well within the Act's capacious definition of "air pollutant," EPA has statutory authority to regulate emission of such gases from new motor vehicles. That definition—which includes "any air pollution agent..., including any physical, chemical, ... substance ... emitted into ... the ambient air ...," §7602(g) (emphasis added)—embraces all airborne compounds of whatever stripe. Moreover, carbon dioxide and other greenhouse gases are undoubtedly "physical [and] chemical ... substance[s]." Ibid. EPA's reliance on postenactment congressional actions and deliberations it views as tantamount to a command to refrain from regulating greenhouse gas emissions is unavailing. Even if postenactment legislative history could shed light on the meaning of an otherwise-unambiguous statute, EPA identifies nothing suggesting that Congress meant to curtail EPA's power to treat greenhouse gases as air pollutants. The Court has no difficulty reconciling Congress' various efforts to promote interagency collaboration and research to better understand climate change with the agency's preexisting mandate to regulate "any air pollutant" that may endanger the public welfare. FDA v. Brown & Williamson Tobacco Corp., 529 U.S. 120, 133, distinguished. Also unpersuasive is EPA's argument that its regulation of motor-vehicle carbon dioxide emissions would require it to tighten mileage standards, a job (according to EPA) that Congress has assigned to the Department of Transportation. The fact that DOT's mandate to promote energy efficiency by setting mileage standards may overlap with EPA's environmental responsibilities in no way licenses EPA to shirk its duty to protect the public "health" and "welfare," §7521(a)(1). Pp. 25–30.

4. EPA's alternative basis for its decision—that even if it has statutory authority to regulate greenhouse gases, it would be unwise to do so at this time—rests on reasoning divorced from the statutory text. While the statute conditions EPA action on its formation of a "judgment," that judgment must relate to whether an air pollutant "cause[s], or contribute[s] to, air pollution which may reasonably be anticipated to endanger public health or welfare." §7601(a)(1). Under the Act's clear terms, EPA can avoid promulgating regulations only if it determines that greenhouse gases do not contribute to climate change or if it provides some reasonable explanation as to why it cannot or will not exercise its discretion to determine whether they do. It has refused to do so, offering instead a laundry list of reasons not to regulate, including the existence of voluntary Executive Branch programs providing a response to global warming and impairment of the President's ability to negotiate with developing nations to reduce emissions. These policy judgments

have nothing to do with whether greenhouse gas emissions contribute to climate change and do not amount to a reasoned justification for declining to form a scientific judgment. Nor can EPA avoid its statutory obligation by noting the uncertainty surrounding various features of climate change and concluding that it would therefore be better not to regulate at this time. If the scientific uncertainty is so profound that it precludes EPA from making a reasoned judgment, it must say so. The statutory question is whether sufficient information exists for it to make an endangerment finding. Instead, EPA rejected the rulemaking petition based on impermissible considerations. Its action was therefore "arbitrary, capricious, or otherwise not in accordance with law," §7607(d)(9). On remand, EPA must ground its reasons for action or inaction in the statute. Pp. 30–32.
415 F. 3d 50, reversed and remanded.

STEVENS, J., delivered the opinion of the Court, in which KENNEDY, SOUTER, GINSBURG, and BREYER, JJ., joined. ROBERTS, C. J., filed a dissenting opinion, in which SCALIA, THOMAS, and ALITO, JJ., joined. SCALIA, J., filed a dissenting opinion, in which ROBERTS, C. J., and THOMAS and ALITO, JJ., joined.

Analysis

The Clean Air Act (CAA), as amended in 1990, was intended to address the most pressing threats to the environment that existed at the time, such as acid rain and the accumulation of ozone in urban areas. Although the role of greenhouse gases (GHGs) as a cause of global warming was already perceived in some quarters as a growing issue, they were not specifically mentioned in the list of pollutants in the CAA that posed a threat to the air. To the officials from President George W. Bush's administration that had overseen the Environmental Protection Agency (EPA) since 2001, this meant that they were not authorized by the CAA to regulate GHGs. A number of states, including Massachusetts, disagreed with that assessment. After determining that the Bush administration had little interest in utilizing the EPA to reduce GHG emissions, they turned to the courts to force the EPA to act.

The case, *Massachusetts et al. v. Environmental Protection Agency et al.*, had its origins in October 1999 when the International Center for Technology requested that the EPA regulate carbon dioxide and three other GHGs that were being emitted into the atmosphere by motor vehicles. The EPA declined the request claiming they lacked the authority to regulate GHGs. The agency also added that even if they did have that authority under the CAA, they still would not have agreed to the petition. With that precedent set, the George W. Bush administration saw no need to expand the EPA's regulatory authority when they came to power.

California, Connecticut, Illinois, Maine, Massachusetts, New Jersey, New Mexico, New York, Oregon, Rhode Island, Vermont, and Washington, along with three cities and a number of environmental groups, filed suit at the United States Court of Appeals for the D.C. Circuit alleging that the request by the International Center for Technology should have been honored. The U.S. Court of Appeals upheld the decision by the EPA, but the United States Supreme Court decided to review the decision at the request of Massachusetts.

The two questions before the Supreme Court Justices in the case were whether GHGs were pollutants as defined by the CAA and, if so, did they pose a threat to the health and welfare of the general populace. Oral arguments were held on November 29, 2006, with a final decision rendered on April 2, 2007. In a 5–4 ruling, the Justices determined that GHGs were a pollutant and posed a real risk to the health and welfare of the citizens of the United States. The EPA was thus was required to regulate GHG emissions. In response to the ruling, the EPA in 2009 officially determined that GHGs posed a threat to human health and thus issued rules describing how the agency would regulate GHGs into the future.

H. Res. 593

Congratulating scientists F. Sherwood Rowland, Mario Molina, and Paul Crutzen for their work in atmospheric chemistry, particularly concerning the formation and decomposition of ozone, that led to the development of the Montreal Protocol on Substances that Deplete the Ozone Layer

Date: September 17, 2007
Location: Washington, D.C.
Significance: The resolution identifies how the pioneering work of American scientists F. Sherwood Rowland and Mario Molina helped lead to the Montreal Protocol.
Source: United States House of Representatives, 110th Congress. 2007. *H. Res. 593*. https://www.congress.gov/110/bills/hres593/BILLS-110hres593eh.pdf (accessed August 22, 2015).

Whereas in 1973, on the University of California, Irvine campus, chemists F. Sherwood Rowland and Mario Molina began researching the depletion of stratospheric ozone by the chlorofluorocarbon gases then used worldwide as refrigerants and aerosol propellants;

Whereas on June 28, 1974, F. Sherwood Rowland and Mario Molina published in the scientific journal Nature, their path-breaking article, "Stratospheric Sink for Chlorofluoromethanes: Chlorine Atom-Catalysed Destruction of Ozone";

Whereas in 1976, the work of F. Sherwood Rowland and Mario Molina connecting chlorofluorocarbons and atmospheric ozone depletion was confirmed by the National Academy of Sciences;

Whereas in 1978, the United States banned chlorofluorocarbons as propellants in aerosol cans;

Whereas in 1987, because of the research of F. Sherwood Rowland, Mario Molina, Paul Crutzen, and many other scientists, the international community acted through the adoption of the Montreal Protocol on Substances that Deplete the Ozone Layer ("Montreal Protocol");

Whereas the Montreal Protocol created the Multilateral Fund for the Implementation of the Montreal Protocol which provides funds to help developing countries to phase out the use of ozone-depleting substances;

Whereas the Multilateral Fund for Implementation of the Montreal Protocol was the first financial mechanism to be created under an international treaty;

Whereas the Montreal Protocol recognized that world-wide emissions of certain substances can significantly deplete and otherwise modify the ozone layer in a manner that is likely to result in adverse effects on human health and the environment;

Whereas because of the adoption of the Montreal Protocol the levels of chlorofluorocarbon gases in the Earth's atmosphere have decreased;

Whereas on September 17, 1987, the Montreal Protocol was open for signatures;

Whereas to date, 191 nations have signed the Montreal Protocol;

Whereas F. Sherwood Rowland, Mario Molina, and Paul Crutzen were awarded the Nobel Prize for Chemistry in 1995 for their work in atmospheric chemistry, particularly concerning the formation and decomposition of ozone; and

Whereas September 17, 2007, marks the twentieth anniversary of the signing of the Montreal Protocol: Now, therefore, be it

Resolved, That the House of Representatives—

(1) congratulates scientists F. Sherwood Rowland, Mario Molina, and Paul Crutzen for their work in atmospheric chemistry, particularly concerning the formation and decomposition of ozone, that led to the development of the Montreal Protocol on Substances that Deplete the Ozone Layer; and
(2) encourages the continued research of the interaction of humans and their actions with the Earth's ecosystem.

Analysis

The research conducted in the early 1970s by Frank Sherwood Rowland and Mario Molina at the University of California, Irvine on the impact of chlorofluorocarbons (CFCs) on the atmosphere's ozone layer is detailed in the analysis of "The Montreal Protocol on Substances that Deplete the Ozone Layer (Excerpts)." Their work culminated in 1974 with the publication of an article in which they detailed their ozone depletion hypothesis to the scientific community. In 1976, the United States Academy of Sciences released a report that supported Rowland and Molina's theory. In response, the federal government banned the use of CFCs as propellants in aerosol cans in 1978.

Despite the potential implications of their research on the future of the ozone layer, the threat posed was largely ignored internationally as it was viewed as a theoretical issue. That began to change when researchers noticed a doughnut-shaped hole in the ozone layer that emerged in October 1978 over Antarctica. By 1985, the hole was roughly the size of North America. Its existence spurred the drafting and passage of the Montreal Protocol. The pioneering work of Sherwood and Molina on CFCs and their impact on the atmosphere resulted in them sharing the 1995 Nobel Prize in Chemistry with Paul Crutzen, from the Netherlands.

It is sobering to consider that at the time the House of Representatives was congratulating two American scientists for their influential work on chlorofluorocarbons; many of its members were at the same time publicly oblivious to the threat posed to the climate by another largely man-made product, namely greenhouse gas emissions.

FURTHER READING

Benedick, Richard Elliott. 1998. *Ozone Diplomacy: New Directions in Safeguarding the Planet*. Cambridge, MA: Harvard University Press.

Grenthe, Ingmar. 1995. *The Nobel Prize in Chemistry 1995*. http://www.nobelprize.org/nobel_prizes/chemistry/laureates/1995/presentation-speech.html (accessed March 30, 2016).

Rowland, F. Sherwood. 1989. "Chlorofluorocarbons and the Depletion of Stratospheric Ozone." *American Scientist* 77: 36–45.

Global Warming Twenty Years Later
Tipping Points Near

Date: June 23, 2008
Location: Washington, D.C.
Significance: Dr. James Hansen, Director of the NASA Goddard Institute for Space Studies, warned that although the existence of climate change was widely accepted, countries like the United States had been slow to respond to the threats it posed. Without immediate action, "tipping points" would be reached, resulting in the reduced ability of human beings to curb climate change.
Source: Hansen, James. 2008. *Global Warming Twenty Years Later: Tipping Points Near.* http://www.columbia.edu/~jehl/2008/TwentyYearsLater_20080623.pdf (accessed August 6, 2015).

My presentation today is exactly 20 years after my 23 June 1988 testimony to Congress, which alerted the public that global warming was underway. There are striking similarities between then and now, but one big difference.

Again a wide gap has developed between what is understood about global warming by the relevant scientific community and what is known by policymakers and the public. Now, as then, frank assessment of scientific data yields conclusions that are shocking to the body politic. Now, as then, I can assert that these conclusions have a certainty exceeding 99 percent.

The difference is that now we have used up all slack in the schedule for actions needed to defuse the global warming time bomb. The next President and Congress must define a course next year in which the United States exerts leadership commensurate with our responsibility for the present dangerous situation.

Otherwise it will become impractical to constrain atmospheric carbon dioxide, the greenhouse gas produced in burning fossil fuels, to a level that prevents the climate system from passing tipping points that lead to disastrous climate changes that spiral dynamically out of humanity's control.

Changes needed to preserve creation, the planet on which civilization developed, are clear. But the changes have been blocked by special interests, focused on short-term profits, who hold sway in Washington and other capitals.

I argue that a path yielding energy independence and a healthier environment is, barely, still possible. It requires a transformative change of direction in Washington in the next year.

On 23 June 1988 I testified to a hearing, chaired by Senator Tim Wirth of Colorado, that the Earth had entered a long-term warming trend and that human-made greenhouse gases almost surely were responsible. I noted that global warming enhanced both extremes of the water cycle, meaning stronger droughts and forest fires, on the one hand, but also heavier rains and floods.

My testimony two decades ago was greeted with skepticism. But while skepticism is the lifeblood of science, it can confuse the public. As scientists examine a topic from all perspectives, it may appear that nothing is known with confidence. But from such broad openminded study of all data, valid conclusions can be drawn.

My conclusions in 1988 were built on a wide range of inputs from basic physics, planetary studies, observations of on-going changes, and climate models. The evidence was strong enough that I could say it was time to "stop waffling." I was sure that time would bring the scientific community to a similar consensus, as it has.

While international recognition of global warming was swift, actions have faltered. The U.S. refused to place limits on its emissions, and developing countries such as China and India rapidly increased their emissions.

What is at stake? Warming so far, about two degrees Fahrenheit over land areas, seems almost innocuous, being less than day-to-day weather fluctuations. But more warming is already "in the pipeline," delayed only by the great inertia of the world ocean. And climate is nearing dangerous tipping points. Elements of a "perfect storm," a global cataclysm, are assembled.

Climate can reach points such that amplifying feedbacks spur large rapid changes. Arctic sea ice is a current example. Global warming initiated sea ice melt, exposing darker ocean that absorbs more sunlight, melting more ice. As a result, without any additional greenhouse gases, the Arctic soon will be ice-free in the summer.

More ominous tipping points loom. West Antarctic and Greenland ice sheets are vulnerable to even small additional warming. These two-mile-thick behemoths respond slowly at first, but if disintegration gets well underway it will become unstoppable. Debate among scientists is only about how much sea level would rise by a given date. In my opinion, if emissions follow a business-as-usual scenario, sea level rise of at least two meters is likely this century. Hundreds of millions of people would become refugees. No stable shoreline would be reestablished in any time frame that humanity can conceive.

Animal and plant species are already stressed by climate change. Polar and alpine species will be pushed off the planet, if warming continues. Other species attempt to migrate, but as some are extinguished their interdependencies can cause ecosystem collapse. Mass extinctions, of more than half the species on the planet, have occurred several times when the Earth warmed as much as expected if greenhouse gases continue to increase. Biodiversity recovered, but it required hundreds of thousands of years.

The disturbing conclusion, documented in a paper I have written with several of the world's leading climate experts, is that the safe level of atmospheric carbon dioxide is no more than 350 ppm (parts per million) and it may be less. Carbon dioxide amount is already 385 ppm and rising about 2 ppm per year. Stunning corollary: the oft-stated goal to keep global warming less than two degrees Celsius (3.6 degrees Fahrenheit) is a recipe for global disaster, not salvation.

These conclusions are based on paleoclimate data showing how the Earth responded to past levels of greenhouse gases and on observations showing how the world is responding to today's carbon dioxide amount. The consequences of continued increase of greenhouse gases extend far beyond extermination of species and future sea level rise.

Arid subtropical climate zones are expanding poleward. Already an average expansion of about 250 miles has occurred, affecting the southern United States, the Mediterranean region, Australia and southern Africa. Forest fires and drying-up of lakes will increase further unless carbon dioxide growth is halted and reversed.

Mountain glaciers are the source of fresh water for hundreds of millions of people. These glaciers are receding world-wide, in the Himalayas, Andes and Rocky Mountains. They will disappear, leaving their rivers as trickles in late summer and fall, unless the growth of carbon dioxide is reversed. Coral reefs, the rainforest of the ocean, are home for one-third of the species in the sea.

Coral reefs are under stress for several reasons, including warming of the ocean, but especially because of ocean acidification, a direct effect of added carbon dioxide. Ocean life dependent on carbonate shells and skeletons is threatened by dissolution as the ocean becomes more acid.

Such phenomena, including the instability of Arctic sea ice and the great ice sheets at today's carbon dioxide amount, show that we have already gone too far. We must draw down atmospheric carbon dioxide to preserve the planet we know. A level of no more than 350 ppm is still feasible, with the help of reforestation and improved agricultural practices, but just barely—time is running out.

Requirements to halt carbon dioxide growth follow from the size of fossil carbon reservoirs. Coal towers over oil and gas. Phase out of coal use except where the carbon is captured and stored below ground is the primary requirement for solving global warming.

Oil is used in vehicles where it is impractical to capture the carbon. But oil is running out. To preserve our planet we must also ensure that the next mobile energy source is not obtained by squeezing oil from coal, tar shale or other fossil fuels.

Fossil fuel reservoirs are finite, which is the main reason that prices are rising. We must move beyond fossil fuels eventually. Solution of the climate problem requires that we move to carbon-free energy promptly.

Special interests have blocked transition to our renewable energy future. Instead of moving heavily into renewable energies, fossil companies choose to spread doubt about global warming, as tobacco companies discredited the smoking-cancer link. Methods are sophisticated, including funding to help shape school textbook discussions of global warming.

CEOs of fossil energy companies know what they are doing and are aware of long-term consequences of continued business as usual. In my opinion, these CEOs should be tried for high crimes against humanity and nature.

Conviction of ExxonMobil and Peabody Coal CEOs will be no consolation, if we pass on a runaway climate to our children. Humanity would be impoverished by ravages of continually shifting shorelines and intensification of regional climate extremes. Loss of countless species would leave a more desolate planet.

If politicians remain at loggerheads, citizens must lead. We must demand a moratorium on new coal-fired power plants. We must block fossil fuel interests who aim to squeeze every last drop of oil from public lands, off-shore, and wilderness areas. Those last drops are no solution. They yield continued exorbitant profits for a short-sighted self-serving industry, but no alleviation of our addiction or long-term energy source.

Moving from fossil fuels to clean energy is challenging, yet transformative in ways that will be welcomed. Cheap, subsidized fossil fuels engendered bad habits. We import food from halfway around the world, for example, even with healthier products available from nearby fields. Local produce would be competitive if not for fossil fuel subsidies and the fact that climate change damages and costs, due to fossil fuels, are also borne by the public.

A price on emissions that cause harm is essential. Yes, a carbon tax. Carbon tax with 100 percent dividend is needed to wean us off fossil fuel addiction. Tax and dividend allows the marketplace, not politicians, to make investment decisions.

Carbon tax on coal, oil and gas is simple, applied at the first point of sale or port of entry. The entire tax must be returned to the public, an equal amount to each adult, a half-share for children. This dividend can be deposited monthly in an individual's bank account.

Carbon tax with 100 percent dividend is non-regressive. On the contrary, you can bet that low and middle income people will find ways to limit their carbon tax and come out ahead. Profligate energy users will have to pay for their excesses.

Demand for low-carbon high-efficiency products will spur innovation, making our products more competitive on international markets. Carbon emissions will plummet as energy efficiency and renewable energies grow rapidly. Black soot, mercury and other fossil fuel emissions will decline. A brighter, cleaner future, with energy independence, is possible.

Washington likes to spend our tax money line-by-line. Swarms of high-priced lobbyists in alligator shoes help Congress decide where to spend, and in turn the lobbyists' clients provide "campaign" money.

The public must send a message to Washington. Preserve our planet, creation, for our children and grandchildren, but do not use that as an excuse for more tax-and-spend. Let this be our motto: "One hundred percent dividend or fight!"

The next President must make a national low-loss electric grid an imperative. It will allow dispersed renewable energies to supplant fossil fuels for power generation. Technology exists for direct-current high-voltage buried transmission lines. Trunk lines can be completed in less than a decade and expanded analogous to interstate highways.

Government must also change utility regulations so that profits do not depend on selling ever more energy, but instead increase with efficiency. Building code and vehicle efficiency requirements must be improved and put on a path toward carbon neutrality.

The fossil-industry maintains its strangle-hold on Washington via demagoguery, using China and other developing nations as scapegoats to rationalize inaction. In fact, we produced most of the excess carbon in the air today, and it is to our advantage as a nation to move smartly in developing ways to reduce emissions. As with the ozone problem, developing countries can be allowed limited extra time to reduce emissions. They will cooperate: they have much to lose from climate change and much to gain from clean air and reduced dependence on fossil fuels.

We must establish fair agreements with other countries. However, our own tax and dividend should start immediately. We have much to gain from it as a nation, and other countries will copy our success. If necessary, import duties on products from uncooperative countries can level the playing field, with the import tax added to the dividend pool.

Democracy works, but sometimes churns slowly. Time is short. The 2008 election is critical for the planet. If Americans turn out to pasture the most brontosaurian congressmen, if Washington adapts to address climate change, our children and grandchildren can still hold great expectations.

Analysis

On the twentieth anniversary of his testimony before the United States Senate's Commission on Energy and Natural Resources, Dr. James Hansen, Director of the NASA Goddard Institute for Space Studies, gave a presentation before the United States Senate's Select Committee on Energy Independence and Global Warming entitled *Global Warming Twenty Years Later: Tipping Points Near*. The intervening years between the presentations had been challenging for Hansen. His willingness to challenge federal officials who did not believe in the man-made causes of global climate change made him a political target. This was especially true during President George W. Bush's years in office. His administration's efforts to shape NASA's climate change research findings through censoring reports and appointing unqualified individuals to positions of authority within the agency led Hansen to publicly complain about the executive branch's interference in the ability of NASA to conduct its scientific research.

Hansen began his 2008 presentation by briefly revisiting his 1988 testimony. He then turned to the prevalence of carbon dioxide in the atmosphere at levels that already posed a threat to living organisms around the world. Blame was placed on politicians for the increasing problem of climate change, as they had been warned that the climate was warming radically twenty years before and had done little to mitigate the problem. Hansen appealed to the nation's citizens to vote out of office politicians who refused to address global warming and climate change. He also urged citizens to quit waiting for politicians to come to a consensus on climate change and instead work locally to protect their future environment. Failure to do so would result in tipping points being reached, which was the point that human intervention would yield reduced impacts against accelerating climate changes.

FURTHER READING

Besel, Richard D. 2013. "Accommodating Climate Change Science: James Hansen and the Rhetorical / Political Emergence of Global Warming." *Science in Context* 26: 137–152.

Bowen, Mark. 2008. *Censoring Science: Inside the Political Attack on Dr. James Hansen and the Truth of Global Warming*. New York: Dutton.

Hansen, James E. 2009. *Storms of my Grandchildren: The Truth About the Coming Climate Catastrophe and Our Last Chance to Save Humanity*. New York: Bloomsbury.

Secretary Chu's Remarks at the Harvard University Commencement—As Prepared for Delivery

Date: June 4, 2009
Location: Cambridge, Massachusetts
Significance: Nobel Prize Winner and United States Secretary of Energy Steven Chu issued a "call to arms" to address the threat of global climate warming.
Source: Chu, Steven. 2009. *Secretary Chu's Remarks at the Harvard University Commencement–As Prepared for Delivery.* http://energy.gov/articles/secretary-chus-remarks-harvard-university-commencement-prepared-delivery (accessed May 1, 2016).

Remarks of U.S. Secretary Steven Chu, Harvard Commencement Address, Thursday, June 4, 2009

Madame President Faust, members of the Harvard Corporation and the Board of Overseers, faculty, family, friends, and, most importantly, today's graduates, thank you for letting me share this wonderful day with you.

I am not sure I can live up to the high standards of Harvard Commencement speakers. Last year, JK Rowling, the billionaire novelist, who started as a classics student, graced this podium. The year before, Bill Gates, the mega-billionaire philanthropist and computer nerd, stood here. Today, sadly, you have me. I am not a billionaire, but at least I am a nerd.

I am grateful to receive an honorary degree from Harvard, an honor that means more to me than you might imagine. You may have heard this morning that I was the academic failure of my family. Both of my brothers have degrees from Harvard. My older brother, Gilbert, has an M.D.–Ph.D. from Harvard. My younger brother, Morgan, who was just named today to the Board of Overseers, has a law degree. I was awarded a Nobel Prize. I thought my mother would be pleased. Not so. When I called her on the morning of the announcement, she replied, "That's nice, but when are you going to visit me next." Now, as the last brother with a degree from Harvard, maybe, at last, she will be satisfied.

Another difficulty with giving a Harvard commencement address is that some students may disapprove of the fact that I am borrowing material from my previous speeches and from other authors. I ask that you forgive me for two reasons.

First, in order to be heard, it is important to deliver the same message more than once.

Second, authors who borrow from others are following in the footsteps of the best. Ralph Waldo Emerson, who graduated from Harvard at the age of 18, noted "All my best thoughts were stolen by the ancients." Picasso declared "Good artists borrow. Great artists steal." Why should commencement speakers be held to a higher standard?

I also want to point out the irony of speaking to graduates of an institution that would have rejected me, had I the chutzpah to apply. I am married to "Dean Jean," a former Dean of Admissions at Stanford. She assures me that she would have rejected me, given the chance. When I showed her a draft of this speech, she objected strongly to my use of the word "rejected." She never rejected applicants; her letters stated that "we are unable to offer you admission." I have great difficulty understanding the difference. After all, Deans of Admissions of highly selective schools are in reality, "Deans of Rejection." Clearly, I have a lot to learn about marketing.

My address will follow the classical sonata form of commencement addresses. The first movement, just presented, were light-hearted remarks. This next movement consists of unsolicited advice, which is rarely valued, seldom remembered, never followed. As Oscar Wilde said, "The only thing to do with good advice is to pass it on. It is never of any use to oneself."

Here comes the advice.

First, every time you celebrate an achievement, be thankful to those who made it possible. Thank your parents and friends who supported you, thank your professors who were inspirational, and especially thank the other professors whose less-than-brilliant lectures forced you to teach yourself. Going forward, the ability to teach yourself is the hallmark of a great liberal arts education and will be the key to your success. To your fellow students who have added immeasurably to your education during those late night discussions, hug them.

Second, in your future life, cultivate a generous spirit. In all negotiations, don't bargain for the last, little advantage. Leave the change on the table. In your collaborations, always remember that "credit" is not a conserved quantity. In a successful collaboration, everybody gets 90% of the credit.

Jimmy Stewart, as Elwood P. Dowd in the movie Harvey, got it exactly right. "Years ago my mother used to say to me, 'in this world, Elwood, you must be .she always used to call me Elwood .in this world Elwood, you must be Oh so smart, or Oh so pleasant.' Well, for years I was smart.... I recommend pleasant. You may quote me."

My third piece of advice is as follows. As you begin this new stage of your lives, follow your passion. If you don't have a passion, don't be satisfied until you find one. Life is too short to go through it without caring deeply about something. When I was your age, I was incredibly single-minded in my goal to be a physicist. After college, I spent eight years as a graduate student and post doc at Berkeley, and then nine years at Bell Labs. During that my time, my central focus and professional joy was physics.

Here is my final advice. Pursuing a personal passion is important, but it should not be your only goal. When you are old and gray, and look back on your life, you will want to be proud of what you have done. The source of that pride won't be the things you have acquired or the recognition you have received. It will be the lives you have touched and the difference you have made.

After nine years at Bell labs, I decided to leave the warm, cozy ivory tower for what I considered to be the "real world": a university. Bell Labs, to quote what was said of Mary Poppins, was "practically perfect in every way," but I wanted to leave behind more than articles in scientific journals. I wanted to teach and give birth to my own set of scientific children.

Ted Geballe, a friend and distinguished colleague of mine at Stanford who also went from Berkeley to Bell Labs to Stanford years earlier, described our motives best:

"The best part of working at a university is the students. They come in fresh, enthusiastic, open to ideas, unscarred by the battles of life. They don't realize it, but they're the recipients of the best our society can offer. If a mind is ever free to be creative, that's the time. They come in believing textbooks are authoritative, but eventually they figure out that textbooks and professors don't know everything, and then they start to think on their own. Then, I begin learning from them."

My students, post doctoral fellows, and the young researchers who worked with me at Bell Labs, Stanford, and Berkeley have been extraordinary. Over 30 former group members are now professors, many at the best research institutions in the world including Harvard. I have learned much from them. Even now, in rare moments on weekends, the remaining members of my biophysics group meet with me in the ether world of cyberspace. I began teaching with the idea of giving back, but I received more that I gave.

This brings me to the final movement of this speech. It begins with a story about an extraordinary scientific discovery and the new dilemma it poses. It's a call to arms and about making a difference.

In the past several decades, our climate has been changing. Climate change is not new: the Earth went through six ice ages in the past 600,000 years. However, recent measurements show that the climate has begun to change rapidly. The size of the North polar ice cap in the month of September is only half the size it was fifty years ago. The sea level has been rising since direct measurements began in 1870, but the rate since 1990 is five times faster than it was since the beginning of recorded measurements.

Here is the remarkable scientific discovery: For the first time in human history, science is now making predictions of how our actions will affect the world fifty and a hundred years from now. These changes are due to the increase in carbon dioxide put into the atmosphere since the beginning of the industrial revolution.

The Earth has warmed up by roughly 0.8 degrees Celsius since the beginning of the industrial revolution. There is already approximately a one degree rise built into the system, even if we stop all greenhouse gas emissions today. Why? It will take decades to warm up the deep oceans before the temperature reaches a new equilibrium.

If the world continues on a business-as-usual path, the Intergovernmental Panel on Climate Change predicts that there is a fifty-fifty chance the temperature will exceed five degrees by the end of this century. This increase may not sound like much, but let me remind you that during the last ice age, the world was only 6 degrees colder. During this time, most of Canada and the U.S. down to Ohio and Pennsylvania were covered year round by a glacier. A world five degrees warmer will be a very different place. The change will be so rapid, that many species, including us, will have a hard time adapting.

We also face the specter of non-linear "tipping points" that may cause much more severe changes. An example of a tipping point is the thawing of the permafrost. The permafrost contains immense amounts of frozen organic matter that have been accumulating for millennia. If the soil melts, microbes will spring to life and cause this debris to rot. The difference in biological activity below freezing and above freezing is something we are very familiar with. Frozen food remains edible for a very long time in the freezer, but once thawed, it spoils quickly. How much methane and carbon dioxide might be released from the rotting permafrost? Even if only a fraction of the carbon is released, it could be greater than all the greenhouse gases we have released to date. Once started, a runaway effect could begin.

The climate problem is an unintended consequence of our success. We depend on fossil energy to keep our homes warm in the winter, cool in the summer, and lit at night; we use it to travel across town and across continents. Energy is a fundamental reason for the prosperity we enjoy, and we will not surrender this prosperity. The United States has four and a half percent of the world's population, but we consume twenty five percent of the energy. By contrast, there are 1.6 billion people who don't have access to electricity. Hundreds of millions of people still cook with twigs or dung. The life we enjoy may not be within the easy reach of many in the developing world, but it is within sight, and they want what we have.

Here is the dilemma. How much are we willing to invest, as a world society, to mitigate the consequences of climate change that will not be realized for at least 100 years? Deeply rooted in all cultures, is the notion of generational responsibility. Parents work hard so that their children will have a better life. Climate change will affect the entire world, but our natural focus is on the welfare of our immediate families. Can we, as a world society, meet our responsibility to future generations?

While I am worried, I am hopeful we will solve this problem. I became the director of the Lawrence Berkeley National Laboratory in part because I wanted to enlist some of the best scientific minds to help battle against climate change. I was there only four and a half years, the shortest serving director in the 78 year history of the Lab, but when I left to become Secretary of Energy, a number of very exciting energy institutes at the Berkeley Lab and UC Berkeley had been established.

I am extremely privileged to be part of the Obama Administration. If there ever was a time to help steer America and the world towards a path of sustainable energy, now is the time. The message the President is delivering is not one of doom and gloom, but of optimism and opportunity. I share this optimism. The task ahead is daunting, but we can and will succeed.

We know some of the answers already. There are immediate and significant savings in energy efficiency and conservation. Energy efficiency is not just low-hanging fruit; it is fruit lying on the ground. We have the potential to make buildings 80% more efficient with investments that will pay for themselves in less than fifteen years. Buildings consume 40% of the energy we use, and a transition to energy efficient buildings will cut our carbon emissions by one third.

We are revving up the remarkable American innovation engine that will be the basis of a new prosperity. We will invent much improved methods to harness the sun, the wind, nuclear power, and capture and sequester the carbon dioxide emitted from our power plants. Advanced bio-fuels and the electrification of personal vehicles make us less dependent on foreign oil.

In the coming decades, we will almost certainly face higher oil prices and a carbon-constrained economy. We have the opportunity to lead in the development of a new, industrial revolution. The great hockey player, Wayne Gretzky, was asked how he positions himself on the ice. He replied, he skates "to where the puck is going to be, not where it's been." America must do the same.

The Obama Administration is laying a new foundation for a prosperous and sustainable energy future, but we don't have all of the answers. That's where you come in. I am asking you, the Harvard graduates, to join us. As our future intellectual leaders, take the time to learn more about what's at stake, and then act on that knowledge. As future scientists and engineers, I ask you to give us better technology solutions. As future economists and political scientists, I ask you to create better policy options. As future business leaders, I ask that you make sustainability an integral part of your business.

Finally, as humanists, I ask that you speak to our common humanity. One of the cruelest ironies about climate change is that the ones who will be hurt the most are the most innocent: the world's poorest and those yet to be born.

The coda to this last movement is borrowed from two humanists.

The first quote is from Martin Luther King. Although he spoke on ending the war in Vietnam in1967, his message seems so fitting for today's climate crisis:

"This call for a worldwide fellowship that lifts neighborly concern beyond one's tribe, race, class, and nation is in reality a call for an all-embracing and unconditional love for all mankind. This oft misunderstood, this oft misinterpreted concept, so readily dismissed by the Nietzsches of the world as a weak and cowardly force, has now become an absolute necessity for the survival of man ..We are now faced with the fact, my friends, that tomorrow is today. We are confronted with the fierce urgency of now. In this unfolding conundrum of life and history, there is such a thing as being too late."

The final message is from William Faulkner. On December 10th, 1950, his Nobel Prize banquet speech was about the role of humanists in a world facing potential nuclear holocaust:

"I believe that man will not merely endure: he will prevail. He is immortal, not because he alone among creatures has an inexhaustible voice, but because he has a soul, a spirit capable of compassion and sacrifice and endurance. The poet's, the writer's, duty is to write about these things. It is his privilege to help man endure by lifting his heart, by reminding him of the courage and honor and hope and pride and compassion and pity and sacrifice which have been the glory of his past."

Graduates, you have an extraordinary role to play in our future. As you pursue your private passions, I hope you will also develop the passion and the voice to help the world in ways both large and small. Nothing will give you greater satisfaction.

Please accept my warmest congratulations. May you prosper and may you help save our planet for your children and for all the future children of the world.

Analysis

Dr. Steven Chu was uniquely qualified to serve as United States Secretary of Energy when he assumed the position in President Barack Obama's administration since he was a 2007 recipient of the Nobel Prize in Physics. Through the commencement speeches he delivered in 2009, most notably at Harvard University and the California Institute of Technology, he provided valuable insights into how the Obama administration viewed the threats posed by ongoing global climate change. Chu essentially set the public tone in what eventually became one of the key initiatives of Obama's presidency.

Further Reading

Chu, Steven. 2009. *Secretary Chu's Remarks at the California Institute of Technology Commencement—As Prepared for Delivery.* http://energy.gov/articles/secretary-chus-remarks-california-institute-technology-commencement-prepared-delivery.

Goodell, Jeff. 2009. "The Secretary of Saving the Planet." *Rolling Stone* 1081: 58–87.

Hand, Eric. 2009. "Newsmaker of the Year: The Power Player." *Nature* 462: 978–983.

Global Climate Change Impacts in the United States (Executive Summary)

Date: June 2009
Location: Washington, D.C.
Significance: The second National Assessment of the Potential Consequences of Climate Variability produced by the United States Global Change Research Program provides a detailed overview of how the federal government perceived the present and future impacts of climate change on both the country as a whole and its individual geographic regions in 2009.
Source: U.S. Global Change Research Program. 2009. *Global Climate Change Impacts in the United States.* http://downloads.globalchange.gov/usimpacts/pdfs/climate-impacts-report.pdf (accessed October 3, 2015).

Observations show that warming of the climate is unequivocal. The global warming observed over the past 50 years is due primarily to human-induced emissions of heat-trapping gases. These emissions come mainly from the burning of fossil fuels (coal, oil, and gas), with important contributions from the clearing of forests, agricultural practices, and other activities.

Warming over this century is projected to be considerably greater than over the last century. The global average temperature since 1900 has risen by about 1.5°F. By 2100, it is projected to rise another 2 to 11.5°F. The U.S. average temperature has risen by a comparable amount and is very likely to rise more than the global average over this century, with some variation from place to place. Several factors will determine future temperature increases. Increases at the lower end of this range are more likely if global heat-trapping gas emissions are cut substantially. If emissions continue to rise at or near current rates, temperature increases are more likely to be near the upper end of the range. Volcanic eruptions or other natural variations could temporarily counteract some of the human-induced warming, slowing the rise in global temperature, but these effects would only last a few years.

Reducing emissions of carbon dioxide would lessen warming over this century and beyond. Sizable early cuts in emissions would significantly reduce the pace and the overall amount of climate change. Earlier cuts in emissions would have a greater effect in reducing climate change than comparable reductions made later. In addition, reducing emissions of some shorter-lived heat-trapping gases, such as methane, and some types of particles, such as soot, would begin to reduce warming within weeks to decades.

Climate-related changes have already been observed globally and in the United States. These include increases in air and water temperatures, reduced frost days, increased frequency and intensity of heavy downpours, a rise in sea level, and reduced snow cover, glaciers, permafrost, and sea ice. A longer ice-free period on lakes and rivers, lengthening of the growing season, and increased water vapor in the atmosphere have also been observed. Over the past 30 years, temperatures have risen faster in winter than in any other season, with average winter temperatures in the Midwest and northern Great Plains increasing more than 7°F. Some of the changes have been faster than previous assessments had suggested.

These climate-related changes are expected to continue while new ones develop. Likely future changes for the United States and surrounding coastal waters include more intense hurricanes with related increases in wind, rain, and storm surges (but not necessarily an increase in the number of these storms that make landfall), as well as drier conditions in the Southwest and Caribbean. These changes will affect human health, water supply, agriculture, coastal areas, and many other aspects of society and the natural environment.

This report synthesizes information from a wide variety of scientific assessments and recently published research to summarize what is known about the observed and projected consequences of climate change on the United States. It combines analysis of impacts on various sectors such as energy, water, and transportation at the national level with an assessment of key impacts on specific regions of the United States. For example, sea-level rise will increase risks of erosion, storm surge damage, and flooding for coastal communities, especially in the Southeast and parts of Alaska. Reduced snowpack and earlier snow melt will alter the timing and amount of water supplies, posing significant challenges for water resource management in the West.

Society and ecosystems can adjust to some climatic changes, but this takes time. The projected rapid rate and large amount of climate change over this century will challenge the ability of society and natural systems to adapt. For example, it is difficult and expensive to alter or replace infrastructure designed to last for decades (such as buildings, bridges, roads, airports, reservoirs, and ports) in response to continuous and/or abrupt climate change.

Impacts are expected to become increasingly severe for more people and places as the amount of warming increases. Rapid rates of warming would lead to particularly large impacts on natural ecosystems and the benefits they provide to humanity. Some of the impacts of climate change will be irreversible, such as species extinctions and coastal land lost to rising seas.

Unanticipated impacts of increasing carbon dioxide and climate change have already occurred and more are possible in the future. For example, it has recently been observed that the increase in atmospheric carbon dioxide concentration is causing an increase in ocean acidity. This reduces the ability of corals and other sea life to build shells and skeletons out of calcium carbonate. Additional impacts in the future might stem from unforeseen changes in the climate system, such as major alterations in oceans, ice, or storms; and unexpected consequences of ecological changes, such as massive dislocations of species or pest outbreaks. Unexpected social or economic changes, including major shifts in wealth, technology, or societal priorities would also affect our ability to respond to climate change. Both anticipated and unanticipated impacts become more challenging with increased warming.

Projections of future climate change come from careful analyses of outputs from global climate models run on the world's most advanced computers. The model simulations analyzed in this report used plausible scenarios of human activity that generally lead to further increases in heat-trapping emissions. None of the scenarios used in this report assumes adoption of policies explicitly designed to address climate change. However, the level of emissions varies among scenarios because of differences in assumptions about population, economic activity, choice of energy technologies, and other factors. Scenarios cover a range of emissions of heat-trapping gases, and the associated climate projections illustrate that lower emissions result in less climate change and thus reduced impacts over this century and beyond. Under all scenarios considered in this report, however, relatively large and sustained changes in many aspects of climate are projected by the middle of this century, with even larger changes by the end of this century, especially under higher emissions scenarios.

In projecting future conditions, there is always some level of uncertainty. For example, there is a high degree of confidence in projections that future temperature increases will be greatest in the Arctic and in the middle of continents. For precipitation, there is high confidence in projections of continued increases in the Arctic and sub-Arctic (including Alaska) and decreases in the regions just outside the tropics, but the precise location of the transition between these is less certain. At local to regional scales and on time frames up to a few years, natural climate variations can be relatively large and can temporarily mask the progressive nature of global climate change. However, the science of making skillful projections at these scales has progressed considerably, allowing useful information to be drawn from regional climate studies such as those highlighted in this report.

This report focuses on observed and projected climate change and its impacts on the United States. However, a discussion of these issues would be incomplete without mentioning some of the actions society can take to respond to the climate challenge. The two major categories are "mitigation" and "adaptation." Mitigation refers to options for limiting climate change by, for example, reducing heat-trapping emissions such as carbon dioxide, methane, nitrous oxide, and halocarbons, or removing some of the heat-trapping gases from the atmosphere. Adaptation refers to changes made to better respond to present or future climatic and other environmental conditions, thereby reducing harm or taking advantage of opportunity. Effective mitigation measures reduce the need for adaptation. Mitigation and adaptation are both essential parts of a comprehensive climate change response strategy.

Carbon dioxide emissions are a primary focus of mitigation strategies. These include improving energy efficiency, using energy sources that do not produce carbon dioxide or produce less of it, capturing and storing carbon dioxide from fossil fuel use, and so on. Choices made about emissions reductions now and over the next few decades will have far reaching consequences for climate-change impacts. The importance of mitigation is clear in comparisons of impacts resulting from higher versus lower emissions scenarios considered in this report. Over the long term, lower emissions will lessen both the magnitude of climate-change impacts and the rate at which they appear. Smaller climate changes that come more slowly make the adaptation challenge more tractable.

However, no matter how aggressively heat-trapping emissions are reduced, some amount of climate change and resulting impacts will continue due to the effects of gases that have already been released. This is true for several reasons. First, some of these gases are very long-lived and the levels of atmospheric heat-trapping gases will remain elevated for hundreds of years or more. Second, the Earth's vast oceans have absorbed much of the heat added to the climate system due to the increase in heat-trapping gases, and will retain that heat for many decades. In addition, the factors that determine emissions, such as energy-supply systems, cannot be changed overnight. Consequently, there is also a need for adaptation.

Adaptation can include a wide range of activities. Examples include a farmer switching to growing a different crop variety better suited to warmer or drier conditions; a company relocating key business centers away from coastal areas vulnerable to sea-level rise and hurricanes; and a community altering its zoning and building codes to place fewer structures in harm's way and making buildings less vulnerable to damage from floods, fires, and other extreme events. Some adaptation options that are currently being pursued in various regions and sectors to deal with climate change and/or other environmental issues are identified in this report.

However, it is clear that there are limits to how much adaptation can achieve.

Humans have adapted to changing climatic conditions in the past, but in the future, adaptations will be particularly challenging because society won't be adapting to a new steady state but rather to a rapidly moving target. Climate will be continually changing, moving at a relatively rapid rate, outside the range to which society has adapted in the past. The precise amounts and timing of these changes will not be known with certainty.

In an increasingly interdependent world, U.S. vulnerability to climate change is linked to the fates of other nations. For example, conflicts or mass migrations of people resulting from food scarcity and other resource limits, health impacts, or environmental stresses in other parts of the world could threaten U.S. national security. It is thus difficult to fully evaluate the impacts of climate change on the United States without considering the consequences of climate change elsewhere. However, such analysis is beyond the scope of this report.

Finally, this report identifies a number of areas in which inadequate information or understanding hampers our ability to estimate future climate change and its impacts. For example, our knowledge of changes in tornadoes, hail, and ice storms is quite limited, making it difficult to know if and how such events have changed as climate has warmed, and how they might change in the future. Research on ecological responses to climate change is also limited, as is our understanding of social responses. The section titled *An Agenda for Climate Impacts Science* at the end of this report offers some thoughts on the most important ways to improve our knowledge. Results from such efforts would inform future assessments that continue building our understanding of humanity's impacts on climate, and climate's impacts on us.

Key Findings

1. **Global warming is unequivocal and primarily human-induced.** Global temperature has increased over the past 50 years. This observed increase is due primarily to human-induced emissions of heat-trapping gases.
2. **Climate changes are underway in the United States and are projected to grow.** Climate-related changes are already observed in the United States and its coastal waters. These include increases in heavy downpours, rising temperature and sea level, rapidly retreating glaciers, thawing permafrost, lengthening growing seasons, lengthening ice-free seasons in the ocean and on lakes and rivers, earlier snowmelt, and alterations in river flows. These changes are projected to grow.

3. **Widespread climate-related impacts are occurring now and are expected to increase.** Climate changes are already affecting water, energy, transportation, agriculture, ecosystems, and health. These impacts are different from region to region and will grow under projected climate change.
4. **Climate change will stress water resources.** Water is an issue in every region, but the nature of the potential impacts varies. Drought, related to reduced precipitation, increased evaporation, and increased water loss from plants, is an important issue in many regions, especially in the West. Floods and water quality problems are likely to be amplified by climate change in most regions. Declines in mountain snowpack are important in the West and Alaska where snowpack provides vital natural water storage.
5. **Crop and livestock production will be increasingly challenged.** Many crops show positive responses to elevated carbon dioxide and low levels of warming, but higher levels of warming often negatively affect growth and yields. Increased pests, water stress, diseases, and weather extremes will pose adaptation challenges for crop and livestock production.
6. **Coastal areas are at increasing risk from sea-level rise and storm surge.** Sea-level rise and storm surge place many U.S. coastal areas at increasing risk of erosion and flooding, especially along the Atlantic and Gulf Coasts, Pacific Islands, and parts of Alaska. Energy and transportation infrastructure and other property in coastal areas are very likely to be adversely affected.
7. **Risks to human health will increase.** Harmful health impacts of climate change are related to increasing heat stress, waterborne diseases, poor air quality, extreme weather events, and diseases transmitted by insects and rodents. Reduced cold stress provides some benefits. Robust public health infrastructure can reduce the potential for negative impacts.
8. **Climate change will interact with many social and environmental stresses.** Climate change will combine with pollution, population growth, overuse of resources, urbanization, and other social, economic, and environmental stresses to create larger impacts than from any of these factors alone.
9. **Thresholds will be crossed, leading to large changes in climate and ecosystems.** There are a variety of thresholds in the climate system and ecosystems. These thresholds determine, for example, the presence of sea ice and permafrost, and the survival of species, from fish to insect pests, with implications for society. With further climate change, the crossing of additional thresholds is expected.
10. **Future climate change and its impacts depend on choices made today.** The amount and rate of future climate change depend primarily on current and future human-caused emissions of heat-trapping gases and airborne particles. Responses involve reducing emissions to limit future warming, and adapting to the changes that are unavoidable.

Analysis

In the Global Change Research Act of 1990 (GCRA), Congress tasked the United States Global Change Research Program (USGCRP) with producing a report for the President of the United States on a four-year cycle that identified the current impacts of climate change on the United States and posited what those changes portended for the next 25 years. The first National Assessment of the Potential Consequences of Climate Variability was produced in 2000 by the United States Global Change Research Program. The next would not be published until 2009, after the conclusion of President George W. Bush's term in office. His administration had questioned the veracity of climate science and thus was unwilling to conduct the assessment as required by the GCRA. Since combating global climate change was a cause that would help define his presidency, President Barack Obama made the completion of the second national assessment a priority during his first year in office. The third iteration of the assessment was conducted during Obama's second term in office.

FURTHER READING

U.S. Global Change Research Program. 2009. *Global Climate Change Impacts in the United States: A State of Knowledge Report.* New York: Cambridge University Press.

Safe Climate Act (Excerpt)

Date: July 7, 2009
Location: Washington, D.C.
Significance: The Safe Climate Act, which was Title III of the American Clean Energy and Security Act of 2009, included the cap-and-trade plan to control global warming that was advocated by President Barack Obama during his first campaign for the presidency.
Source: United States Senate, 111th Congress, First Session. 2009. H.R. 2454. https://www.gpo.gov/fdsys/pkg/BILLS-111hr2454pcs/pdf/BILLS-111hr2454pcs.pdf (accessed January 6, 2016).

Title III—Reducing Global Warming Pollution

Sec. 301. Short Title.

This title, and sections 112, 116, 221, 222, 223, and 401 of this Act, and the amendments made by this title and those sections, may be cited as the "Safe Climate Act."

Subtitle A—Reducing Global Warming Pollution

Sec. 311. Reducing Global Warming Pollution.

The Clean Air Act (42 U.S.C. and following) is amended by adding after title VI the following new title:

"Title VII—Global Warming Pollution Reduction Program: "Part A—Global Warming Pollution Reduction Goals And Targets

"Sec. 701. Findings and Purpose.

"(a) FINDINGS.—The Congress finds as follows:

"(1) Global warming poses a significant threat to the national security, economy, public health and welfare, and environment of the United States, as well as of other nations.

"(2) Reviews of scientific studies, including by the Intergovernmental Panel on Climate Change and the National Academy of Sciences, demonstrate that global warming is the result of the combined anthropogenic greenhouse gas emissions from numerous sources of all types and sizes. Each increment of emission, when combined with other emissions, causes or contributes materially to the acceleration and extent of global warming and its adverse effects for the lifetime of such gas in the atmosphere. Accordingly, controlling emissions in small as well as large amounts is essential to prevent, slow the pace of, reduce the threats from, and mitigate global warming and its adverse effects.

"(3) Because they induce global warming, greenhouse gas emissions cause or contribute to injuries to persons in the United States, including—

"(A) adverse health effects such as disease and loss of life;

"(B) displacement of human populations;

"(C) damage to property and other interests related to ocean levels, acidification, and ice changes;

"(D) severe weather and seasonal changes;

"(E) disruption, costs, and losses to business, trade, employment, farms, subsistence, aesthetic enjoyment of the environment, recreation, culture, and tourism;

"(F) damage to plants, forests, lands, and waters;

"(G) harm to wildlife and habitat;

"(H) scarcity of water and the decreased abundance of other natural resources;

"(I) worsening of tropospheric air pollution;

"(J) substantial threats of similar damage; and

"(K) other harm.

"(4) That many of these effects and risks of future effects of global warming are widely shared does not minimize the adverse effects individual persons have suffered, will suffer, and are at risk of suffering because of global warming.

"(5) That some of the adverse and potentially catastrophic effects of global warming are at risk of occurring and not a certainty does not negate the harm persons suffer from actions that increase the likelihood, extent, and severity of such future impacts.

"(6) Nations of the world look to the United States for leadership in addressing the threat of and harm from global warming. Full implementation of the Safe Climate Act is critical to engage other nations in an international effort to mitigate the threat of and harm from global warming.

"(7) Global warming and its adverse effects are occurring and are likely to continue and increase in magnitude, and to do so at a greater and more harmful rate, unless the Safe Climate Act is fully implemented and enforced in an expeditious manner.

"(b) PURPOSE.—It is the general purpose of the Safe Climate Act to help prevent, reduce the pace of, mitigate, and remedy global warming and its adverse effects. To fulfill such purpose, it is necessary to—

"(1) require the timely fulfillment of all governmental acts and duties, both substantive and procedural, and the prompt compliance of covered entities with the requirements of the Safe Climate Act;

"(2) establish and maintain an effective, transparent, and fair market for emission allowances and preserve the integrity of the cap on emissions and of offset credits;

"(3) advance the production and deployment of clean energy and energy efficiency technologies; and

"(4) ensure effective enforcement of the Safe Climate Act by citizens, States, Indian tribes, and all levels of government because each violation of the Safe Climate Act is likely to result in an additional increment of greenhouse gas emission and will slow the pace of implementation of the Safe Climate Act and delay the achievement of the goals set forth in section 702, and cause or contribute to global warming and its adverse effects.

"Sec. 702. Economy-Wide Reduction Goals.

"The goals of the Safe Climate Act are to reduce steadily the quantity of United States greenhouse gas emissions such that—

"(1) in 2012, the quantity of United States greenhouse gas emissions does not exceed 97 percent of the quantity of United States greenhouse gas emissions in 2005;

"(2) in 2020, the quantity of United States greenhouse gas emissions does not exceed 80 percent of the quantity of United States greenhouse gas emissions in 2005;

"(3) in 2030, the quantity of United States greenhouse gas emissions does not exceed 58 percent of the quantity of United States greenhouse gas emissions in 2005; and

"(4) in 2050, the quantity of United States greenhouse gas emissions does not exceed 17 percent of the quantity of United States greenhouse gas emissions in 2005.

"Sec. 703. Reduction Targets for Specified Sources.

"(a) IN GENERAL.—The regulations issued under section 721 shall cap and reduce annually the greenhouse gas emissions of capped sources each calendar year beginning in 2012 such that—

"(1) in 2012, the quantity of greenhouse gas emissions from capped sources does not exceed 97 percent of the quantity of greenhouse gas emissions from such sources in 2005;

"(2) in 2020, the quantity of greenhouse gas emissions from capped sources does not exceed 83 percent of the quantity of greenhouse gas emissions from such sources in 2005;

"(3) in 2030, the quantity of greenhouse gas emissions from capped sources does not exceed 58 percent of the quantity of greenhouse gas emissions from such sources in 2005; and

"(4) in 2050, the quantity of greenhouse gas emissions from capped sources does not exceed 17 percent of the quantity of greenhouse gas emissions from such sources in 2005.

"(b) DEFINITION.—For purposes of this section, the term 'greenhouse gas emissions from such sources in 2005 means emissions to which section 722 would have applied if the requirements of this title for the specified year had been in effect for 2005.

"Sec. 704. Supplemental Pollution Reductions.

"For the purposes of decreasing the likelihood of catastrophic climate change, preserving tropical forests, building capacity to generate offset credits, and facilitating international action on global warming, the Administrator shall set aside the percentage specified in section 781 of the quantity of emission allowances established under section 721(a) for each year, to be used to achieve a reduction of greenhouse gas emissions from deforestation in developing countries in accordance with part E. In 2020, activities supported under part E shall provide greenhouse gas reductions in an amount equal to an additional 10 percentage points of reductions from United States greenhouse gas emissions in 2005. The Administrator shall distribute these allowances with respect to activities in countries that enter into and implement agreements or arrangements relating to reduced deforestation as described in section 754(a)(2).

Analysis

While pursuing the presidency, Senator Barack Obama proposed that the United States adopt a cap-and-trade program in order to reduce the emissions of greenhouse gases that were fueling climate change. In theory, the way that cap-and-trade economic systems operate is that a country is granted a limited amount of credits which allow it to emit greenhouses gases up to a specified ceiling, hence the cap. More credits could be acquired by trading for them from another country that had an excess inventory or by an act that helped the environment for the long-term, such as the protection of large swaths of forested lands. In order to reduce the amount of pollutants being emitted over time, mechanisms are included that periodically reduce the amount of credits available for use, thereby forcing countries to reduce their harmful emissions. Obama envisioned that such an initiative, combined with an investment by the federal government of $150 billion in energy-saving technologies, would lead to an 80% reduction in the emissions of carbon dioxide and other greenhouse gases within the United States by 2050.

President Obama's cap-and-trade system was included in Title III of the American Clean Energy and Security Act of 2009. The administration, with the assistance of key Congressional leaders like Congressman Henry Waxman of California, was able to successfully steer the legislation through the United States House of Representatives. Within the Senate, the Act faced entrenched opposition. Critics alleged that passage of the legislation was going to severely hurt the economy and would result in the loss of millions of jobs. The argument held sway as Democrats in that body were unable to find enough votes to advance the legislation.

FURTHER READING

Accordino, Megan H., and Deepak Rajagopal. 2015. "When a National Cap-and-Trade Policy with a Carve-Out Provision May Be Preferable to a National CO2 Tax." *Energy Journal* 36: 189–207.
Hoffman, Matthew J. 2011. *Climate Governance at the Crossroads: A Global Response After Kyoto*. New York: Oxford University Press.
Kollmuss, Anja, et al. 2010. *Handbook of Carbon Offset Programs: Trading Systems, Funds, Protocols and Standards*. Washington, D.C.: Earthscan.
Rabe, Barry G. 2016. "The Durability of Carbon Cap-and-Trade Policy." *Governance* 29: 103–119.
Sinden, Amy. 2010. "Shifting the Domestic and International Logjams on Climate Change: A New Defense of Cap and Dividend." *Tulane Journal of International and Comparative Law* 19: 79–93.

Remarks by the President at the Morning Plenary Session of the United Nations Climate Change Conference

Date: December 18, 2009
Location: Copenhagen, Denmark
Significance: President Barack Obama challenged his counterparts from developed countries around the world to work to substantially reduce greenhouse gas (GHG) emissions within their respective borders.
Source: The White House, Office of the Press Secretary. 2009. *Remarks by the President at the Morning Plenary Session of the United Nations Climate Change Conference.* https://www.whitehouse.gov/the-press-office/remarks-president-morning-plenary-session-united-nations-climate-change-conference (accessed August 5, 2015).

THE PRESIDENT: Good morning. It is an honor for me to join this distinguished group of leaders from nations around the world. We come here in Copenhagen because climate change poses a grave and growing danger to our people. All of you would not be here unless you—like me—were convinced that this danger is real. This is not fiction, it is science. Unchecked, climate change will pose unacceptable risks to our security, our economies, and our planet. This much we know.

The question, then, before us is no longer the nature of the challenge—the question is our capacity to meet it. For while the reality of climate change is not in doubt, I have to be honest, as the world watches us today, I think our ability to take collective action is in doubt right now, and it hangs in the balance.

I believe we can act boldly, and decisively, in the face of a common threat. That's why I come here today—not to talk, but to act. (Applause.)

Now, as the world's largest economy and as the world's second largest emitter, America bears our responsibility to address climate change, and we intend to meet that responsibility. That's why we've renewed our leadership within international climate change negotiations. That's why we've worked with other nations to phase out fossil fuel subsidies. That's why we've taken bold action at home—by making historic investments in renewable energy; by putting our people to work increasing efficiency in our homes and buildings; and by pursuing comprehensive legislation to transform to a clean energy economy.

These mitigation actions are ambitious, and we are taking them not simply to meet global responsibilities. We are convinced, as some of you may be convinced, that changing the way we produce and use energy is essential to America's economic future—that it will create millions of new jobs, power new industries, keep us competitive, and spark new innovation. We're convinced, for our own self-interest, that the way we use energy, changing it to a more efficient fashion, is essential to our national security, because it helps to reduce our dependence on foreign oil, and helps us deal with some of the dangers posed by climate change.

So I want this plenary session to understand, America is going to continue on this course of action to mitigate our emissions and to move towards a clean energy economy, no matter what happens here in Copenhagen. We think it is good for us, as well as good for the world. But we also believe that we will all be stronger, all be safer, all be more secure if we act together. That's why it is in our mutual interest to achieve a global accord in which we agree to certain steps, and to hold each other accountable to certain commitments.

After months of talk, after two weeks of negotiations, after innumerable side meetings, bilateral meetings, endless hours of discussion among negotiators, I believe that the pieces of that accord should now be clear.

First, all major economies must put forward decisive national actions that will reduce their emissions, and begin to turn the corner on climate change. I'm pleased that many of us have already done so. Almost all the major economies have put forward legitimate targets, significant targets, ambitious targets. And I'm confident that America will fulfill the commitments that we have made: cutting our emissions in the range of 17 percent by 2020, and by more than 80 percent by 2050 in line with final legislation.

Second, we must have a mechanism to review whether we are keeping our commitments, and exchange this information in a transparent manner. These measures need not be intrusive, or infringe upon sovereignty. They must, however, ensure that an accord is credible, and that we're living up to our obligations. Without such accountability, any agreement would be empty words on a page.

I don't know how you have an international agreement where we all are not sharing information and ensuring that we are meeting our commitments. That doesn't make sense. It would be a hollow victory.

Number three, we must have financing that helps developing countries adapt, particularly the least developed and most vulnerable countries to climate change. America will be a part of fast-start funding that will ramp up to $10 billion by 2012. And yesterday, Secretary Hillary Clinton, my Secretary of State, made it clear that we will engage in a global effort to mobilize $100 billion in financing by 2020, if—and only if—it is part of a broader accord that I have just described.

Mitigation. Transparency. Financing. It's a clear formula—one that embraces the principle of common but differentiated responses and respective capabilities. And it adds up to a significant accord—one that takes us farther than we have ever gone before as an international community.

I just want to say to this plenary session that we are running short on time. And at this point, the question is whether we will move forward together or split apart, whether we prefer posturing to action. I'm sure that many consider this an imperfect framework that I just described. No country will get everything that it wants. There are those developing countries that want aid with no strings attached, and no obligations with respect to transparency. They think that the most advanced nations should pay a higher price; I understand that. There are those advanced nations who think that developing countries either cannot absorb this assistance, or that will not be held accountable effectively, and that the world's fastest-growing emitters should bear a greater share of the burden.

We know the fault lines because we've been imprisoned by them for years. These international discussions have essentially taken place now for almost two decades, and we have very little to show for it other than an increased acceleration of the climate change phenomenon. The time for talk is over. This is the bottom line: We can embrace this accord, take a substantial step forward, continue to refine it and build upon its foundation. We can do that, and everyone who is in this room will be part of a historic endeavor—one that makes life better for our children and our grandchildren.

Or we can choose delay, falling back into the same divisions that have stood in the way of action for years. And we will be back having the same stale arguments month after month, year after year, perhaps decade after decade, all while the danger of climate change grows until it is irreversible.

Ladies and gentlemen, there is no time to waste. America has made our choice. We have charted our course. We have made our commitments. We will do what we say. Now I believe it's the time for the nations and the people of the world to come together behind a common purpose.

We are ready to get this done today—but there has to be movement on all sides to recognize that it is better for us to act than to talk; it's better for us to choose action over inaction; the future over the past—and with courage and faith, I believe that we can meet our responsibility to our people, and the future of our planet. Thank you very much.

Analysis

After the United States failed to ratify the Kyoto Protocol to the United Nations Framework Convention on Climate Change, it lost the credibility necessary to lead on the international stage the effort to address climate change. While pursing the presidency, Senator Barack Obama declared his intention to reclaim the United States' leadership role. This speech marked the opening salvo in that effort.

President Obama stated that the United States' status as the second-largest emitter of greenhouse gases (GHG) in the world required the country to lead the effort to address the causes and consequences of global climate change. Towards that end, Obama pledged that the United States would reduce GHG emissions by 17% by 2020, and by 80% by 2050. In addition, he promised that the United States would heavily invest in a variety of clean energy initiatives. He challenged the leaders of other countries to launch their own programs, as collective action would be required in order to achieve meaningful results on a global scale.

Further Reading

Parker, Charles F., et al. 2012. "Fragmented Climate Change Leadership: Making Sense of the Ambiguous Outcome of COP-15." *Environmental Politics* 21: 268–286.

Szarka, Joseph. 2012. "The EU, the USA and the Climate Divide: Reappraising Strategic Choices." *European Political Science* 11: 31–40.

Endangerment and Cause or Contribute Findings for Greenhouse Gases Under Section 202(a) of the Clean Air Act
Final Rule (Introduction)

Date: December 15, 2009
Location: Washington, D.C.
Significance: In response to the decision by the United States Supreme Court in *Massachusetts et al. v. Environmental Protection Agency*, the Environmental Protection Agency issued rules to regulate the emission of greenhouse gases by stationary and mobile sources.
Source: Environmental Protection Agency. 2009. *Environmental Protection Agency: 40 CFR Chapter I: Endangerment and Cause or Contribute Findings for Greenhouse Gases Under Section 202(a) of the Clean Air Act: Final Rule.* https://www3.epa.gov/climatechange/Downloads/endangerment/Federal_Register-EPA-HQ-OAR-2009-0171-Dec.15-09.pdf (accessed May 1, 2016).

A. Overview

Pursuant to CAA section 202(a), the Administrator finds that greenhouse gases in the atmosphere may reasonably be anticipated both to endanger public health and to endanger public welfare. Specifically, the Administrator is defining the "air pollution" referred to in CAA section 202(a) to be the mix of six long-lived and directly-emitted greenhouse gases: carbon dioxide (CO_2), methane (CH_4), nitrous oxide (N_2O), hydrofluorocarbons (HFCs), perfluorocarbons (PFCs), and sulfur hexafluoride (SF_6). In this document, these six greenhouse gases are referred to as "well-mixed greenhouse gases" in this document (with more precise meanings of "long lived" and "well mixed" provided in Section IV.A).

The Administrator has determined that the body of scientific evidence compellingly supports this finding. The major assessments by the U.S. Global Climate Research Program (USGCRP), the Intergovernmental Panel on Climate Change (IPCC), and the National Research Council (NRC) serve as the primary scientific basis supporting the Administrator's endangerment finding. The Administrator reached her determination by considering both observed and projected effects of greenhouse gases in the atmosphere, their effect on climate, and the public health and welfare risks and impacts associated with such climate change. The Administrator's assessment focused on public health and public welfare impacts within the United States. She also examined the evidence with respect to impacts in other world regions, and she concluded that these impacts strengthen the case for endangerment to public health and welfare because impacts in other world regions can in turn adversely affect the United States.

The Administrator recognizes that human-induced climate change has the potential to be far-reaching and multidimensional, and in light of existing knowledge, that not all risks and potential impacts can be quantified or characterized with uniform metrics. There is variety not only in the nature and potential magnitude of risks and impacts, but also in our ability to characterize, quantify and project such impacts into the future. The Administrator is using her judgment, based on existing science, to weigh the threat for each of the identifiable risks, to weigh the potential benefits where relevant, and ultimately to assess whether these risks and effects, when viewed in total, endanger public health or welfare.

The Administrator has considered how elevated concentrations of the well-mixed greenhouse gases and associated climate change affect public health by evaluating the risks associated with changes in air quality, increases in temperatures, changes in extreme weather events, increases in food- and water-borne pathogens, and changes in aeroallergens. The evidence concerning adverse air quality impacts provides strong and clear support for an endangerment finding. Increases in ambient ozone are expected to occur over broad areas of the country, and they are expected to increase serious adverse health effects in large population areas that are and may continue to be in nonattainment. The evaluation of the potential risks associated with increases in ozone in attainment areas also supports such a finding.

The impact on mortality and morbidity associated with increases in average temperatures, which increase the likelihood of heat waves, also provides support for a public health endangerment finding. There are uncertainties over the net health impacts of a temperature increase due to decreases in cold-related mortality, but some recent evidence suggests that the net impact on mortality is more likely to be adverse, in a context where heat is already the leading cause of weather-related deaths in the United States.

The evidence concerning how human-induced climate change may alter extreme weather events also clearly supports a finding of endangerment, given the serious adverse impacts that can result from such events and the increase in risk, even if small, of the occurrence and intensity of events such as hurricanes and floods. Additionally, public health is expected to be adversely affected by an increase in the severity of coastal storm events due to rising sea levels.

There is some evidence that elevated carbon dioxide concentrations and climate changes can lead to changes in aeroallergens that could increase the potential for allergenic illnesses. The evidence on pathogen borne disease vectors provides directional support for an endangerment finding. The Administrator acknowledges the many uncertainties in these areas. Although these adverse effects provide some support for an endangerment finding, the Administrator is not placing primary weight on these factors.

Finally, the Administrator places weight on the fact that certain groups, including children, the elderly, and the poor, are most vulnerable to these climate-related health effects.

The Administrator has considered how elevated concentrations of the well-mixed greenhouse gases and associated climate change affect public welfare by evaluating numerous and far-ranging risks to food production and agriculture, forestry, water resources, sea level rise and coastal areas, energy, infrastructure, and settlements, and ecosystems and wildlife. For each of these sectors, the evidence provides support for a finding of endangerment to public welfare. The evidence concerning adverse impacts in the areas of water resources and sea level rise and coastal areas provides the clearest and strongest support for an endangerment finding, both for current and future generations. Strong support is also found in the evidence concerning infrastructure and settlements, as well ecosystems and wildlife. Across the sectors, the potential serious adverse impacts of extreme events, such as wildfires, flooding, drought, and extreme weather conditions, provide strong support for such a finding.

Water resources across large areas of the country are at serious risk from climate change, with effects on water supplies, water quality, and adverse effects from extreme events such as floods and droughts. Even areas of the country where an increase in water flow is projected could face water resource problems from the supply and water quality problems associated with temperature increases and precipitation variability, as well as the increased risk of serious adverse effects from extreme events, such as floods and drought. The severity of risks and impacts is likely to increase over time with accumulating greenhouse gas concentrations and associated temperature increases and precipitation changes.

Overall, the evidence on risk of adverse impacts for coastal areas provides clear support for a finding that greenhouse gas air pollution endangers the welfare of current and future generations. The most serious potential adverse effects are the increased risk of storm surge and flooding in coastal areas from sea level rise and more intense storms. Observed sea level rise is already increasing the risk of storm surge and flooding in some coastal areas. The conclusion in the assessment literature that there is the potential for hurricanes to become more intense (and even some evidence that Atlantic hurricanes have already become more intense) reinforces the judgment that coastal communities are now endangered by human-induced climate change, and may face substantially greater risk in the future. Even if there is a low probability of raising the destructive power of hurricanes, this threat is enough to support a finding that coastal communities are endangered by greenhouse gas air pollution. In addition, coastal areas face other adverse impacts from sea level rise such as land loss due to inundation, erosion, wetland submergence, and habitat loss. The increased risk associated with these adverse impacts also endangers public welfare, with an increasing risk of greater adverse impacts in the future.

Strong support for an endangerment finding is also found in the evidence concerning energy, infrastructure, and settlements, as well ecosystems and wildlife. While the impacts on net energy demand may be viewed as generally neutral for purposes of making an endangerment determination, climate change is expected to result in an increase in electricity production, especially supply for peak demand. This may be exacerbated by the potential for adverse impacts from climate change on hydropower resources as well as the potential risk of serious adverse effects on energy infrastructure from extreme events. Changes in extreme weather events threaten energy, transportation, and water resource infrastructure. Vulnerabilities of industry, infrastructure, and settlements to climate change are generally greater in high-risk locations, particularly coastal and riverine areas, and areas whose economies are closely linked with climate-sensitive resources. Climate change will likely interact with and possibly exacerbate ongoing environmental change and environmental pressures in settlements, particularly in Alaska where indigenous communities are facing major environmental and cultural impacts on their historic lifestyles. Over the 21st century, changes in climate will cause some species to shift north and to higher elevations and fundamentally rearrange U.S. ecosystems. Differential capacities for range shifts and constraints from development, habitat fragmentation, invasive species, and broken ecological connections will likely alter ecosystem structure, function, and services, leading to predominantly negative consequences for biodiversity and the provision of ecosystem goods and services.

There is a potential for a net benefit in the near term for certain crops, but there is significant uncertainty about whether this benefit will be achieved given the various potential adverse impacts of climate change on crop yield, such as the increasing risk of extreme weather events. Other aspects of this sector may be adversely affected by climate change, including livestock management and irrigation requirements, and there is a risk of adverse effect on a large segment of the total crop market. For the near term, the concern over the potential for adverse effects in certain parts of the agriculture sector appears generally comparable to the potential for benefits for certain crops. However, the body of evidence points towards increasing risk of net adverse impacts on U.S. food production and agriculture over time, with the potential for significant disruptions and crop failure in the future.

For the near term, the Administrator finds the beneficial impact on forest growth and productivity in certain parts of the country from elevated carbon dioxide concentrations and temperature increases to date is offset by the clear risk from the observed increases in wildfires, combined with risks from the spread of destructive pests and disease. For the longer term, the risk from adverse effects increases over time, such that overall climate change presents serious adverse risks for forest productivity. There is compelling reason to find that the support for a positive endangerment finding increases as one considers expected future conditions where temperatures continue to rise.

Looking across all of the sectors discussed above, the evidence provides compelling support for finding that greenhouse gas air pollution endangers the public welfare of both current and future generations. The risk and the severity of adverse impacts on public welfare are expected to increase over time.

The Administrator also finds that emissions of well-mixed greenhouse gases from the transportation sources covered under CAA section 202(a) 3 contribute to the total greenhouse gas air pollution, and thus to the climate change problem, which is reasonably anticipated to endanger public health and welfare. The Administrator is defining the air pollutant that contributes to climate change as the aggregate group of the well-mixed greenhouse gases. The definition of air pollutant used by the Administrator is based on the similar attributes of these substances. These attributes include the fact that they are sufficiently long-lived to be well mixed globally in the atmosphere, that they are directly emitted, and that they exert a climate warming effect by trapping outgoing, infrared heat that would otherwise escape to space, and that they are the focus of climate change science and policy.

In order to determine if emissions of the well-mixed greenhouse gases from CAA section 202(a) source categories contribute to the air pollution that endangers public health and welfare, the Administrator compared the emissions from these CAA section 202(a) source categories to total global and total U.S. greenhouse gas emissions, finding that these source categories are responsible for about 4 percent of total global well-mixed greenhouse gas emissions and just over 23 percent of total U.S. well-mixed greenhouse gas emissions. The Administrator found that these comparisons, independently and together, clearly establish that these emissions contribute to greenhouse gas concentrations. For example, the emissions of well-mixed greenhouse gases from CAA section 202(a) sources are larger in magnitude than the total well-mixed greenhouse gas emissions from every other individual nation with the exception of China, Russia, and India, and are the second largest emitter within the United States behind the electricity generating sector. As the Supreme Court noted, "[j]udged by any standard, U.S. motor-vehicle emissions make a meaningful contribution to greenhouse gas concentrations and hence, * * * to global warming." Massachusetts v. EPA, 549 U.S. 497, 525 (2007).

The Administrator's findings are in response to the Supreme Court's decision in Massachusetts v. EPA. That case involved a 1999 petition submitted by the International Center for Technology Assessment and 18 other environmental and renewable energy industry organizations requesting that EPA issue standards under CAA section 202(a) for the emissions of carbon dioxide, methane, nitrous oxide, and hydrofluorocarbons from new motor vehicles and engines. The Administrator's findings are in response to this petition and are for purposes of CAA section 202(a).

B. Background Information Helpful To Understand These Findings

This section provides some basic information regarding greenhouse gases and the CAA section 202(a) source categories, as well as the ongoing joint-rulemaking on greenhouse gases by EPA and the Department of Transportation. Additional technical and legal background, including a summary of the Supreme Court's Massachusetts v. EPA decision, can be found in the Proposed Endangerment and Contribution Findings (74 FR 18886, April 24, 2009).

1. Greenhouse Gases and Transportation Sources Under CAA Section 202(a)

Greenhouse gases are naturally present in the atmosphere and are also emitted by human activities. Greenhouse gases trap the Earth's heat that would otherwise escape from the atmosphere, and thus form the greenhouse effect that helps keep the Earth warm enough for life. Human activities are intensifying the naturally-occurring greenhouse effect by adding greenhouse gases to the atmosphere. The primary greenhouse gases of concern that are directly emitted by human activities include carbon dioxide, methane, nitrous oxide, hydrofluorocarbons, perfluorocarbons, and sulfur hexafluoride. Other pollutants (such as aerosols) and other human activities, such as land use changes that alter the reflectivity of the Earth's surface, also cause climatic warming and cooling effects. In these Findings, the term "climate change" generally refers to the global warming effect plus other associated changes (e.g., precipitation effects, sea level rise, changes in the frequency and severity of extreme weather events) being induced by human activities, including activities that emit greenhouse gases. Natural causes also, contribute to climate change and climatic changes have occurred throughout the Earth's history. The concern now, however, is that the changes taking place in our atmosphere as a result of the well-documented buildup of greenhouse gases due to human activities are changing the climate at a pace and in a way that threatens human health, society, and the natural environment. Further detail on the state of climate change science can be found in Section III of these

Findings as well as the technical support document (TSD) that accompanies this action (www.epa.gov/climatechange/ endangerment.html).

The transportation sector is a major source of greenhouse gas emissions both in the United States and in the rest of the world. The transportation sources covered under CAA section 202(a)—the section of the CAA under which these Findings occur—include passenger cars, light- and heavy-duty trucks, buses, and motorcycles. These transportation sources emit four key greenhouse gases: carbon dioxide, methane, nitrous oxide, and hydrofluorocarbons. Together, these transportation sources are responsible for 23 percent of total annual U.S. greenhouse gas emissions, making this source the second largest in the United States behind electricity generation. Further discussion of the emissions data supporting the Administrator's cause or contribute finding can be found in Section V of these Findings, and the detailed greenhouse gas emissions data for section 202(a) source categories can be found in Appendix B of EPA's TSD.

2. Joint EPA and Department of Transportation Proposed Greenhouse Gas Rule

On September 15, 2009, EPA and the Department of Transportation's National Highway Safety Administration (NHTSA) proposed a National Program that would dramatically reduce greenhouse gas emissions and improve fuel economy for new cars and trucks sold in the United States. The combined EPA and NHTSA standards that make up this proposed National Program would apply to passenger cars, light duty trucks, and medium-duty passenger vehicles, covering model years 2012 through 2016. They proposed to require these vehicles to meet an estimated combined average emissions level of 250 grams of carbon dioxide per mile, equivalent to 35.5 miles per gallon (MPG) if the automobile industry were to meet this carbon dioxide level solely through fuel economy improvements. Together, these proposed standards would cut carbon dioxide emissions by an estimated 950 million metric tons and 1.8 billion barrels of oil over the lifetime of the vehicles sold under the program (model years 2012–2016). The proposed rulemaking can be viewed at (74 FR 49454, September 28, 2009).

C. Public Involvement

In response to the Supreme Court's decision, EPA has been examining the scientific and technical basis for the endangerment and cause or contribute decisions under CAA section 202(a) since 2007. The science informing the decision-making process has grown stronger since our work began. EPA's approach to evaluating the science, including comments submitted during the public comment period, is further discussed in Section III.A of these Findings. Public review and comment has always been a major component of EPA's process.

1. EPA's Initial Work on Endangerment

As part of the Advance Notice of Proposed Rulemaking: Regulating Greenhouse Gas Emissions under the Clean Air Act (73 FR 44353) published in July 2008, EPA provided a thorough discussion of the issues and options pertaining to endangerment and cause or contribute findings under the CAA. The Agency also issued a TSD providing an overview of all the major scientific assessments available at the time and emission inventory data relevant to the contribution finding (Docket ID No. EPA-HQ-OAR-

2008–0318). The comment period for that Advance Notice was 120 days, and it provided an opportunity for EPA to hear from the public with regard to the issues involved in endangerment and cause or contribute findings as well as the supporting science. EPA received, reviewed and considered numerous comments at that time and this public input was reflected in the Findings that the Administrator proposed in April 2009. In addition, many comments were received on the TSD released with the Advance Notice and reflected in revisions to the TSD released in April 2009 to accompany the Administrator's proposal. All public comments on the Advance Notice are contained in the public docket for this action (Docket ID No. EPA–HQ–OAR–2008–0318) accessible through www.regulations.gov.

2. Public Involvement Since the April 2009 Proposed Endangerment Finding

The Proposed Endangerment and Cause or Contribute Findings for Greenhouse Gases (Proposed Findings) was published on April 24, 2009 (74 FR 18886). The Administrator's proposal was subject to a 60-day public comment period, which ended June 23, 2009, and also included two public hearings. Over 380,000 public comments were received on the Administrator's proposed endangerment and cause or contribute findings, including comments on the elements of the Administrator's April 2009 proposal, the legal issues pertaining to the Administrator's decisions, and the underlying TSD containing the scientific and technical information.

A majority of the comments (approximately 370,000) were the result of mass mail campaigns, which are defined as groups of comments that are identical or very similar in form and content. Overall, about two-thirds of the mass-mail comments received are supportive of the Findings and generally encouraged the Administrator both to make a positive endangerment determination and implement greenhouse gas emission regulations. Of the mass mail campaigns in disagreement with the Proposed Findings most either oppose the proposal on economic grounds (e.g., due to concern for regulatory measures following an endangerment finding) or take issue with the proposed finding that atmospheric greenhouse gas concentrations endanger public health and welfare. Please note that for mass mailer campaigns, a representative copy of the comment is posted in the public docket for this Action (Docket ID No. EPA–HQ–OAR–2009–0171) at www.regulations.gov.

Approximately 11,000 other public comments were received. These comments raised a variety of issues related to the scientific and technical information EPA relied upon in making the Proposed Findings, legal and procedural issues, the content of the Proposed Findings, and the implications of the Proposed Findings.

In light of the very large number of comments received and the significant overlap between many comments, EPA has not responded to each comment individually. Rather, EPA has summarized and provided responses to each significant argument, assertion and question contained within the totality of the comments. EPA's responses to some of the most significant comments are provided in these Findings. Responses to all significant issues raised by the comments are contained in the 11 volumes of the Response to Comments document, organized by subject area (found in docket EPA–HQ–OAR–2009–0171).

3. Issues Raised Regarding the Rulemaking Process

EPA received numerous comments on process-related issues, including comments urging the Administrator to delay issuing the final findings, arguing that it was improper for the Administrator to sever the endangerment and cause or contribute findings from the attendant section 202(a) standards, arguing the final decision was preordained by the President's May vehicle announcement, and questioning the adequacy of the comment period. Summaries of key comments and EPA's responses are discussed in this section. Additional and more detailed responses can be found in the Response to Comments document, Volume 11. As noted in the Response to Comments document, EPA also received comments supporting the overall process.

a. It Is Reasonable for the Administrator To Issue the Endangerment and Cause or Contribute Findings Now

Though the Supreme Court did not establish a specific deadline for EPA to act, more than two and a half years have passed since the remand from the Supreme Court, and it has been 10 years since EPA received the original petition requesting that EPA regulate greenhouse gas emissions from new motor vehicles. EPA has a responsibility to respond to the Supreme Court's decision and to fulfill its obligations under current law, and there is good reason to act now given the urgency of the threat of climate change and the compelling scientific evidence.

Many commenters urge EPA to delay making final findings for a variety of reasons. They note that the Supreme Court did not establish a deadline for EPA to act on remand. Commenters also argue that the Supreme Court's decision does not require that EPA make a final endangerment finding, and thus that EPA has discretionary power and may decline to issue an endangerment finding, not only if the science is too uncertain, but also if EPA can provide "some reasonable explanation" for exercising its discretion. These commenters interpret the Supreme Court decision not as rejecting all policy reasons for declining to undertake an endangerment finding, but rather as dismissing solely the policy reasons EPA set forth in 2003. Some commenters cite language in the Supreme Court decision regarding EPA's discretion regarding "the manner, timing, content, and coordination of its regulations," and the Court's declining to rule on "whether policy concerns can inform EPA's actions in the event that it makes" a CAA section 202(a) finding to support their position.

Commenters then suggest a variety of policy reasons that EPA can and should make to support a decision not to undertake a finding of endangerment under CAA section 202(a)(1). For example, they argue that a finding of endangerment would trigger several other regulatory programs—such as the Prevention of Significant Deterioration (PSD) provisions—that would impose an unreasonable burden on the economy and government, without providing a benefit to the environment. Some commenters contend that EPA should defer issuing a final endangerment finding while Congress considers legislation. Many commenters note the ongoing international discussions regarding climate change and state their belief that unilateral EPA action would interfere with those negotiations. Others suggest deferring the EPA portion of the joint U.S. Department of Transportation (DOT)/EPA rulemaking because they argue that the new Corporate Average Fuel Economy (CAFE) standards will effectively result in lower greenhouse gas emissions from new motor vehicles, while avoiding the inevitable problems and concerns of regulating greenhouse gases under the CAA.

Other commenters argue that the endangerment determination has to be made on the basis of scientific considerations only. These commenters state that the Court was clear that "[t]he statutory question is whether sufficient information exists to make an endangerment finding," and thus, only if "the scientific uncertainty is so profound that it precludes EPA from making a reasoned judgment as to whether greenhouse gases contribute to global warming," may EPA avoid making a positive or negative endangerment finding. Many commenters urge EPA to take action quickly. They note that it has been 10 years since the original petition requesting that EPA regulate greenhouse gas emissions from motor vehicles was submitted to EPA. They argue that climate change is a serious problem that requires immediate action.

EPA agrees with the commenters who argue that the Supreme Court decision held that EPA is limited to consideration of science when undertaking an endangerment finding, and that we cannot delay issuing a finding due to policy concerns if the science is sufficiently certain (as it is here). The Supreme Court stated that "EPA can avoid taking further action only if it determines that greenhouse gases do not contribute to climate change or if it provides some reasonable explanation as to why it cannot or will not exercise its discretion to determine whether they do" 549 U.S. at 533. Some commenters point to this last provision, arguing that the policy reasons they provide are a "reasonable explanation" for not moving forward at this time. However, this ignores other language in the decision that clearly indicates that the Court interprets the statute to allow for the consideration only of science. For example, in rejecting the policy concerns expressed by EPA in its 2003 denial of the rulemaking petition, the Court noted that "it is evident [the policy considerations] have nothing to do with whether greenhouse gas emissions contribute to climate change. Still less do they amount to a reasoned justification for declining to form a scientific judgment" Id. at 533–34 (emphasis added).

Moreover, the Court also held that "[t]he statutory question is whether sufficient information exists to make an endangerment finding" Id. at 534. Taken as a whole, the Supreme Court's decision clearly indicates that policy reasons do not justify the Administrator avoiding taking further action on the question here.

We also note that the language many commenters quoted from the Supreme Court decision about EPA's discretion regarding the manner, timing and content of Agency actions, and the ability to consider policy concerns, relate to the motor vehicle standards required in the event that EPA makes a positive endangerment finding, and not the finding itself. EPA has long taken the position that it does have such discretion in the standard-setting step under CAA section 202(a).

b. The Administrator Reasonably Proceeded With the Endangerment and Cause or Contribute Findings Separate From the CAA Section 202(a) Standard Rulemaking

As discussed in the Proposed Findings, typically endangerment and cause or contribute findings have been proposed concurrently with proposed standards under various sections of the CAA, including CAA section 202(a). EPA received numerous comments on its decision to propose the endangerment and cause or contribute findings separate from any standards under CAA section 202(a).

Commenters argue that EPA has no authority to issue an endangerment determination under CAA section 202(a) separate and apart from the rulemaking to establish emissions standards under CAA section 202(a). According to these commenters, CAA section 202(a) provides only one reason to issue an endangerment determination, and that is as the basis for promulgating emissions standards for new motor vehicles; thus, it does not authorize such a stand-alone endangerment finding, and EPA may not create its own procedural rules completely divorced from the statutory text. They continue by stating that while CAA section 202(a) says EPA may issue emissions standards conditioned on such a finding, it does not say EPA may first issue an endangerment determination and then issue emissions standards. In addition, they contend, the endangerment proposal and the emissions standards proposal need to be issued together so commenters can fully understand the implications of the endangerment determination. Failure to do so, they argue, deprives the commenters of the opportunity to assess the regulations that will presumably follow from an endangerment finding. They also argue that the expected overlap between reductions in emissions of greenhouse gases from CAA section 202(a) standards issued by EPA and CAFE standards issued by DOT calls into question the basis for the CAA section 202(a) standards and the related endangerment finding, and that EPA is improperly motivated by an attempt to trigger a cascade of regulations under the CAA and/or to promote legislation by Congress.

EPA disagrees with the commenters' claims and arguments. The text of CAA section 202(a) is silent on this issue. It does not specify the timing of an endangerment finding, other than to be clear that emissions standards may not be issued unless such a determination has been made. EPA is exercising the procedural discretion that is provided by CAA section 202(a)'s lack of specific direction. The text of CAA section 202(a) envisions two separate actions by the Administrator: (1) A determination on whether emissions from classes or categories of new motor vehicles cause or contribute to air pollution that may reasonably be anticipated to endanger, and (2) a separate decision on issuance of appropriate emissions standards for such classes or categories. The procedure followed in this rulemaking, and the companion rulemaking involving emissions standards for light duty motor vehicles, is consistent with CAA section 202(a). EPA will issue final emissions standards for new motor vehicles only if affirmative findings are made concerning contribution and endangerment, and such emissions standards will not be finalized prior to making any such determinations. While it would also be consistent with CAA section 202(a) to issue the greenhouse gas endangerment and contribution findings and emissions standards for new light-duty vehicles in the same rulemaking, e.g., a single proposal covering them and a single final rule covering them, nothing in CAA section 202(a) requires such a procedural approach, and nothing in the approach taken in this case violates the text of CAA section 202(a). Since Congress was silent on this issue, and more than one procedural approach may accomplish the requirements of CAA section 202(a), EPA has the discretion to use the approach considered appropriate in this case. Once the final affirmative contribution and endangerment findings are made, EPA has the authority to issue the final emissions standards for new light-duty motor vehicles; however, as the Supreme Court has noted, the agency has 'significant latitude as to the manner, timing, [and] content * * * of its regulations . * * *' Massachusetts v. EPA, 549 U.S. at 533. That includes the discretion to issue them in a separate rulemaking.

Commenters' argument would also lead to the conclusion that EPA could not make an endangerment finding for the entire category of new motor vehicles, as it is doing here, unless EPA also conducted a rulemaking that set emissions standards for all the classes and categories of new motor vehicles at the same time. This narrow procedural limitation would improperly remove discretion that CAA section 202(a) provides to EPA.

EPA has the discretion under CAA section 202(a) to consider classes or categories of new motor vehicles separately or together in making a contribution and endangerment determination. This discretion would be removed under commenters' interpretation, by limiting this to only those cases in which EPA was also ready to issue emissions standards for all of the classes or categories covered by the endangerment finding. However, nothing in the text of CAA section 202(a) places such a limit on EPA's discretion in determining how to group classes or categories of new motor vehicles for purposes of the contribution and endangerment findings. This limitation would not be appropriate, because the issues of contribution and endangerment are separate and distinct from the issues of setting emissions standards. EPA, in this case, is fully prepared to go forward with the contribution and endangerment determination, while it is not ready to proceed with rulemaking for each and every category of new motor vehicles in the first rulemaking to set emissions standards. Section 202(a) of the CAA provides EPA discretion with regard to when and how it conducts its rulemakings to make contribution and endangerment findings, and to set emissions standards, and the text of CAA section 202(a) does not support commenters attempt to limit such discretion.

Concerns have been raised that the failure to issue the proposed endangerment finding and the proposed emissions standard together preclude commenters from assessing and considering the implications of the endangerment finding and the regulations that would likely flow from such a finding. However, commenters have failed to explain how this interferes in any way with their ability to comment on the endangerment finding. In fact it does not interfere, because the two proposals address separate and distinct issues. The endangerment finding concerns the contribution of new motor vehicles to air pollution and the effect of that air pollution on public health or welfare. The emissions standards, which have been proposed (74 FR 49454, September 28, 2009), concern the appropriate regulatory emissions standards if affirmative findings are made on contribution and endangerment. These two proposals address different issues. While commenters have the opportunity to comment on the proposed emissions standards in that rulemaking, they have not shown, and cannot show, that they need to have the emissions standards proposal before them in order to provide relevant comments on the proposed contribution or endangerment findings. Further discussion of this issue can be found in Section II of these Findings, and discussion of the timing of this action and its relationship to other CAA provisions and Congressional action can be found in Section III of these Findings and Volume 11 of the Response to Comments document.

c. The Administrator's Final Decision Was Not Preordained by the President's May Vehicle Announcement

EPA received numerous comments arguing that the President's announcement of a new "National Fuel Efficiency Policy" on May 19, 2009, seriously undermines EPA's ability to provide objective consideration of and a legally adequate response to comments objecting to the previously proposed endangerment findings.

Commenters' conclusion is based on the view that the President's announced policy requires EPA to promulgate greenhouse gas emissions standards under CAA section 202(a), that the President's and Administrator Jackson's announcement indicated that the endangerment rulemaking was but a formality and that a final endangerment finding was a fait accompli. Commenters argue that this means the result of this rulemaking has been preordained and the merits of the issues have been prejudged.

EPA disagrees. Commenters' arguments wholly exaggerate and mischaracterize the circumstances. In the April 24, 2009, endangerment proposal EPA was clear that the two steps in the endangerment provision have to be satisfied in order for EPA to issue emissions standards for new motor vehicles under CAA section 202(a) (74 FR at 18888,

April 24, 2009). This was repeated when EPA issued the Notice of Upcoming Joint Rulemaking to Establish Vehicle GHG Emissions and CAFE Standards (74 FR 24007 May 22, 2009) (Notice of Intent or NOI). This was repeated again when EPA issued proposed greenhouse gas emissions standards for certain new motor vehicles (74 FR 49454, September 28, 2009). EPA has consistently made it clear that issuance of new motor vehicle standards requires and is contingent upon satisfaction of the two-part endangerment test.

On May 19, 2009, EPA issued the joint Notice of Intent, which indicated EPA's intention to propose new motor vehicle standards. All of the major motor vehicle manufacturers, their trade associations, the State of California, and several environmental organizations announced their full support for the upcoming rulemaking. Not surprisingly, on the same day the President also announced his full support for this action. Commenters, however, erroneously equate this Presidential support with a Presidential directive that requires EPA to prejudge and preordain the result of this rulemaking. The only evidence they point to are simply indications of Presidential support. Commenters point to a press release, which unsurprisingly refers to the Agency's announcement as delivering on the President's commitment to enact more stringent fuel economy standards, by bringing "all stakeholders to the table and [coming] up with a plan" for solving a serious problem. The plan that was announced, of course, was a plan to conduct notice and comment rulemaking. The press release itself states that President Obama "set in motion a new national policy," with the policy "aimed" at reducing greenhouse gas emissions for new cars and trucks. What was "set in motion" was a notice and comment rulemaking described in the NOI issued by EPA on the same day. Neither the President nor EPA announced a final rule or a final direction that day, but instead did no more than announce a plan to go forward with a notice and comment rulemaking. That is how the plan "delivers on the President's commitment" to enact more stringent standards. The announcement was that a notice and comment rulemaking would be initiated with the aim of adopting certain emissions standards.

That is no different from what EPA or any other agency states when it issues a notice of proposed rulemaking. It starts a process that has the aim of issuing final regulations if they are deemed appropriate at the end of the public process. The fact that an Agency proposes a certain result, and expects that a final rule will be the result of setting such a process in motion, is the ordinary course of affairs in notice and comment rulemakings. This does not translate into prejudging the final result or having a preordained result that de facto negates the public comment process. The President's press release of May 19, 2009, was a recognition that this notice and comment rulemaking process would be set in motion, as well as providing his full support for the Agency to go forward in this direction; it was no more than that.

The various stakeholders who announced their support for the plan that had been set in motion all recognized that full notice and comment rulemaking was part of the plan, and they all reserved their rights to participate in such notice and comment rulemaking. For example, see the letter of support from Ford Motor Company, which states that "Ford fully supports proposal and adoption of such a National Program, which we understand will be subject to full notice-and comment rulemaking, affording all interested parties including Ford the right to participate fully, comment, and submit information, the results of which are not pre-determined but depend upon processes set by law."

d. The Notice and Comment Period Was Adequate

Many commenters argue that the 60-day comment period was inadequate. Commenters claim that a 60-day period was insufficient time to fully evaluate the science and other information that informed the Administrator's proposal. Some commenters assert that because the comment period for the Proposed Finding substantially overlapped with the comment period for the Mandatory Greenhouse Gas Reporting Rule, as well as Congress' consideration of climate legislation, their ability to fully participate in the notice and comment period was "seriously compromised." Moreover, they continue, because EPA had not yet proposed CAA section 202(a) standards, there was no valid reason to fail to extend the comment period. Several commenters and other entities had also requested that EPA extend the comment period.

Some commenters assert that the notice provided by this rulemaking was "defective" because the **Federal Register** notice announcing the proposal had an error in the e-mail address for the docket. At least one commenter suggests that this error deprives potential commenters of their Due Process under the Fifth Amendment of the Constitution, citing Armstrong v. Manzo, 380 U.S. 545, 552 (1965), and that failure to "correct" the minor typographical error in the e-mail address and extend the comment period would make the rule "subject to reversal" in violation of the CAA, Administrative Procedure Act (APA), the Due Process clause of the Constitution, and EO 12866.

Finally, for many of the same reasons that commenters argue a 60-day comment period was inadequate, several commenters request that EPA reopen and/or extend the comment period. One commenter requests that the comment period be reopened because there was new information regarding data used by EPA in the Proposed Findings. In particular, the commenter alleges that it recently became aware that one of the sources of global climate data had destroyed the raw data for its data set of global surface temperatures. The commenter argues that this alleged destruction of raw data violates scientific standards, calls into question EPA's reliance on that data in these Findings, and necessitates a reopening of the proceedings. Other commenters request that the comment period be extended and/or reopened due to the release of a Federal government document on the impact of climate change in the United States near the end of the comment period, as well as the release of an internal EPA staff document discussing the science.

The official public comment period on the proposed rule was adequate. First, a 60-day comment period satisfies the procedural requirements of CAA section 307 of the CAA, which requires a 30-day comment period, and that the docket be kept open to receive rebuttal or supplemental information as follow-up to any hearings for 30 days following the hearings. EPA met those obligations here—the comment period opened on April 24, 2009, the last hearing was on May 21, 2009, and the comment period closed June 23, 2009.

Second, as explained in letters denying requests to extend the comment period, a very large part of the information and analyses for the Proposed Findings had been previously released in July 30, 2008, as part of the Advance Notice of Proposed Rulemaking: Regulating Greenhouse Gas Emissions under the Clean Air Act (ANPR) (73 FR 44353). The public comment period for the ANPR is discussed above in Section I.C.1 of these Findings. The Administrator explained that the comment period for that ANPR was 120 days and that the major recent scientific assessments that EPA relied upon in the TSD released with the ANPR had previously each gone through their own public review processes and have been publicly available for some time. In other words, EPA has provided ample time for review, particularly with regard to the technical support for the

Findings. See, for example, EPA Letter to Congressman Issa dated June 17, 2009, a copy of which is available at http://epa.gov/ climatechange/endangerment.html.

Moreover, the comment period was not rendered insufficient merely because other climate-related proceedings were occurring simultaneously.

While one commenter suggests that the convergence of several different climate-related activities has "seriously compromised" their ability to participate in the comment process, that commenter was able to submit an 89 page comment on this proposal alone. Moreover, it is hardly rare that more than one rule is out for comment at the same time. As noted above, EPA has received a substantial number of significant comments on the Proposed Findings, and has thoroughly considered and responded to significant comments.

EPA finds no evidence that a typographical error in the docket e-mail address of the **Federal Register** notice announcing the proposal prevented the public from having a meaningful opportunity to comment, and therefore deprived them of due process. Although the minor error—which involved a word processing auto-correction that turned a short dash into a long dash—appeared in the FR version of the Proposed Findings, the e-mail address is correct in the signature version of the Proposed Findings posted on EPA's Web site until publication in the Federal Register, and in the "Instructions for Submitting Written Comments" document on the Web site for the rulemaking. EPA has received over 190,000 e-mails to the docket e-mail address to date, so the minor typographical error appearing in only one location has not been an impediment to interested parties' e-mailing comments. Moreover, EPA provided many other avenues for interested parties to submit comments in addition to the docket e-mail address, including via www.regulations.gov, mail, and fax; each of these options have been utilized by many commenters. EPA is confident that the minor typographical error did not prevent anyone from submitting written comments, by e-mail or otherwise, and that the public was provided "meaningful participation in the regulatory process" as mentioned in EO 12866.

Our response regarding the request to reopen the comment period due to concerns about alleged destruction of raw global surface data is discussed more fully in the Response to Comments document, Volume 11. The commenter did not provide any compelling reason to conclude that the absence of these data would materially affect the trends in the temperature records or conclusions drawn about them in the assessment literature and reflected in the TSD. The Hadley Centre/Climate Research Unit (CRU) temperature record (referred to as HadCRUT) is just one of three global surface temperature records that EPA and the assessment literature refer to and cite. National Oceanic and Atmospheric Administration (NOAA) and National Aeronautics and Space Administration (NASA) also produce temperature records, and all three temperature records have been extensively peer reviewed. Analyses of the three global temperature records produce essentially the same long-term trends as noted in the Climate Change Science Program (CCSP) (2006) report "Temperature Trends in the Lower Atmosphere," IPCC (2007), and NOAA's study "State of the Climate in 2008". Furthermore, the commenter did not demonstrate that the allegedly destroyed data would materially alter the HadCRUT record or meaningfully hinder its replication. The raw data, a small part of which has not been public (for reasons described at: https:// www.uea.ac.uk/mac/comm/media/press/2009/nov/CRUupdate), are available in a quality-controlled (or homogenized, value-added) format and the methodology for developing the quality-controlled data is described in the peer reviewed literature (as documented at http:// www.cru.uea.ac.uk/cru/data/temperature/).

The release of the U.S. Global Climate Research Program (USGCRP) report on impacts of climate change in the United States in June 2009 also did not necessitate extending the comment period. This report was issued by the USGCRP, formerly the Climate Change Science Program (CCSP), and synthesized information contained in prior CCSP reports and other synthesis reports, many of which had already been published (and were included in the TSD for the Proposed Findings). Further, the USGCRP report itself underwent notice and comment before it was finalized and released.

Regarding the internal EPA staff paper that came to light during the comment period, several commenters submitted a copy of the EPA staff paper with their comments; EPA's response to the issues raised by the staff paper are discussed in the Response to Comments document, Volume 1. The fact that some internal agency deliberations were made public during the comment period does not in and of itself call into question those deliberations. As our responses to comments explain, EPA considered the concerns noted in the staff paper during the proposal stage, as well as when finalizing the Findings. There was nothing about those internal comments that required an extension or reopening of the comment period.

Thus, the opportunity for comment fully satisfies the CAA and Constitutional requirement of Due Process. Cases cited by commenters do not indicate otherwise. The comment period and thorough response to comment documents in the docket indicate that EPA has given people an opportunity to be heard in a "meaningful time and a meaningful matter." Armstrong v. Manzo, 380 U.S. 545, 552 (1965). Interested parties had full notice of the rulemaking proceedings and a significant opportunity to participate through the comment process and multiple hearings.

For all the above reasons, EPA's denial of the requests for extension or reopening of the comment period was entirely reasonable in light of the extensive opportunity for public comment and heavy amount of public participation during the comment period. EPA has fully complied with all applicable public participation requirements for this rulemaking.

e. These Findings Did Not Necessitate a Formal Rulemaking Under the Administrative Procedure Act

One commenter, with the support of others, requests that EPA undertake a formal rulemaking process for the Findings, on the record, in accordance with the procedures described in sections 556–557 of the Administrative Procedure Act (APA). The commenter requests a multi-step process, involving additional public notice, an on-the-record proceeding (e.g., formal administrative hearing) with the right of appeal, utilization of the Clean Air Scientific Advisory Committee (CASAC) and its advisory proceedings, and designation of representatives from other executive branch agencies to participate in the formal proceeding and any CASAC advisory proceeding.

The commenter asserts that while EPA is not obligated under the CAA to undertake these additional procedures, the Agency nonetheless has the legal authority to engage in such a proceeding. The commenter believes this proceeding would show that EPA is "truly committed to scientific integrity and transparency." The commenter cites several cases to argue that refusal to proceed on the record would be "arbitrary and capricious" or would be an "abuse of discretion." The allegation at the core of the commenter's argument is that profound and wide-ranging scientific uncertainties exist in the Proposed Findings and in the impacts on health and welfare discussed in the TSD. To support this argument, the commenter provides lengthy criticisms of the science. The commenter also argues that the regulatory cascade that would be "unleashed" by a positive endangerment finding warrants the more formal proceedings.

Finally, the commenter suggests that EPA engage in "formal rulemaking" procedures in part due to the Administrative Conference of the United States' (ACUS) recommended factors for engaging in formal rulemaking. The commenter argues that the current action is "complex," "open-ended," and the costs that errors in the action may pose are "significant."

EPA is denying the request to undertake an "on the record" formal rulemaking. EPA is under no obligation to follow the extraordinarily rarely used formal rulemaking provisions of the APA. First, CAA section 307(d) of the CAA clearly states that the rulemaking provisions of CAA section 307(d), not APA sections 553 through 557, apply to certain specified actions, such as this one. EPA has satisfied all the requirements of CAA section 307(d). Indeed, the commenter itself "is not asserting that the Clean Air Act expressly requires" the additional procedures it requests. Moreover, the commenter does not discuss how the suggested formal proceeding would fit into the informal rulemaking requirements of CAA section 307(d) that do apply.

Formal rulemaking is very rarely used by Federal agencies. The formal rulemaking provisions of the APA are only triggered when the statute explicitly calls for proceedings "on the record after opportunity for an agency hearing." United States v. Florida East Coast Ry. Co., 410 U.S. 224, 241 (1973). The mere mention of the word "hearing" does not trigger the formal rulemaking provisions of the APA. Id. The CAA does not include the statutory phrase required to trigger the formal rulemaking provisions of the APA (and as noted above the APA does not apply in the first place). Congress specified that certain rulemakings under the CAA follow the rulemaking procedures outlined in CAA section 307(d) rather than the APA "formal rulemaking" commenter suggests.

Despite the inapplicability of the formal rulemaking provisions to this action, commenters suggest that to refuse to voluntarily undertake rulemaking provisions not preferred by Congress would make EPA's rulemaking action an "abuse of discretion." EPA disagrees with this claim, and cases cited by the commenter do not indicate otherwise. To support the idea that an agency decision to engage in informal rulemaking could be an abuse of discretion, commenter cites Ford Motor Co. v. FTC, 673 F.2d 1008 (9th Cir. 1981). In Ford Motor Co., the court ruled that the FTC's decision regarding an automobile dealership should have been resolved through a rulemaking rather than an individualized adjudication. Id. at 1010. In that instance, the court favored "rulemaking" over adjudication—not "formal rulemaking" over the far more common "informal rulemaking." The case stands only for the non-controversial proposition that sometimes agency use of adjudications may rise to an abuse of discretion where a rulemaking would be more appropriate—whether formal or informal. The Commenter does not cite a single judicial opinion stating that an agency abused its discretion by following the time-tested and Congressionally-favored informal rulemaking provisions of the CAA or the APA instead of the rarely used formal APA rulemaking provisions.

The commenter also alludes to the possibility that the choice of informal rulemaking may be "arbitrary and capricious. EPA disagrees that the choice to follow the frequently used, and CAA required, informal rulemaking procedures is arbitrary and capricious. The commenter cites Vermont Yankee Nuclear Power Corp. v. NRDC, 435 U.S. 519 (1978) for the proposition that "extremely compelling circumstances" could lead to a court overturning agency action for declining to follow extraneous procedures. As the commenter notes, in Vermont Yankee the Supreme Court overturned a lower court decision for imposing additional requirements not required by applicable statutes. Even if the dicta in Vermont Yankee could be applied contrary to the holding of the case in the way the commenter suggests, EPA's decision to follow frequently used informal rulemaking procedures for this action is highly reasonable.

As for the ACUS factors the commenter cites in support of its request, as the commenter notes, the ACUS factors are mere recommendations. While EPA certainly respects the views of ACUS, the recommendations are not binding on the Agency. In addition, EPA has engaged in a thorough, traditional rulemaking process that ensures that any concerns expressed by the commenter have been addressed. EPA has fully satisfied all applicable law in their consideration of this rulemaking.

Finally, as explained in Section III of these Findings and the Response to Comments document, EPA's approach to evaluating the evidence before it was entirely reasonable, and did not require a formal hearing. EPA relied primarily on robust synthesis reports that have undergone peer review and comment. The Agency also carefully considered the comments received on the Proposed Findings and TSD, including review of attached studies and documents. The public has had ample opportunity to provide its views on the science, and the record supporting these final findings indicates that EPA carefully considered and responded to significant public comments. To the extent the commenter's concern is that a formal proceeding will help ensure the right action in response to climate change is taken, that is not an issue for these Findings. As discussed in Section III of these Findings, this science-based judgment is not the forum for considering the potential mitigation options or their impact.

Analysis

Although ordered by the United States Supreme Court in the 2007 decision in *Massachusetts et al. v. Environmental Protection Agency* to regulate the emission of greenhouse gases into the environment, the Environmental Protection Agency (EPA) under the George W. Bush administration proved reticent to act. In contrast, it took less than a year into President Barack Obama's presidency for the EPA to produce extensive rules covering carbon dioxide, methane, nitrous oxide, hydrofluorocarbons, perfluorocarbons, and sulfur hexafluoride. Congressional critics alleged that the EPA utilized its loss in *Massachusetts et al. v. Environmental Protection Agency* to acquire far more authority over the regulation of GHGs than the Supreme Court ever intended. It is obvious in the "Introduction" to the **Endangerment and Cause or Contribute Findings for Greenhouse Gases Under Section 202(a) of the Clean Air Act: Final Rule** that the EPA's administrators anticipated the criticism, as they justified extensively in the document what granted them the authority to take such far-reaching steps to prevent the escalation of global warming through the emission of six specific types of GHGs.

Further Reading

Carlson, Ann, and Megan Herzog. 2014. "EPA Greenhouse Gas Rules at Stake in U.S. Supreme Court." *Trends* 45, no. 4: 2–6.

Westmoreland, Joshua K. 2010. "Global Warming and Originalism: The Role of the EPA in the Obama Administration" *Boston College Environmental Law Review* 37: 225–256.

Copenhagen Accord

Date: December 18, 2009
Location: Copenhagen, Denmark
Significance: The attendees at the Fifteenth Conference of the Parties to the United Nations Framework Convention on Climate Change (COP-15) agreed to significantly reduce greenhouse emissions in order to ensure that the temperature globally does not rise above 2 degrees Celsius from 1900 levels.
Source: United Nations Framework Convention on Climate Change. 2009. *Report of the Conference of the Parties on its Fifteenth Session, Held in Copenhagen from 7 to 19 December 2009.* http://unfccc.int/resource/docs/2009/cop15/eng/11a01.pdf (accessed January 6, 2016).

The Heads of State, Heads of Government, Ministers, and other heads of the following delegations present at the United Nations Climate Change Conference 2009 in Copenhagen: Albania, Algeria, Armenia, Australia, Austria, Bahamas, Bangladesh, Belarus, Belgium, Benin, Bhutan, Bosnia and Herzegovina, Botswana, Brazil, Bulgaria, Burkina Faso, Cambodia, Canada, Central African Republic, Chile, China, Colombia, Congo, Costa Rica, Côte d'Ivoire, Croatia, Cyprus, Czech Republic, Democratic Republic of the Congo, Denmark, Djibouti, Eritrea, Estonia, Ethiopia, European Union, Fiji, Finland, France, Gabon, Georgia, Germany, Ghana, Greece, Guatemala, Guinea, Guyana, Hungary, Iceland, India, Indonesia, Ireland, Israel, Italy, Japan, Jordan, Kazakhstan, Kiribati, Lao People's Democratic Republic, Latvia, Lesotho, Liechtenstein, Lithuania, Luxembourg, Madagascar, Malawi, Maldives, Mali, Malta, Marshall Islands, Mauritania, Mexico, Monaco, Mongolia, Montenegro, Morocco, Namibia, Nepal, Netherlands, New Zealand, Norway, Palau, Panama, Papua New Guinea, Peru, Poland, Portugal, Republic of Korea, Republic of Moldova, Romania, Russian Federation, Rwanda, Samoa, San Marino, Senegal, Serbia, Sierra Leone, Singapore, Slovakia, Slovenia, South Africa, Spain, Swaziland, Sweden, Switzerland, the former Yugoslav Republic of Macedonia, Tonga, Trinidad and Tobago, Tunisia, United Arab Emirates, United Kingdom of Great Britain and Northern Ireland, United Republic of Tanzania, United States of America, Uruguay and Zambia,

In pursuit of the ultimate objective of the Convention as stated in its Article 2,

Being guided by the principles and provisions of the Convention,

Noting the results of work done by the two Ad hoc Working Groups,

Endorsing decision 1/CP.15 on the Ad hoc Working Group on Long-term Cooperative Action and decision 1/CMP.5 that requests the Ad hoc Working Group on Further Commitments of Annex I Parties under the Kyoto Protocol to continue its work,

Have agreed on this Copenhagen Accord which is operational immediately.

1. We underline that climate change is one of the greatest challenges of our time. We emphasise our strong political will to urgently combat climate change in accordance

with the principle of common but differentiated responsibilities and respective capabilities. To achieve the ultimate objective of the Convention to stabilize greenhouse gas concentration in the atmosphere at a level that would prevent dangerous anthropogenic interference with the climate system, we shall, recognizing the scientific view that the increase in global temperature should be below 2 degrees Celsius, on the basis of equity and in the context of sustainable development, enhance our long-term cooperative action to combat climate change. We recognize the critical impacts of climate change and the potential impacts of response measures on countries particularly vulnerable to its adverse effects and stress the need to establish a comprehensive adaptation programme including international support.

2. We agree that deep cuts in global emissions are required according to science, and as documented by the IPCC Fourth Assessment Report with a view to reduce global emissions so as to hold the increase in global temperature below 2 degrees Celsius, and take action to meet this objective consistent with science and on the basis of equity. We should cooperate in achieving the peaking of global and national emissions as soon as possible, recognizing that the time frame for peaking will be longer in developing countries and bearing in mind that social and economic development and poverty eradication are the first and overriding priorities of developing countries and that a low-emission development strategy is indispensable to sustainable development.

3. Adaptation to the adverse effects of climate change and the potential impacts of response measures is a challenge faced by all countries. Enhanced action and international cooperation on adaptation is urgently required to ensure the implementation of the Convention by enabling and supporting the implementation of adaptation actions aimed at reducing vulnerability and building resilience in developing countries, especially in those that are particularly vulnerable, especially least developed countries, small island developing States and Africa. We agree that developed countries shall provide adequate, predictable and sustainable financial resources, technology and capacity-building to support the implementation of adaptation action in developing countries.

4. Annex I Parties commit to implement individually or jointly the quantified economy-wide emissions targets for 2020, to be submitted in the format given in Appendix I by Annex I Parties to the secretariat by 31 January 2010 for compilation in an INF document. Annex I Parties that are Party to the Kyoto Protocol will thereby further strengthen the emissions reductions initiated by the Kyoto Protocol. Delivery of reductions and financing by developed countries will be measured, reported and verified in accordance with existing and any further guidelines adopted by the Conference of the Parties, and will ensure that accounting of such targets and finance is rigorous, robust and transparent.

5. Non-Annex I Parties to the Convention will implement mitigation actions, including those to be submitted to the secretariat by non–Annex I Parties in the format given in Appendix II by 31 January 2010, for compilation in an INF document, consistent with Article 4.1 and Article 4.7 and in the context of sustainable development. Least developed countries and small island developing States may undertake actions voluntarily and on the basis of support. Mitigation actions subsequently taken and envisaged by Non-Annex I Parties, including national inventory reports, shall be communicated through national communications consistent with Article 12.1(b)

every two years on the basis of guidelines to be adopted by the Conference of the Parties. Those mitigation actions in national communications or otherwise communicated to the Secretariat will be added to the list in appendix II. Mitigation actions taken by Non-Annex I Parties will be subject to their domestic measurement, reporting and verification the result of which will be reported through their national communications every two years. Non-Annex I Parties will communicate information on the implementation of their actions through National Communications, with provisions for international consultations and analysis under clearly defined guidelines that will ensure that national sovereignty is respected. Nationally appropriate mitigation actions seeking international support will be recorded in a registry along with relevant technology, finance and capacity building support. Those actions supported will be added to the list in appendix II. These supported nationally appropriate mitigation actions will be subject to international measurement, reporting and verification in accordance with guidelines adopted by the Conference of the Parties.

6. We recognize the crucial role of reducing emission from deforestation and forest degradation and the need to enhance removals of greenhouse gas emission by forests and agree on the need to provide positive incentives to such actions through the immediate establishment of a mechanism including REDD-plus, to enable the mobilization of financial resources from developed countries.

7. We decide to pursue various approaches, including opportunities to use markets, to enhance the cost-effectiveness of, and to promote mitigation actions. Developing countries, especially those with low emitting economies should be provided incentives to continue to develop on a low emission pathway.

8. Scaled up, new and additional, predictable and adequate funding as well as improved access shall be provided to developing countries, in accordance with the relevant provisions of the Convention, to enable and support enhanced action on mitigation, including substantial finance to reduce emissions from deforestation and forest degradation (REDD-plus), adaptation, technology development and transfer and capacity-building, for enhanced implementation of the Convention. The collective commitment by developed countries is to provide new and additional resources, including forestry and investments through international institutions, approaching USD 30 billion for the period 2010–2012 with balanced allocation between adaptation and mitigation. Funding for adaptation will be prioritized for the most vulnerable developing countries, such as the least developed countries, small island developing States and Africa. In the context of meaningful mitigation actions and transparency on implementation, developed countries commit to a goal of mobilizing jointly USD 100 billion dollars a year by 2020 to address the needs of developing countries. This funding will come from a wide variety of sources, public and private, bilateral and multilateral, including alternative sources of finance. New multilateral funding for adaptation will be delivered through effective and efficient fund arrangements, with a governance structure providing for equal representation of developed and developing countries. A significant portion of such funding should flow through the Copenhagen Green Climate Fund.

9. To this end, a High Level Panel will be established under the guidance of and accountable to the Conference of the Parties to study the contribution of the potential sources of revenue, including alternative sources of finance, towards meeting this goal.

10. We decide that the Copenhagen Green Climate Fund shall be established as an

operating entity of the financial mechanism of the Convention to support projects, programme, policies and other activities in developing countries related to mitigation including REDD-plus, adaptation, capacity-building, technology development and transfer.
11. In order to enhance action on development and transfer of technology we decide to establish a Technology Mechanism to accelerate technology development and transfer in support of action on adaptation and mitigation that will be guided by a country-driven approach and be based on national circumstances and priorities.
12. We call for an assessment of the implementation of this Accord to be completed by 2015, including in light of the Convention's ultimate objective. This would include consideration of strengthening the long-term goal referencing various matters presented by the science, including in relation to temperature rises of 1.5 degrees Celsius.

Analysis

At the Copenhagen meeting, attendees identified 2 degrees Celsius over 1900 levels to be the threshold that, if crossed, irreparable damage to the climate would occur. At that point, locales like the Marshall Islands would be underwater. This would have severe consequences for all, even those nations located far from the oceans, as those displaced by rising waters would need to be resettled. Although the entire world was threatened by the rising global temperatures, most of the countries involved exempted themselves from having to make cuts to the emissions of greenhouse gases (GHGs) within their borders. This was justified because they were not responsible for the rise in greenhouse gases globally. The blame for that problem was given to the industrialized countries, namely Brazil, China, European Union, India, Japan, Russia, and the United States. Those countries were charged with making the required reductions. Although the cuts would negatively impact their economies, the industrialized countries agreed to the desires of the whole. In truth, their agreement to do what was necessary provided the respective leaders a public relations coup and nothing else. All that they really committed to was the intent to cut emissions. Even that did not have to be honored because the agreement was non-binding.

One success that had long-term benefit for the "developing" countries was the establishment of a $100 billion fund that was financed by their industrialized counterparts. That fund would be used between 2010 and 2020 to help the poorest countries acquire technologies that would help wean them from substances like coal that contributed to the increase in the amount of GHGs.

Further Reading

Dimitrov, Radoslav S. 2010. "Inside UN Climate Change Negotiations: The Copenhagen Conference." *Review of Policy Research* 27: 795–821.
Ferrey, Steven. 2012. "Changing Venue of International Governance and Finance: Exercising Legal Control Over the $100 Billion Per Year Climate Fund?" *Wisconsin International Law Journal* 30: 26–111.
Parker, Charles F., et al. 2012. "Fragmented Climate Change Leadership: Making Sense of the Ambiguous Outcome of COP-15." *Environmental Politics* 21: 268–286.
Szarka, Joseph. 2012. "The EU, the USA and the Climate Divide: Reappraising Strategic Choices." *European Political Science* 11: 31–40.

Richard A. Muller
Statement to the Committee on Science, Space and Technology of the United States House of Representatives

Date: March 31, 2011
Location: Washington, D.C.
Significance: Dr. Richard Muller, a known skeptic of man-made global warming, testified that the preliminary data generated from the Berkeley Earth Surface Temperature project suggested that there was scientific evidence that man was responsible for the rise in surface temperatures over time.
Source: United States House of Representatives. 2011. *Statement to the Committee on Science, Space and Technology of the United States House of Representatives.* https://science.house.gov/sites/republicans.science.house.gov/files/documents/hearings/Muller%20Testimony%20rev2.pdf (accessed August 23, 2015).

Richard A. Muller
Professor of Physics
University of California, Berkeley
Chair, Berkeley Earth Surface Temperature Project
31 March 2011

Executive Summary

The Berkeley Earth Surface Temperature project was created to make the best possible estimate of global temperature change using as complete a record of measurements as possible and by applying novel methods for the estimation and elimination of systematic biases. It was organized under the auspices of Novim, a non-profit public interest group. Our approach builds on the prior work of the groups at NOAA, NASA, and in the UK (Hadley Center—Climate Research Unit, or HadCRU).

Berkeley Earth has assembled 1.6 billion temperature measurements, and will soon make these publicly available in a relatively easy to use format.

The difficult issues for understanding global warming are the potential biases. These can arise from many technical issues, including data selection, substandard temperature station quality, urban vs rural effects, station moves, and changes in the methods and times of measurement.

We have done an initial study of the station selection issue. Rather than pick stations with long records (as done by the prior groups) we picked stations randomly from the complete set. This approach eliminates station selection bias. Our results are shown in the Figure; we see a global warming trend that is very similar to that previously reported by the other groups.

We have also studied station quality. Many U.S. stations have low quality rankings according to a study led by Anthony Watts. However, we find that the warming seen in the "poor" stations is virtually indistinguishable from that seen in the "good" stations.

We are developing statistical methods to address the other potential biases.

I suggest that Congress consider the creation of a Climate-ARPA to facilitate the study of climate issues.

Based on the preliminary work we have done, I believe that the systematic biases that are the cause for most concern can be adequately handled by data analysis techniques. The world temperature data has sufficient integrity to be used to determine global temperature trends.

Testimony of Richard A. Muller

Thank you Chairman Hall and Ranking Member Johnson for this opportunity to testify before the Committee.

I am a Professor of Physics at UC Berkeley and Faculty Senior Scientist at the Lawrence Berkeley Laboratory. I founded the Berkeley Earth Surface Temperature project under the auspices of Novim, a non-profit public interest group. My testimony represents my personal views and not those of the above organizations.

[[Italic part for written statement only, not to be read aloud]]

I've published papers on climate change in Science, Nature, and other refereed journals; I am the author of a technical book on the subject.

My papers on climate change have appeared in Nature, Science, Paleoceanography, and the Journal of Geophysical Research. I wrote a technical book on the Earth's past temperature changes: "Ice Ages and Astronomical Causes," Springer 2000. I am the author of "Physics for Future Presidents," a popular book which describes many misuses of data in climate. I was a cited referee on the report of the NRC on the hockey stick controversy. For two years I wrote an online column for MIT's Technology Review. My major awards for scientific achievement include the Alan T. Waterman Award of the National Science Foundation, the Texas Instruments Founders Prize, a MacArthur Prize Fellowship, and election to the American Academy of Arts and Sciences and to the California Academy of Sciences.

The Berkeley Earth Surface Temperature study has received a total of $623,087 in financial support from:

The Lee and Juliet Folger Fund ($20,000)
Lawrence Berkeley National Laboratory ($188,587)
William K. Bowes, Jr. Foundation ($100,000)
Fund for Innovative Climate and Energy Research (created by Bill Gates) ($100,000)
Charles G. Koch Charitable Foundation ($150,000)
The Ann & Gordon Getty Foundation ($50,000)
We have also received funding from a number of private individuals, totaling $14,500.
For more information on Berkeley Earth, see www.BerkeleyEarth.org
For more information on Novim, see www.Novim.org

I begin by talking about **Global Warming**

Prior groups at NOAA, NASA, and in the UK *(HadCRU)* estimate about a 1.2 degree C land temperature rise from the early 1900s to the present. This 1.2 degree rise is what we call **global warming**. Their work is excellent, and the Berkeley Earth project strives to build on it.

Human caused global warming is somewhat smaller. According to the most recent IPCC report (2007), the human component became apparent only after 1957, and it amounts to "most" of the 0.7 degree rise since then. **Let's assume the human-caused warming is 0.6 degrees.**

The magnitude of this temperature rise is a key scientific and public policy concern. A 0.2 degree uncertainty puts the human component between 0.4 and 0.8 degrees—a factor of two uncertainty. Policy depends on this number. It needs to be improved.

Berkeley Earth is working to improve on the accuracy of this key number by using a more complete set of data, and by looking at biases in a new way.

The project has already merged 1.6 billion land surface temperature measurements from 16 sources, most of them publicly available, and is putting them in a simple format to allow easy use by scientists around the world. By using all the data and new statistical approaches that can handle short records, and by using novel approaches to estimation and avoidance of systematic biases, we expect to improve on the accuracy of the estimate of the Earth's temperature change.

I'll now talk about potential **Bias in Data Selection**

Prior groups *(NOAA, NASA, HadCRU)* selected for their analysis 12% to 22% of the roughly 39,000 available stations. *(The number of stations they used varied from 4,500 to a maximum of 8,500.)*

They believe their station selection was unbiased. Outside groups have questioned that, and claimed that the selection picked records with large temperature increases. Such bias could be inadvertent, for example, a result of choosing long continuous records. *(A long record might mean a station that was once on the outskirts and is now within a city.)*

To avoid such station selection bias, Berkeley Earth has developed techniques to work with all the available stations. *This requires a technique that can include short and discontinuous records.*

In an initial test, Berkeley Earth chose stations randomly from the complete set of 39,028 stations. Such a selection is free of station selection bias.

In our preliminary analysis of these stations, we found a warming trend that is shown in the figure. It is very similar to that reported by the prior groups: a rise of about 0.7 degrees C since 1957. *(Please keep in mind that the Berkeley Earth curve, in black, does not include adjustments designed to eliminate systematic bias.)*

The Berkeley Earth agreement with the prior analysis surprised us, since our preliminary results don't yet address many of the known biases. When they do, it is possible that the corrections could bring our current agreement into disagreement.

Why such close agreement between our uncorrected data and their adjusted data? One possibility is that the systematic corrections applied by the other groups are small. We don't yet know.

The main value of our preliminary result is that it demonstrates the Berkeley Earth ability to use all records, including those that are short or fragmented. When we apply our approach to the complete data collection, we will largely eliminate the station selection bias, and significantly reduce statistical uncertainties.

Let me now address the problem of **Poor Temperature Station Quality**

Many temperature stations in the U.S. are located near buildings, in parking lots, or close to heat sources. Anthony Watts and his team has shown that most of the current stations in the U.S. *Historical Climatology* Network would be ranked "poor" by NOAA's own standards, with error uncertainties up to 5 degrees C.

Did such poor station quality exaggerate the estimates of global warming? We've studied this issue, and our preliminary answer is **no**.

The Berkeley Earth analysis shows that over the past 50 years the poor stations in the U.S. network do not show greater warming than do the good stations.

Thus, although poor station quality might affect absolute temperature, it does not appear to affect trends, and for global warming estimates, the trend is what is important.

Our key caveat is that our results are preliminary and have not yet been published in a peer reviewed journal. We have begun that process of submitting a paper to the Bulletin of the American Meteorological Society, and we are preparing several additional papers for publication elsewhere.

NOAA has already published a similar conclusion—that station quality bias did not affect estimates of global warming—based on a smaller set of stations, and Anthony Anthony Watts and his team have a paper submitted, which is in late stage peer review, using over 1000 stations, but it has not yet been accepted for publication and I am not at liberty to discuss their conclusions and how they might differ. We have looked only at average temperature changes, and additional data needs to be studied, to look at (for example) changes in maximum and minimum temperatures.

In fact, in our preliminary analysis the good stations report more warming in the U.S. than the poor stations by 0.009 ± 0.009 degrees per decade, opposite to what might be expected, but also consistent with zero. We are currently checking these results and performing the calculation in several different ways. But we are consistently finding that there is no enhancement of global warming trends due to the inclusion of the poorly ranked U.S. stations.

Berkeley Earth hopes to complete its analysis including systematic bias avoidance in the next few weeks. We are now studying new approaches to reducing biases from:

1. *Urban heat island effects.* Some stations in cities show more rapid warming than do stations in rural areas.
2. *Time of observation bias.* When the time of recording temperature is changed, stations will typically show different mean temperatures than they did previously. This is sometimes corrected in the processes used by existing groups. But this cannot be done easily for remote stations or those that do not report times of observations.
3. *Station moves.* If a station is relocated, this can cause a "jump" in its temperatures. This is typically corrected in the adjustment process used by other groups. Is the correction introducing another bias? The corrections are sometimes done by hand, making replication difficult.
4. *Change of instrumentation.* When thermometer type is changed, there is often an offset introduced, which must be corrected.

Potential Legislation

I was asked what legislation could advance our knowledge of climate change. After some consideration, I felt that the creation of a Climate Advanced Research Project Agency, or Climate-ARPA, could help.

Without the efforts of Anthony Watts and his team, we would have only a series of anecdotal images of poor temperature stations, and we would not be able to evaluate the integrity of the data.

This is a case in which scientists receiving no government funding did work crucial to understanding climate change. Similarly for the work done by Steve McIntyre. Their "amateur" science is not amateur in quality; it is true science, conducted with integrity and high standards.

Government policy needs to encourage such work. **Climate-ARPA** could be an organization that provides quick funding to worthwhile projects without regard to whether they support or challenge current understanding.

In Summary

Despite potential biases in the data, methods of analysis can be used to reduce bias effects well enough to enable us to measure long-term Earth temperature changes. Data integrity is adequate. Based on our initial work at Berkeley Earth, I believe that some of the most worrisome biases are less of a problem than I had previously thought.

Analysis

Prior to his testimony before the Committee on Science, Space and Technology of the United States House of Representatives, Dr. Richard A. Muller was skeptical of claims that man was responsible for global warming. This was due to his belief that most of the "evidence" for the theory was derived from non-scientists, such as Al Gore. What scientific evidence did exist was derived from researchers who had obvious biases. He thus decided, through the Berkeley Earth Surface Temperature project, to use all of the scientific evidence for and against global warming to see if a trend for or against man-made global warming emerged. After adjusting for factors like bias and location of the temperature sensors, the preliminary data determined that there had been an increase in global temperatures of 0.7 degrees C since 1957. This was a similar conclusion to studies that he had derided in the past.

When he arrived to testify before the Committee on Science, Space and Technology of the United States House of Representatives, it was expected that he was going to buttress the claims of climate skeptics and deniers, but instead presented extensive evidence that man-made global warming was occurring and posed a threat to human existence. He urged the Congressmen to work to curb the threat to mankind. What made his testimony especially compelling and influential was that it was based solely on scientific evidence, rather than advancing a political agenda.

FURTHER READING

Lemonick, Michael D. 2011. "'I Stick to the Science.'" *Scientific American* 304: 84–87.

2013 Highlights of Progress
Responses to Climate Change by the National Water Program (Excerpt)

Date: April 24, 2014
Where: Washington, D.C.
Significance: The report provides an overview of how the Environmental Protection Agency views the threat posed by climate change. It also details the specific steps that the agency is undertaking to ensure that the nation's water needs are met into the future.
Source: United: U.S. Environmental Protection Agency. 2014. *2013 Highlights of Progress: Responses to Climate Change by the National Water Program.* http://water.epa.gov/scitech/climatechange/upload/Final-2013-NWP-Climate-Highlights-Report.pdf (accessed November 30, 2015).

National Program Highlights

Vision Area 1: Water Infrastructure
   Vision: In the face of a changing climate, resilient and adaptable drinking water, wastewater and stormwater utilities (water sector) ensure clean and safe water to protect the nation's public health and environment by making smart investment decisions to improve the sustainability of their infrastructure and operations and the communities they serve, while reducing greenhouse gas emissions through greater energy efficiency.

1. **Address Climate Change in Clean Water State Revolving Fund:** In 2013, the Clean Water State Revolving Fund (CWSRF) program developed a comprehensive list of CWSRF-eligible projects to increase climate/weather-related resilience at water utilities to implement the Disaster Relief Appropriations Act of 2013. Climate/weather-related eligibilities were also discussed at the fall Council for Infrastructure Financing Authorities conference, where the CWSRF program delivered a presentation to State and Regional counterparts regarding the CWSRF's ability to promote climate/weather-related resilience in the water sector.

    The CWSRF program also revised its Annual Review Checklist to incorporate several questions on resilience to climate change and extreme weather and participated with the DWSRF in the development of a draft guide for small utilities that want to become more resilient to flooding.

2. **Expand Climate Ready Water Utilities Program Outreach:** Through the Climate Ready Water Utilities (CRWU) initiative, EPA has provided 15 workshops and webinars and reached over 2000 people. Key 2013 activities included:

- convening a working group to develop version 3.0 of the Climate Resilience Evaluation and Awareness Tool (CREAT) including utilities, academia, associations and other Federal partners;
- holding two-day emergency response workshops with demonstrations of tools and resources to aid utilities, review climate impacts, and discuss planning options with different sector stakeholders, such as local governments, first responders, and community leaders; and
- hosting webinars with the Water Utility Climate Alliance to help utilities plan for climate change and other threats.

Ten webinars were held in 2013 and there are plans for at least four more over 2014. Each webinar is recorded and archived on EPA's website at www.epa.gov/climatereadyutilities.

3. **Publish Reports of Water Utility Extreme Weather Case Studies:** EPA worked with partners to organize workshops in six communities with a focus on areas that have already experienced extreme events, including drought, flooding, wildfires, sea level rises, and heat waves. The communities where workshops were held included:
   - Georgia: Upper Apalachicola–Chattahoochee–Flint River Basin;
   - California: Russian River Watershed;
   - Virginia: Tidewater Area;
   - Washington DC: National Capital Area;
   - Kansas/Missouri: Lower Missouri River Basin; and
   - Texas: Central Region.

   Fact sheets and reports on the lessons learned from these case studies were published throughout 2013. Fact sheets are available at: http://cpo.noaa.gov/ClimatePrograms/ClimateandSocietalInteractions/SARPProgram/ExtremeEventsCaseStudies.aspx. Partners included the National Oceanic and Atmospheric Administration, Water Environment Research Foundation, Water Research Foundation, Concurrent Technologies Corporation, and Noblis.

4. **Expand WaterSense to Commercial Kitchen Products:** In September 2013, EPA finalized the first WaterSense specification for a commercial kitchen product. Pre-rinse spray valves—which remove excess food waste from dishes prior to dishwashing—are now eligible to earn the WaterSense label and help food service establishments save water, energy, and money. Pre-rinse spray valves can account for nearly one-third of the water used in a typical commercial kitchen. If every U.S. commercial food service establishment installed and used a WaterSense labeled pre-rinse spray valve, we could save more than 10 billion gallons of water, and more than $225 million in water and energy costs annually across the country. Because kitchens use hot water to rinse dishes, installing a WaterSense labeled pre–rinse spray valve can also reduce a commercial kitchen's annual natural gas use by more than 6,400 cubic feet per year.

5. **Expand WaterSense Partners**: In 2013, the number of WaterSense partners across the country continued to grow, increasing by close to 120 to a total of 1,474 partners, which includes water utilities, state and local governments, manufacturers, retailers, and builders

Vision Area 2: Watersheds and Wetlands

Vision: Watersheds are protected, maintained and restored to ensure climate resilience and to preserve the social and economic benefits they provide; and the nation's wetlands

are maintained and improved using integrated approaches that recognize their inherent value as well as their role in reducing the impacts of climate change.

6. **Build State and Local Capacity to Protect Healthy Watersheds and Enhance Climate Change Resiliency:** EPA's Healthy Watersheds Program (HWP) is working to build state and local capacity to identify and protect healthy watersheds using a systems approach that recognizes watersheds as dynamic, interconnected ecosystems. Natural, intact watersheds are better equipped to withstand, recover from, and adapt to natural and man-made disturbances, including climate change. Implementing strategies to maintain and protect healthy watersheds is key toward enhancing climate change resiliency. In 2013, HWP worked to build state and local capacity to identify and protect healthy watersheds at a variety of scales and locations:
    - Identification and Protection of Kansas's Healthy Watersheds;
    - California Integrated Assessment of Watershed Health;
    - Aquatic Ecosystem Protection in Minnesota's Snake River Watershed;
    - Establishing Temperature Regime Characteristics of High Quality Streams in Connecticut;
    - Sustaining West Virginia's Natural Capital: A Framework for Green Infrastructure; and
    - Green Infrastructure Practitioners Guide and Ulster County New York Case Study.

    For more information on the Healthy Watershed Program see: www.epa.gov/healthywatersheds.

7. **Sign Joint Memorandum of Understanding to Promote Healthy Watershed Protection:** On February 22, 2013, EPA, The Nature Conservancy (TNC), and the Association of Clean Water Administrators (ACWA) jointly signed the Memorandum of Understanding (MOU) to promote EPA's Healthy Watersheds Program (HWP: www.epa.gov/healthywatersheds). This MOU formalizes a mutual collaboration between these groups as they strive to develop and implement healthy watersheds programs in states and with regional aquatic ecosystem programs. These programs include working with states and other partners to identify healthy watersheds statewide and to implement healthy watershed protection plans, to integrate such protection into EPA programs and to increase awareness and understanding of the importance of protecting our remaining healthy watersheds. TNC, EPA, and ACWA recognize that healthy, intact watersheds can offset the potential impacts of climate change in a variety of ways including maintenance of baseflow during periods of drought, native vegetation that provides cooling during heat waves, carbon storage in native vegetation and soils, and enhanced stormwater infiltration capacity that mitigates downstream flooding.

    The partners will promote data gathering/data sharing and evaluation of conservation and environmental outcomes resulting from the implementation of state and regional healthy watershed programs. See: http://water.epa.gov/polwaste/nps/watershed/hwi-mou.cfm.

8. **Assess Climate Change in 20 Watersheds:** The EPA Office of Research and Development released in 2013 a report that evaluated streamflow, nitrogen, phosphorus, and sediment in 20 different watersheds across the U.S. for the periods 1970–2000 and 2041–2070 to examine the effects of six different scenarios of climate change and urban/residential development. Additional scenarios were evaluated for five of

the watersheds to examine implications of using different methodological choices for this and similar studies. The results indicate that different conditions by mid-21st century are possible for many watersheds, with larger differences likely where development is concentrated. The results also showed sensitivity to the methodological choices, such as use of different watershed models and approaches to downscaling results from global-scale models. (See EPA/600/R-12/058F). (http://www.epa.gov/ncea).

Vision Area 3: Coastal and Ocean Waters

Vision: Adverse effects of climate change and unintended adverse consequences of responses to climate change have been successfully prevented or reduced in the ocean and coastal environment. Federal, tribal, state, and local agencies, organizations, and institutions are working cooperatively; and information necessary to integrate climate change considerations into ocean and coastal management is produced, readily available, and used.

9. **Publish for Peer Review *Being Prepared for Climate Change* workbook**: A public draft of the Climate Ready Estuaries (CRE) Program workbook titled *Being Prepared for Climate Change* was sent out for peer review in September and posted on the CRE website in October 2013 for public comment. The workbook applies a risk management methodology for climate change adaptation and helps organizations prepare vulnerability assessments and action plans. The vulnerability assessment methodology of the *Being Prepared for Climate Change* workbook was shared with staff from NEPs, EPA Regions, EPA headquarters, and other federal partners. For more information see http://water.epa.gov/type/oceb/cre/news.cfm.

10. **Hold Climate Change Vulnerability Workshop:** On February 25, 2013, EPA held a climate change vulnerability assessment workshop with two main goals:
    - to share the vulnerability assessment methodology of the *Being Prepared for Climate Change* workbook; and
    - to hear from NEP staff (a main audience for the workbook) about what they want to know in order to prepare high-level, risk-based vulnerability assessments.

    The workshop provided a step-by-step walk through of the vulnerability assessment steps in the CRE workbook. The San Juan Bay Estuary Program (SJBEP) also shared some of its experience working on a climate change vulnerability assessment. SJBEP was part of a 2012 Climate Ready Estuaries pilot project to use an early version of the workbook for their assessment.

11. **Publish Study of Climate Change Impacts on Salmon Populations**: EPA's Office of Research and Development used empirically based simulation modeling of 48 sockeye salmon populations to examine how reliably alternative monitoring designs and fish stock assessment methods can estimate the relative contribution of climate compared to non-climatic factors. The study covered a range of scenarios for ocean conditions, salmon productivity, and human-induced changes and found that distinguishing climate-related effects on salmon productivity from non-climate sources will be difficult, especially if climatic changes occur rapidly and concurrently with major anthropogenic disturbances. Better understanding of the mechanisms underlying the relationship between climate and salmon productivity may be essential to avoid undesirable management outcomes. *Fisheries Research* 147:10–23.

12. **Publish Research on Sea Surface Temperatures in Pacific Northwest:** A remotely-sensed dataset was used to focus on the nearshore environment of the North Pacific to identify and describe broad-scale sea surface temperature (SST) patterns. Satellite remotely-sensed mean, monthly SST data were used to create a 29-year nearshore (< 20 km offshore) time series of SST along the North Pacific coastline. The scalable nature of the methodology is useful to both broader-scale and more focused analyses, and puts an environmental factor of primary importance (SST) into the hands of researchers studying nearshore environments by providing web-based access to it. Reference: Payne, M. C., Reusser, D., Brown, C. A., and Lee II, H. (2012) "Ecoregional analysis of nearshore sea-surface temperature in the North Pacific." *PLoS ONE,* 7(1):12 pages.

Vision Area 4: Water Quality
Vision: The Nation's surface water, drinking water, and ground water quality are protected, and the risks of climate change to human health and the environment are diminished, through a variety of adaptation and mitigation strategies.

13. **Develop Climate Change Extension for the Stormwater Calculator:** The Stormwater Calculator (SWC) is a desktop tool intended to help users at individual sites manage stormwater by reducing runoff through infiltration and retention (i.e., green infrastructure). The SWC was launched in 2013 and uses EPA's Stormwater Management Model (SWMM) as its computational engine. A climate change extension to the SWC was developed in 2013 and released in final form early in 2014. This extension allows users to apply different future climate change scenarios that modify the historical precipitation events and evaporation rates normally used by the calculator. This climate change extension will help site owners, developers, and planners design more robust stormwater management solutions in the face of uncertain future climatic conditions.

14. **Develop Improved Monitoring of Water Temperature and Flow:** EPA's Office of Research and Development released an external review draft report of technical "best practices" describing sensor deployment for and data collection of continuous temperature and flow at ungaged sites in wadeable streams. The draft report addresses questions related to equipment needs; configuration, placement, and installation of equipment; and data retrieval and processing. (See EPA/600/R-13/170). (http://cfpub.epa.gov/ncea/global/recordisplay.cfm?deid=261911).

15. **Publish Study of Climate Impacts on Nitrogen in Water:** EPA's Office of Research and Development investigated the effects of projected changes in land cover and climate (precipitation, temperature and atmospheric CO_2 concentrations) on simulated NO_3 and organic nitrogen discharge for two watersheds within the Neuse River Basin, NC for years 2010 to 2070. Results showed nitrogen discharges were most sensitive to changes in precipitation and temperature, with sensitivities to CO_2 and land cover only one-tenth as much. With nitrogen discharge showing high sensitivity to P+T change, this study suggests more emphasis should be placed on investigating impacts of climate change on nutrient transport compared to land cover change in the Neuse River Basin. (See "Relative Sensitivity of Simulated Nitrogen Discharge to Projected Changes in Climate and Land Cover for Two Watersheds in North Carolina, USA," presented at AGU Fall Meeting 2013, San Francisco, CA, December 09–13).

Vision Area 5: Working with Tribes

Vision: Tribes are able to preserve, adapt, and maintain the viability of their culture, traditions, natural resources, and economies in the face of a changing climate.

16. **Support EPA's Tribal-Focused Environmental Risk and Sustainability Tool (Tribal-FERST).** The EPA Office of Water is working with the Office of Research and Development to develop and implement Tribal-FERST, which is a web-based geospatial decision support tool designed to serve as a research framework to provide tribes with easy access to the best available human health and ecological science.

    Tribes and partners throughout the United States are providing input on the design and content of Tribal-FERST. The United South and Eastern Tribes (USET) is partnering with EPA to develop the Tribal-FERST guidance document and connect its water quality exchange database and data transfer network with Tribal-FERST. The Pleasant Point Passamaquoddy Tribe of Maine is currently piloting Tribal-FERST as part of its sustainable and healthy community effort. In addition, water programs in EPA's regional offices are working with Tribes to assist them in responding to a range of climate change related issues. These activities, described in greater detail in the next Part of this report, include:
    - Region 2 is maintaining a dialog with the Tribal nations regarding **climate change adaptation and Traditional Ecological Knowledge (TEK). A climate change grant** to the Nations was extended through September 2013 to support vulnerability assessment of nation lands and planning climate adaptation strategies.
    - Region 4 initiated collaboration with United South and Eastern Tribes, Inc., (USET), which serves 26 Tribes from Texas to Maine and is located in Nashville, TN. Region 4 is working with USET to **build their capacity to provide energy management assistance to Tribal water utilities.**
    - Region 6 held a **climate change workshop in Albuquerque, NM for Tribal and Environmental Justice Communities vulnerable to climate change impacts.**
    - Region 7 tribes are incorporating climate change science into their CWA 106 programs addressing **water quality monitoring.**

Vision Area 6: Cross-cutting Program Support

17. **Develop Office of Water Draft *Climate Change Adaptation Implementation Plan:*** The Office of Water worked with EPA Regional Water Division staff to draft a *Climate Change Adaptation Implementation Plan*. The draft *Plan* was organized on the template adopted by EPA and is comparable to each of the 16 other national program office and Regional office climate change adaptation implementation plans. The Office of Water draft *Plan* was released for public comment in September 2013. A final *Plan* will be published in the fall of 2014. More information is available at: http://www.epa.gov/water/climatechange.

18. **Co-Chair Climate Change Adaptation and Water Stakeholder Group:** In 2013, the Office of Water staff served as co-chair of a newly established Climate Change Workgroup of the Advisory Committee on Water Information (ACWI). The Workgroup includes 40 representatives from federal agencies and stakeholder organizations and provides advice and comment to federal agencies on a range of climate change and water resources issues, including the progress in implementing

the *National Action Plan: Priorities for Managing Freshwater Resources in a Changing Climate*. More information is available at www.acwi.org.
19. **Contribute to Federal Interagency Climate Adaptation Projects:** National Water Program staff also participated in a range of workgroups within EPA and among other federal agencies working to adapt to a changing climate including the:
    - EPA Cross-Agency Climate Change Adaptation Workgroup;
    - Interagency Council on Climate Resilience and Preparedness;
    - Water Resources Workgroup of the Interagency Council on Climate Change Resilience and Preparedness;
    - Interagency Joint Working Group implementing the final *Fish Wildlife and Plants Climate Adaptation Strategy*;
    - *National Ocean Policy* Implementation Plan workgroup on climate change;
    - Interagency Ocean Acidification Working Group; and
    - Coral Reef Task Force.

The Office of Water also has an interagency agreement with NOAA in which climate adaptation is a joint focus.

Analysis

The Environmental Protection Agency's Global Water Program publishes an annual report that includes six vision statements which guide its work related to global warming. Four of the statements relate directly to water. The other two address administrative functions. The report also details the projects that are completed in a given year and charts the direction for future research.

Further Reading

Christian-Smith, Juliet, and Peter H. Gleick. 2012. *A Twenty-First Century U.S. Water Policy*. New York: Oxford University Press.

Environmental Protection Agency, Office of Water. 2014. *National Water Program Guidance Addendum: Fiscal Year 2015*. http://water.epa.gov/resource_performance/planning/upload/FY-2015-NWPG-4-22-2014-Narrative-with-covers.pdf (accessed September 17, 2014).

Ingram, B. Lynn, and Frances Malamud-Roam. 2013. *The West Without Water: What Past Floods, Droughts, and Other Climatic Clues Tell Us About Tomorrow*. Berkeley: University of California Press.

Climate Change Impacts in the United States (Overview and Report Findings)

Date: May 2014
Location: Washington, D.C.
Significance: Produced by the United States Global Research Program, the National Climate Assessment identifies how climate change is impacting the nation and what effects can be anticipated in the future.
Source: Melillo, Jerry M., Terese Richmond, and Gary W. Yohe, eds., 2014. *Climate Change Impacts in the United States*. Washington, D.C.: U.S. Government Printing Office. http://s3.amazonaws.com/nca2014/high/NCA3_Highlights_HighRes.pdf?download=1 (accessed October 3, 2015).

Climate change is already affecting the American people in far reaching ways. Certain types of extreme weather events with links to climate change have become more frequent and/or intense, including prolonged periods of heat, heavy downpours, and, in some regions, floods and droughts. In addition, warming is causing sea level to rise and glaciers and Arctic sea ice to melt, and oceans are becoming more acidic as they absorb carbon dioxide. These and other aspects of climate change are disrupting people's lives and damaging some sectors of our economy.

Climate Change: Present and Future

Evidence for climate change abounds, from the top of the atmosphere to the depths of the oceans. Scientists and engineers from around the world have meticulously collected this evidence, using satellites and networks of weather balloons, thermometers, buoys, and other observing systems. Evidence of climate change is also visible in the observed and measured changes in location and behavior of species and functioning of ecosystems. Taken together, this evidence tells an unambiguous story: the planet is warming, and over the last half century, this warming has been driven primarily by human activity.

Multiple lines of independent evidence confirm that human activities are the primary cause of the global warming of the past 50 years. The burning of coal, oil, and gas, and clearing of forests have increased the concentration of carbon dioxide in the atmosphere by more than 40% since the Industrial Revolution, and it has been known for almost two centuries that this carbon dioxide traps heat. Methane and nitrous oxide emissions from agriculture and other human activities add to the atmospheric burden of heat-trapping gases. Data show that natural factors like the sun and volcanoes cannot have caused the warming observed over the past 50 years. Sensors on satellites have measured the sun's output with great accuracy and found no overall increase during the past half century. Large volcanic eruptions during this period, such as Mount Pinatubo in 1991, have exerted a short term cooling influence. In fact, if not for human activities, global climate would actually have cooled slightly over the past 50 years. The pattern of temperature change through the layers of the atmosphere, with warming near the surface and cooling higher up in the stratosphere, further confirms that it is the buildup of heat-trapping gases (also known as "greenhouse gases") that has caused most of the Earth's warming over the past half century.

Because human-induced warming is superimposed on a background of natural variations in climate, warming is not uniform over time. Short-term fluctuations in the long-term upward trend are thus natural and expected. For example, a recent slowing in the rate of surface air temperature rise appears to be related to cyclic changes in the oceans and in the sun's energy output, as well as a series of small volcanic eruptions and other factors. Nonetheless, global temperatures are still on the rise and are expected to rise further.

U.S. average temperature has increased by 1.3°F to 1.9°F since 1895, and most of this increase has occurred since 1970. The most recent decade was the nation's and the world's hottest on record, and 2012 was the hottest year on record in the continental United States. All U.S. regions have experienced warming in recent decades, but the extent of warming has not been uniform. In general, temperatures are rising more quickly in the north. Alaskans have experienced some of the largest increases in temperature between 1970 and the present. People living in the Southeast have experienced some of the smallest temperature increases over this period.

Temperatures are projected to rise another 2°F to 4°F in most areas of the United States over the next few decades. Reductions in some short-lived human-induced emissions that contribute to warming, such as black carbon (soot) and methane, could reduce some of the projected warming over the next couple of decades, because, unlike carbon dioxide, these gases and particles have relatively short atmospheric lifetimes.

The amount of warming projected beyond the next few decades is directly linked to the cumulative global emissions of heat-trapping gases and particles. By the end of this century, a roughly 3°F to 5°F rise is projected under a lower emissions scenario, which would require substantial reductions in emissions (referred to as the "B1 scenario"), and a 5°F to 10°F rise for a higher emissions scenario assuming continued increases in emissions, predominantly from fossil fuel combustion (referred to as the "A2 scenario"). These projections are based on results from 16 climate models that used the two emissions scenarios in a formal inter-model comparison study. The range of model projections for each emissions scenario is the result of the differences in the ways the models represent key factors such as water vapor, ice and snow reflectivity, and clouds, which can either dampen or amplify the initial effect of human influences on temperature. The net effect of these feedbacks is expected to amplify warming. More information about the models and scenarios used in this report can be found in Appendix 5 of the full report.

Prolonged periods of high temperatures and the persistence of high nighttime temperatures have increased in many locations (especially in urban areas) over the past half century. High nighttime temperatures have widespread impacts because people, livestock, and wildlife get no respite from the heat. In some regions, prolonged periods of high temperatures associated with droughts contribute to conditions that lead to larger wildfires and longer fire seasons. As expected in a warming climate, recent trends show that extreme heat is becoming more common, while extreme cold is becoming less common. Evidence indicates that the human influence on climate has already roughly doubled the probability of extreme heat events such as the record-breaking summer heat experienced in 2011 in Texas and Oklahoma. The incidence of record-breaking high temperatures is projected to rise.

Human-induced climate change means much more than just hotter weather. Increases in ocean and freshwater temperatures, frost-free days, and heavy downpours have all been documented. Global sea level has risen, and there have been large reductions in snow-cover extent, glaciers, and sea ice. These changes and other climatic changes have affected and will continue to affect human health, water supply, agriculture, transportation, energy, coastal areas, and many other sectors of society, with increasingly adverse impacts on the American economy and quality of life.

Some of the changes discussed in this report are common to many regions. For example, large increases in heavy precipitation have occurred in the Northeast, Midwest, and Great Plains, where heavy downpours have frequently led to runoff that exceeded the capacity of storm drains and levees, and caused flooding events and accelerated erosion. Other impacts, such as those associated with the rapid thawing of permafrost in Alaska, are unique to a particular U.S. region.

Permafrost thawing is causing extensive damage to infrastructure in our nation's largest state.

Some impacts that occur in one region ripple beyond that region. For example, the dramatic decline of summer sea ice in the Arctic—a loss of ice cover roughly equal to half the area of the continental United States—exacerbates global warming by reducing the reflectivity of Earth's surface and increasing the amount of heat absorbed. Similarly, smoke from wildfires in one location can contribute to poor air quality in faraway regions, and evidence suggests that particulate matter can affect atmospheric properties and therefore weather patterns. Major storms and the higher storm surges exacerbated by sea level rise that hit the Gulf Coast affect the entire country through their cascading effects on oil and gas production and distribution.

Water expands as it warms, causing global sea levels to rise; melting of land-based ice also raises sea level by adding water to the oceans. Over the past century, global average sea level has risen by about 8 inches. Since 1992, the rate of global sea level rise measured by satellites has been roughly twice the rate observed over the last century, providing evidence of acceleration. Sea level rise, combined with coastal storms, has increased the risk of erosion, storm surge damage, and flooding for coastal communities, especially along the Gulf Coast, the Atlantic seaboard, and in Alaska. Coastal infrastructure, including roads, rail lines, energy infrastructure, airports, port facilities, and military bases, are increasingly at risk from sea level rise and damaging storm surges. Sea level is projected to rise by another 1 to 4 feet in this century, although the rise in sea level in specific regions is expected to vary from this global average for a number of reasons. A wider range of scenarios, from 8 inches to more than 6 feet by 2100, has been used in risk-based analyses in this report. In general, higher emissions scenarios that lead to more warming would be expected to lead to higher amounts of sea level rise. The stakes are high, as nearly five million Americans and hundreds of billions of dollars of property are located in areas that are less than four feet above the local high-tide level.

In addition to causing changes in climate, increasing levels of carbon dioxide from the burning of fossil fuels and other human activities have a direct effect on the world's oceans.

Carbon dioxide interacts with ocean water to form carbonic acid, increasing the ocean's acidity. Ocean surface waters have become 30% more acidic over the last 250 years as they have absorbed large amounts of carbon dioxide from the atmosphere. This ocean acidification makes water more corrosive, reducing the capacity of marine organisms with shells or skeletons made of calcium carbonate (such as corals, krill, oysters, clams, and crabs) to survive, grow, and reproduce, which in turn will affect the marine food chain.

Widespread Impacts

Impacts related to climate change are already evident in many regions and sectors and are expected to become increasingly disruptive across the nation throughout this century and beyond. Climate changes interact with other environmental and societal factors in ways that can either moderate or intensify these impacts.

Some climate changes currently have beneficial effects for specific sectors or regions. For example, current benefits of warming include longer growing seasons for agriculture and longer ice-free periods for shipping on the Great Lakes. At the same time, however, longer growing seasons, along with higher temperatures and carbon dioxide levels, can increase pollen production, intensifying and lengthening the allergy season. Longer ice-free periods on the Great Lakes can result in more lake-effect snowfalls.

Sectors affected by climate changes include agriculture, water, human health, energy, transportation, forests, and ecosystems. Climate change poses a major challenge to U.S. agriculture because of the critical dependence of agricultural systems on climate. Climate change has the potential to both positively and negatively affect the location, timing, and productivity of crop, livestock, and fishery systems at local, national, and global scales. The United States produces nearly $330 billion per year in agricultural commodities. This productivity is vulnerable to direct impacts on crops and livestock from changing climate conditions and extreme weather events and indirect impacts through increasing pressures from pests and pathogens. Climate change will also alter the stability of food supplies and create new food security challenges for the United States as the world seeks to feed nine billion people by 2050. While the agriculture sector has proven to be adaptable to a range of stresses, as evidenced by continued growth in production and efficiency across the United States, climate change poses a new set of challenges.

Water quality and quantity are being affected by climate change. Changes in precipitation and runoff, combined with changes in consumption and withdrawal, have reduced surface and groundwater supplies in many areas. These trends are expected to continue, increasing the likelihood of water shortages for many uses. Water quality is also diminishing in many areas, particularly due to sediment and contaminant concentrations after heavy downpours. Sea level rise, storms and storm surges, and changes in surface and groundwater use patterns are expected to compromise the sustainability of coastal freshwater aquifers and wetlands. In most U.S. regions, water resources managers and planners will encounter new risks, vulnerabilities, and opportunities that may not be properly managed with existing practices.

Climate change affects human health in many ways. For example, increasingly frequent and intense heat events lead to more heat-related illnesses and deaths and, over time, worsen drought and wildfire risks, and intensify air pollution. Increasingly frequent extreme precipitation and associated flooding can lead to injuries and increases in waterborne disease. Rising sea surface temperatures have been linked with increasing levels and ranges of diseases. Rising sea levels intensify coastal flooding and storm surge, and thus exacerbate threats to public safety during storms. Certain groups of people are more vulnerable to the range of climate change related health impacts, including the elderly, children, the poor, and the sick. Others are vulnerable because of where they live, including those in floodplains, coastal zones, and some urban areas. Improving and properly supporting the public health infrastructure will be critical to managing the potential health impacts of climate change.

Climate change also affects the living world, including people, through changes in ecosystems and biodiversity. Ecosystems provide a rich array of benefits and services to humanity, including habitat for fish and wildlife, drinking water storage and filtration, fertile soils for growing crops, buffering against a range of stressors including climate change impacts, and aesthetic and cultural values. These benefits are not always easy to quantify, but they support jobs, economic growth, health, and human well-being. Climate change driven disruptions to ecosystems have direct and indirect human impacts, including reduced water supply and quality, the loss of iconic species and landscapes, effects on food chains and the timing and success of species migrations, and the potential for extreme weather and climate events to destroy or degrade the ability of ecosystems to provide societal benefits.

Human modifications of ecosystems and landscapes often increase their vulnerability to damage from extreme weather events, while simultaneously reducing their natural capacity to moderate the impacts of such events. For example, salt marshes, reefs, mangrove forests, and barrier islands defend coastal ecosystems and infrastructure, such as roads and buildings, against storm surges. The loss of these natural buffers due to coastal development, erosion, and sea level rise increases the risk of catastrophic damage during or after extreme weather events. Although floodplain wetlands are greatly reduced from their historical extent, those that remain still absorb floodwaters and reduce the effects of high flows on river-margin lands. Extreme weather events that produce sudden increases in water flow, often carrying debris and pollutants, can decrease the natural capacity of ecosystems to cleanse contaminants.

The climate change impacts being felt in the regions and sectors of the United States are affected by global trends and economic decisions. In an increasingly interconnected world, U.S. vulnerability is linked to impacts in other nations. It is thus difficult to fully evaluate the impacts of climate change on the United States without considering consequences of climate change elsewhere.

Response Options

As the impacts of climate change are becoming more prevalent, Americans face choices. Especially because of past emissions of long-lived heat-trapping gases, some additional climate change and related impacts are now unavoidable. This is due to the long-lived nature of many of these gases, as well as the amount of heat absorbed and retained by the oceans and other responses within the climate system. The amount of future climate change, however, will still largely be determined by choices society makes about emissions. Lower emissions of heat-trapping gases and particles mean less future warming and less-severe impacts; higher emissions mean more warming and more severe impacts. Efforts to limit emissions or increase carbon uptake fall into a category of response options known as "mitigation," which refers to reducing the amount and speed of future climate change by reducing emissions of heat-trapping gases or removing carbon dioxide from the atmosphere.

The other major category of response options is known as "adaptation," and refers to actions to prepare for and adjust to new conditions, thereby reducing harm or taking advantage of new opportunities. Mitigation and adaptation actions are linked in multiple ways, including that effective mitigation reduces the need for adaptation in the future. Both are essential parts of a comprehensive climate change response strategy. The threat of irreversible impacts makes the timing of mitigation efforts particularly critical. This report includes chapters on Mitigation, Adaptation, and Decision Support that offer an overview of the options and activities being planned or implemented around the country as local, state, federal, and tribal governments, as well as businesses, organizations, and individuals begin to respond to climate change. These chapters conclude that while response actions are under development, current implementation efforts are insufficient to avoid increasingly negative social, environmental, and economic consequences.

Large reductions in global emissions of heat-trapping gases, similar to the lower emissions scenario (B1) analyzed in this assessment, would reduce the risks of some of the worst impacts of climate change. Some targets called for in international climate negotiations to date would require even larger reductions than those outlined in the B1 scenario. Meanwhile, global emissions are still rising and are on a path to be even higher than the high emissions scenario (A2) analyzed in this report. The recent U.S. contribution to annual global emissions is about 18%, but the U.S. contribution to cumulative global emissions over the last century is much higher. Carbon dioxide lasts for a long time in the atmosphere, and it is the cumulative carbon emissions that determine the amount of global climate change. After decades of increases, U.S. CO2 emissions from energy use (which account for 97% of total U.S. emissions) declined by around 9% between 2008 and 2012, largely due to a shift from coal to less CO2-intensive natural gas for electricity production. Governmental actions in city, state, regional, and federal programs to promote energy efficiency have also contributed to reducing U.S. carbon emissions. Many, if not most of these programs are motivated by other policy objectives, but some are directed specifically at greenhouse gas emissions. These U.S. actions and others that might be undertaken in the future are described in the Mitigation chapter of this report. Over the remainder of this century, aggressive and sustained greenhouse gas emission reductions by the United States and by other nations would be needed to reduce global emissions to a level consistent with the lower scenario (B1) analyzed in this assessment.

With regard to adaptation, the pace and magnitude of observed and projected changes emphasize the need to be prepared for a wide variety and intensity of impacts. Because of the growing influence of human activities, the climate of the past is not a good basis for future planning. For example, building codes and landscaping ordinances could be updated to improve energy efficiency, conserve water supplies, protect against insects that spread disease (such as dengue fever), reduce susceptibility to heat stress, and improve protection against extreme events. The fact that climate change impacts are increasing points to the urgent need to develop and refine approaches that enable decision-making and increase flexibility and resilience in the face of ongoing and future impacts. Reducing non-climate-related stresses that contribute to existing vulnerabilities can also be an effective approach to climate change adaptation.

Adaptation can involve considering local, state, regional, national, and international jurisdictional objectives. For example, in managing water supplies to adapt to a changing climate, the implications of international treaties should be considered in the context of managing the Great Lakes, the Columbia River, and the Colorado River to deal with increased drought risk. Both "bottom up" community planning and "top down" national strategies may help regions deal with impacts such as increases in electrical brownouts, heat stress, floods, and wildfires.

Proactively preparing for climate change can reduce impacts while also facilitating a more rapid and efficient response to changes as they happen. Such efforts are beginning at the federal, regional, state, tribal, and local levels, and in the corporate and non-governmental sectors, to build adaptive capacity and resilience to climate change impacts. Using scientific information to prepare for climate changes in advance can provide economic opportunities, and proactively managing the risks can reduce impacts and costs over time.

There are a number of areas where improved scientific information or understanding would enhance the capacity to estimate future climate change impacts. For example, knowledge of the mechanisms controlling the rate of ice loss in Greenland and Antarctica is limited, making it difficult for scientists to narrow the range of expected future sea level rise. Improved understanding of ecological and social responses to climate change is needed, as is understanding of how ecological and social responses will interact.

A sustained climate assessment process could more efficiently collect and synthesize the rapidly evolving science and help supply timely and relevant information to decision-makers. Results from all of these efforts could continue to deepen our understanding of the interactions of human and natural systems in the context of a changing climate, enabling society to effectively respond and prepare for our future.

The cumulative weight of the scientific evidence contained in this report confirms that climate change is affecting the American people now, and that choices we make will affect our future and that of future generations.

Report Findings

These findings distill important results that arise from this National Climate Assessment. They do not represent a full summary of all of the chapters' findings, but rather a synthesis of particularly noteworthy conclusions.

1. **Global climate is changing and this is apparent across the United States in a wide range of observations. The global warming of the past 50 years is primarily due to human activities, predominantly the burning of fossil fuels.** Many independent lines of evidence confirm that human activities are affecting climate in unprecedented ways. U.S. average temperature has increased by 1.3°F to 1.9°F since record keeping began in 1895; most of this increase has occurred since about 1970. The most recent

decade was the warmest on record. Because human-induced warming is superimposed on a naturally varying climate, rising temperatures are not evenly distributed across the country or over time.

2. **Some extreme weather and climate events have increased in recent decades, and new and stronger evidence confirms that some of these increases are related to human activities.** Changes in extreme weather events are the primary way that most people experience climate change. Human-induced climate change has already increased the number and strength of some of these extreme events. Over the last 50 years, much of the United States has seen an increase in prolonged periods of excessively high temperatures, more heavy downpours, and in some regions, more severe droughts.

3. **Human-induced climate change is projected to continue, and it will accelerate significantly if global emissions of heat-trapping gases continue to increase.** Heat-trapping gases already in the atmosphere have committed us to a hotter future with more climate-related impacts over the next few decades. The magnitude of climate change beyond the next few decades depends primarily on the amount of heat-trapping gases that human activities emit globally, now and in the future.

4. **Impacts related to climate change are already evident in many sectors and are expected to become increasingly disruptive across the nation throughout this century and beyond.** Climate change is already affecting societies and the natural world. Climate change interacts with other environmental and societal factors in ways that can either moderate or intensify these impacts. The types and magnitudes of impacts vary across the nation and through time. Children, the elderly, the sick, and the poor are especially vulnerable. There is mounting evidence that harm to the nation will increase substantially in the future unless global emissions of heat-trapping gases are greatly reduced.

5. **Climate change threatens human health and well-being in many ways, including through more extreme weather events and wildfire, decreased air quality, and diseases transmitted by insects, food, and water.** Climate change is increasing the risks of heat stress, respiratory stress from poor air quality, and the spread of waterborne diseases. Extreme weather events often lead to fatalities and a variety of health impacts on vulnerable populations, including impacts on mental health, such as anxiety and post-traumatic stress disorder. Large-scale changes in the environment due to climate change and extreme weather events are increasing the risk of the emergence or reemergence of health threats that are currently uncommon in the United States, such as dengue fever.

6. **Infrastructure is being damaged by sea level rise, heavy downpours, and extreme heat; damages are projected to increase with continued climate change.** Sea level rise, storm surge, and heavy downpours, in combination with the pattern of continued development in coastal areas, are increasing damage to U.S. infrastructure including roads, buildings, and industrial facilities, and are also increasing risks to ports and coastal military installations. Flooding along rivers, lakes, and in cities following heavy downpours, prolonged rains, and rapid melting of snowpack is exceeding the limits of flood protection infrastructure designed for historical conditions. Extreme heat is damaging transportation infrastructure such as roads, rail lines, and airport runways.

7. **Water quality and water supply reliability are jeopardized by climate change in**

a variety of ways that affect ecosystems and livelihoods. Surface and groundwater supplies in some regions are already stressed by increasing demand for water as well as declining runoff and groundwater recharge. In some regions, particularly the southern part of the country and the Caribbean and Pacific Islands, climate change is increasing the likelihood of water shortages and competition for water among its many uses. Water quality is diminishing in many areas, particularly due to increasing sediment and contaminant concentrations after heavy downpours.

8. **Climate disruptions to agriculture have been increasing and are projected to become more severe over this century.** Some areas are already experiencing climate-related disruptions, particularly due to extreme weather events. While some U.S. regions and some types of agricultural production will be relatively resilient to climate change over the next 25 years or so, others will increasingly suffer from stresses due to extreme heat, drought, disease, and heavy downpours. From mid-century on, climate change is projected to have more negative impacts on crops and livestock across the country—a trend that could diminish the security of our food supply.

9. **Climate change poses particular threats to Indigenous Peoples' health, well-being, and ways of life.** Chronic stresses such as extreme poverty are being exacerbated by climate change impacts such as reduced access to traditional foods, decreased water quality, and increasing exposure to health and safety hazards. In parts of Alaska, Louisiana, the Pacific Islands, and other coastal locations, climate change impacts (through erosion and inundation) are so severe that some communities are already relocating from historical homelands to which their traditions and cultural identities are tied. Particularly in Alaska, the rapid pace of temperature rise, ice and snow melt, and permafrost thaw are significantly affecting critical infrastructure and traditional livelihoods.

10. **Ecosystems and the benefits they provide to society are being affected by climate change. The capacity of ecosystems to buffer the impacts of extreme events like fires, floods, and severe storms is being overwhelmed.** Climate change impacts on biodiversity are already being observed in alteration of the timing of critical biological events such as spring bud burst and substantial range shifts of many species. In the longer term, there is an increased risk of species extinction. These changes have social, cultural, and economic effects. Events such as droughts, floods, wildfires, and pest outbreaks associated with climate change (for example, bark beetles in the West) are already disrupting ecosystems. These changes limit the capacity of ecosystems, such as forests, barrier beaches, and wetlands, to continue to play important roles in reducing the impacts of these extreme events on infrastructure, human communities, and other valued resources.

11. **Ocean waters are becoming warmer and more acidic, broadly affecting ocean circulation, chemistry, ecosystems, and marine life.** More acidic waters inhibit the formation of shells, skeletons, and coral reefs. Warmer waters harm coral reefs and alter the distribution, abundance, and productivity of many marine species. The rising temperature and changing chemistry of ocean water combine with other stresses, such as overfishing and coastal and marine pollution, to alter marine-based food production and harm fishing communities.

12. **Planning for adaptation (to address and prepare for impacts) and mitigation (to reduce future climate change, for example by cutting emissions) is becoming**

more widespread, but current implementation efforts are insufficient to avoid increasingly negative social, environmental, and economic consequences. Actions to reduce emissions, increase carbon uptake, adapt to a changing climate, and increase resilience to impacts that are unavoidable can improve public health, economic development, ecosystem protection, and quality of life.

Analysis

In the Global Change Research Act of 1990, Congress tasked the United States Global Change Research Program (USGCRP) with producing a report for the President of the United States on a four-year cycle that identified the current impacts of climate change on the United States and posited what those changes portend for the next 25 years. Due to the minimization of climate change research by Republican presidential administrations, the reports have thus far been published irregularly, which is why the third report was not produced until 2014.

The report features contributions from more than twelve federal entities, including the Department of Commerce, Department of State, Environmental Protection Agency, National Science Foundation, and Smithsonian. Before publication, the data in the report is peer-reviewed by outside experts drawn from academic institutions or private companies.

Further Reading

Ingram, B. Lynn, and Frances Malamud-Roam. 2013. *The West Without Water: What Past Floods, Droughts, and Other Climate Clues Tell Us About Tomorrow*. Berkeley: University of California Press.

Nash, Steven. 2015. *Virginia Climate Fever: How Global Warming Will Transform Our Cities, Shorelines, and Forests*. Charlottesville: University of Virginia Press.

U.S. Global Change Research Program. "Globalchange.gov." http://www.globalchange.gov.

One City Built to Last
Transforming New York City's Buildings for a Low-Carbon Future (Executive Summary)

Date: September 21, 2014
Location: New York, New York
Significance: New York City Mayor Bill de Blasio's administration detailed in their plan how the city would reduce its greenhouse emissions by 80 percent of 2005 levels by 2050.
Source: The City of New York, Mayor Bill de Blasio. 2014. *One City Built to Last: Transforming New York City's Buildings for a Low-Carbon Future.* http://www.nyc.gov/html/builttolast/assets/downloads/pdf/OneCity.pdf (accessed July 26, 2015).

What does it mean for a city to be built to last?

In a city that is built to last, our homes and workplaces will require less energy for heating, cooling, and power. The energy those buildings will need comes from renewable sources that do not pollute our air and water or dangerously increase global temperatures. Residents of a city built to last are protected from rising sea levels and extreme weather.

Global climate change is the challenge of our generation. The stakes are high—for New Yorkers and for the world. In the coming years, New York City will face rising sea levels, increased temperatures and heat waves, and an increasing frequency of the most intense storms. These risks are not remote nor distant. They are here today. The damage caused by Hurricane Sandy in 2012 provided vivid evidence of these risks. Almost two years later, we are still recovering. Globally, climate change is having a devastating impact on people's lives as rising sea levels flood coastlines, droughts disrupt livelihoods, and storms, hurricanes, and other extreme weather events threaten security and economic development.

For this reason, New York City is committed to reducing its greenhouse gas (GHG) emissions by 80 percent by 2050—the level the United Nations projects is needed to avoid the most dangerous impacts of climate change—and will chart a long-term course for a total transition away from fossil fuels to renewable sources of energy.

In New York City, our buildings are responsible for the overwhelming share of our GHG emissions, accounting for nearly three-quarters of our contribution to climate change. We can upgrade our buildings to become more energy efficient and power them with renewable sources of energy, reducing our GHG emissions while also helping to make our homes more affordable and creating new jobs and businesses.

Realizing this vision will require engaged communities, energized leadership, and creative solutions. We will rise to the challenge. New York City is one of the world's leaders in real estate development, architecture and engineering firms, skilled labor unions, financial institutions, and research universities. We are uniquely poised to develop the solutions needed to transform our city, and will share these solutions with the world.

We have the power to begin transforming our buildings for a low-carbon future and the complete transition away from fossil fuels—and we will begin today.

The Global Context

Human activities are increasing fossil fuel combustion and changing land use patterns, which leads to GHG emissions that change the chemical composition of the atmosphere. Over time, this has had a direct and measurable impact on human populations and natural ecosystems.

The United Nations Framework Convention on Climate Change (UNFCCC) projects that by 2050, global GHG emissions must be reduced by 50 percent below 1990 levels to avoid the most dangerous impacts of climate change. Developed countries must reduce their emissions even more aggressively—by up to 80 percent by 2050—to account for their greater contribution to global emissions to date and their higher than average per-capita emissions. If we fail, the impacts of climate change will be far-reaching and felt by all, but with the worst consequences for the world's poorest and most vulnerable populations.

Cities must play a leading role in addressing the problem. More than half the world's population now lives in urban areas, which are responsible for the vast majority of global GHG emissions. Cities on coastlines and in other vulnerable locations face the greatest risks from the impacts of climate change. But cities also have the resources, commitment, and ingenuity to prepare for a changing climate and to take bold action to reduce the harmful emissions that are the cause of climate change.

New York City's Role

New Yorkers will continue to lead the way. With our significant public infrastructure, a world class mass transit system, dense living patterns, and capacity for civic innovation, we are uniquely positioned to become the most sustainable big city in the world.

New York City has demonstrated its commitment to adapt to climate change. Mayor de Blasio created the Office of Recovery and Resiliency (ORR) in March 2014, the first City agency in the country dedicated solely to resiliency. ORR is implementing the strategies laid out in *PlaNYC: A Stronger, More Resilient New York*, which includes 257 initiatives to make the city, its communities, and its infrastructure more resilient.

We are also committed to reducing the GHG emissions that are causing the problem. In line with the UN target, we will put New York City on a pathway to achieve an 80 percent reduction in GHG emissions from 2005 levels by 2050.

New Yorkers have the tools and resources to mitigate our GHG emissions. In fact, New York City has already reduced citywide emissions by 19 percent from 2005 levels.

We made this progress by measuring and reporting our annual GHG emissions, and efforts to make investments to clean our electric grid, expand sustainable transportation options, reduce our solid waste, and enact policies to reduce emissions from the energy used in our buildings.

But the majority of the GHG reductions we have achieved so far were the result of switching from coal and oil to natural gas for electricity generation and other improvements to utility operations. Together, these account for more than 80 percent of the reductions. These strategies cannot be replicated, and future reductions will be much more difficult to achieve.

To reach 80 by 50, further GHG reductions will need to come from additional cleaner power generation, more sustainable modes of transportation, and better management of our solid waste. But the biggest untapped opportunity is to improve the energy efficiency of the city's one million buildings.

The energy used in our buildings contributes nearly three-quarters of all citywide emissions. Because our buildings are expected to last well beyond 2050, increasing the energy efficiency of our existing buildings, in addition to new construction, is the most important step we can take to make deep reductions in our carbon emissions.

Reducing energy use in our buildings can also help address our affordable housing crisis. Increasing utility costs are one of the primary contributors to the growing share of New Yorkers who are becoming rent-burdened. Improving efficiency in our residential buildings can help mitigate rising housing costs. In addition, our public housing has significant untapped energy-saving potential. Investments in efficiency in these buildings would help improve the quality of our public housing stock.

Investments in energy efficiency will also help stimulate economic activity that will create new jobs and opportunities for career advancement for thousands of New Yorkers, and will create a healthier and more sustainable place for all.

Opportunities to Reduce Greenhouse Gas Emissions from Buildings

There are three distinct ways to reduce GHG emissions from the energy used in our buildings. The first is by improving the energy efficiency of building systems and operations, and investing in cleaner on-site power generation. These are improvements that building owners and managers can directly control.

The second is by reducing the energy consumption of a building's tenants and occupants, which can account for anywhere between 40–60 percent of a building's energy use. Tenants can directly lower their energy use, with building owners typically having limited control.

The third opportunity is by reducing emissions from the city's power supply. This requires power suppliers to switch to cleaner energy sources and fuel distributors to offer lower-carbon fuels. Building owners and managers do not have direct influence over the power supply, although they can help grow the market for renewables through power purchase agreements and other mechanisms to buy cleaner energy that is generated off-site.

This plan focuses on the opportunities to improve the efficiency of building systems, equipment, and operations; and dramatically expand on-site renewable energy generation. We will call on building owners and managers to make these investments, but also provide the resources they need to do so. We will create tools to empower New Yorkers to reduce their own energy use, and programs to mobilize communities in this effort. With this plan, we can improve the quality of our built environment, create good jobs, and help protect families from rising housing costs.

Current Sustainable Building and Energy Efficiency Policies

The City has already laid the foundation for improving the efficiency of both our public and private buildings.

For our public buildings, we have undertaken major energy efficiency retrofits, invested in clean energy sources, and piloted leading edge technologies across the City's portfolio of roughly 4,000 buildings.

For our privately-owned buildings, the City has focused on providing information to building decision-makers—which include building owners, managers, superintendents, board members, buyers, sellers, and residents—to help them prioritize investments in energy efficiency. The City requires owners of large buildings over 50,000 square feet in floor area to measure and publicly disclose their energy and water use annually through a process called "benchmarking," conduct energy audits and retrocommissioning once every ten years, install energy sub-meters for large commercial tenants, and upgrade lighting in non-residential buildings.

More than 2.1 billion square feet of floor area in New York City buildings has been benchmarked since 2010, and the City has released three reports analyzing this data. The reports found that multifamily residential buildings present the greatest energy saving opportunity due to their relative size and distribution of energy use, and that there is significant variation in energy use among similar types of buildings—presenting enormous opportunities for energy efficiency and GHG reductions.

The City had enacted laws to phase out the use of heavy fuel oils in buildings, which contribute more soot pollution in New York City than all of our cars and trucks combined. To complement the laws, the City launched the NYC Clean Heat program to provide technical assistance to help accelerate the pace of fuel oil conversions. Mayor de Blasio extended the program through 2015, and more than 3,500 fuel oil conversions have now been completed. Last year the City's air quality was the cleanest it has been in more than 50 years, due in large part to the success of the program, and an estimated 800 lives will be saved each year.

The City has created additional resources to help building owners comply with its local laws and make their buildings more energy efficient. These include the New York City Energy Efficiency Corporation (NYCEEC), a not-for-profit financial services firm that has developed innovative energy efficiency and clean energy financing products, and Green Light New York, an educational resource center to provide trainings, demonstrations, and events on emerging technologies and strategies.

Working with stakeholders, the City also launched a the NYC Carbon Challenge as a voluntary program to encourage deep energy retrofits, incorporated energy efficiency and green building principles into the building codes, and changed zoning laws to encourage sustainable building practices.

Charting a Path to 80 by 50

The Bloomberg administration put the City on track to achieve a 30 percent reduction in GHG emissions by 2030—but it is increasingly clear that this is not enough.

To achieve an 80 percent reduction in emissions by 2050 ("80 by 50"), the City must be on a much more ambitious pathway toward deep carbon reductions.

In 2013 the City studied opportunities to achieve an 80 by 50 reduction based on current technologies. The study found that nearly two-thirds of the GHG reductions that are needed must come from more efficient buildings. More than 80 percent of the building-based measures would also be cost-effective from a societal standpoint, but there are major obstacles to scaling up these investments and transitioning to a low-carbon future.

Nothing short of a dramatic transformation in the way energy is used in buildings is necessary to achieve 80 by 50. By 2050, our buildings will need to become high performance structures powered by low-carbon energy sources. Walls and windows must be insulated, building equipment must become more efficient and intelligent, and building systems must be made ready for renewable energy sources to eventually replace fossil fuels for heating, hot water, and cooking. Residents would need to conscientiously conserve energy and water, and building operators will need to become skilled in the latest energy efficiency technologies. Moreover, achieving 80 by 50 would require the deployment of new and promising—but largely unproven—technologies and strategies.

Despite these challenges, the City can focus on efforts in the short- and medium term to put our buildings on the right trajectory. The City has the tools, expertise, and committed leadership to accelerate carbon reductions and place the city on a pathway to 80 by 50. This plan is the first step.

A Roadmap for New York City's Buildings

To be on a pathway to 80 by 50, we will need to reduce GHG emissions from the energy used to heat, cool, and power our buildings by 30 percent from 2005 levels over the next decade. City government can lead the way, and will commit to achieving a 35 percent reduction in emissions in the next ten years.

To achieve these goals, by 2025 we will:

- Complete efficiency improvements in every City-owned building that has significant energy use and install 100 MW of onsite renewable power.
- Implement leading edge performance standards for new construction that cost effectively achieve highly efficient buildings, looking to Passive House, carbon neutral, or "zero net energy" strategies to inform the standards.
- Develop interim energy performance targets for existing buildings to be met through both voluntary reductions and new regulations, such as performance standards and measure-based mandates, which would be triggered if adequate reductions are not achieved.

These goals set a high bar for what we will expect from our buildings in the future. Though ambitious, they are also designed to be flexible. To rise to this challenge, ensure our measures are technically sound, and determine the ideal process of implementation, we will work closely with New York City's world class real estate industry, architects, engineers, labor unions, affordable housing experts, environmental justice leaders, and academics. The City will convene a task force of these stakeholders and conduct a robust technical analysis that will be used to create programs, policies, and regulations that transition our buildings to a low-carbon future.

There are also initiatives that we will begin right away. We will invest in our public buildings to make them models for sustainability. We will improve energy efficiency in affordable housing. We will catalyze a thriving market for energy efficiency and renewable energy generation. We will continue to improve construction standards to make our new buildings the greenest in the world. We will mobilize business sectors, communities, and neighborhoods across the city to improve energy efficiency and reduce GHG emissions. We will provide new opportunities for skill development in energy services and create jobs. And we will invest in and inspire the next generation of clean tech and energy efficiency innovations right here in New York City.

All together, the initiatives outlined in this plan are expected to reduce GHG emissions from the energy we use to heat, cool, and power our buildings by roughly 3.4 million metric tons of carbon dioxide equivalent annually by 2025. These initiatives will also create an estimated 3,500 construction-related jobs, provide training for more than 7,000 building operators and staff, and generate $8.5 billion in total cost savings for New Yorkers.

Creating Efficient Buildings for All New Yorkers

What we are outlining today is not just a carbon mitigation plan: it is also an affordability plan, an economic development plan, and a public health plan.

By lowering building energy use, we will make it easier for people to afford to live in New York City. Energy efficiency measures can reduce the energy bills for tenants who pay their utilities directly, and free up additional funding that owners can invest in other capital upgrades to improve the quality of our housing stock. For the City, the money saved by reducing operating costs can be redirected to other vital investments.

Stimulating new investments in improving the efficiency of our buildings will create broader economic benefits. Boosting the demand for energy efficiency services and contractors will create new jobs for New Yorkers, growth opportunities for local businesses, and training opportunities for building staff to advance their careers and enhance their earning potential.

This plan will also improve the health of all New Yorkers. Reducing GHG emissions often has the added benefit of reducing air pollutants, which lowers health risks such as asthma, heart disease, and premature death. Reducing GHG emissions also helps to protect vulnerable populations from the increased frequency of heat waves that will be brought on by rising global temperatures.

At every step of the way, we are committed to ensuring that all residents in each and every borough have equal access to the benefits that come from retrofitting our buildings, reducing our GHG emissions, and building a more sustainable and resilient place to live. This is our plan to create One City: Built to Last.

Guiding Strategies

The City adhered to five guiding strategies for the proposals outlined in this plan.

1. **Lead by example.** The City will pave the way by implementing innovative technologies and strategies in City-owned buildings that will set the example for achieving deep carbon reductions. This also means working with the federal government on new strategies to reduce energy use in public housing and mobilizing business leaders and communities in our effort to tackle climate change.
2. **Empower New Yorkers to take action.** The City will create programs and policies to reduce the risks and complexities of retrofitting buildings and empower New Yorkers to act. This includes providing educational resources and expanding financial resources to help cover the costs of projects. The City can also research local applicability of emerging technologies and help bring them to market, particularly where they promote deep energy retrofit projects.
3. **Hold New York City's buildings to the highest energy performance standards.** The City will ensure our buildings meet the highest standards. This includes improving compliance with existing laws, raising standards for energy performance on new construction and renovations, and promoting resiliency improvements during efficiency upgrades.
4. **Ensure benefits are shared by New Yorkers in every neighborhood.** The City will promote energy efficiency and renewable energy across more communities and building sectors, including affordable housing and small and mid-sized buildings. The City will also create new programs so local workers benefit from the job growth and economic activity that result from efficiency investments.
5. **Use data, analysis, and stakeholder feedback to drive the approach.** As initiatives are implemented, the City will take a data-driven approach based on energy use information, real estate market data, engineering analysis, and other sources. We will collaborate with stakeholders along the way, particularly when shaping the approach for specific building sectors and communities. We will hold ourselves accountable by closely tracking and reporting our progress to the public.

Key Policies and Programs

The guiding strategies outlined above provide the policy framework for the initiatives we will begin today.

We will make our public buildings models for sustainability.
    Invest in high value projects in all City-owned buildings. The City's Department of Citywide Administrative Services (DCAS) will fund high value efficiency projects identified by City agencies through a competitive selection process. The City will also expand the funding program to reach many more agencies, support new and innovative projects, and provide the incremental cost of efficiency measures in planned capital construction projects.
    Expand solar power on City rooftops. Solar photovoltaic (PV) installations offset electric grid use with a clean and renewable energy source, and when combined with bat-

tery storage, can also provide backup power during extended blackouts. The City will install 100 megawatts (MW) of solar capacity on more than 300 City-owned rooftops over the next ten years, starting with 24 newly re-roofed schools. The City will also prioritize installations paired with battery storage on the City's emergency shelters to improve the city's emergency preparedness.

Implement deep retrofits in key City facilities. The City will enhance its implementation of comprehensive retrofits in City buildings using new, more streamlined contracts that facilitate deep energy retrofits. The City will also expand implementation of combined heat and power (CHP) projects to generate energy more efficiently and reliably.

Improve building operations and maintenance. The City will improve the operations and maintenance (O&M) of City buildings by expanding its preventative maintenance programs, which includes hiring more staff and enhancing training for the city's building operators. The City will also expand its Expenses for Conservation & Efficiency Leadership (ExCEL) Program, a competitive program to award resources for agency-identified O&M measures such as training, tools, and other energy-saving projects.

Pilot new clean energy technology in City buildings. The City will engage companies with emerging energy technologies to pilot their solutions in City facilities. The expanded program will test the performance of technologies in more facilities, identify opportunities for larger-scale deployment in City buildings, and provide case studies to increase market adoption of promising technologies.

Improve the efficiency and quality of New York City's public housing. The New York City Housing Authority (NYCHA) will undertake a partnership with the U.S. Department of Housing and Urban Development (HUD) and private lenders to develop a multiphase, large-scale Energy Performance Contract (EPC). NYCHA will work with HUD to streamline the EPC process and explore the opportunity to leverage financial incentives from third parties as part of the EPC.

We will create a thriving market for energy efficiency and renewable energy.

Launch an Energy and Water Retrofit Accelerator. The City will create a coordinated outreach and technical assistance program to accelerate energy and water retrofits in privately-owned buildings. The "Retrofit Accelerator" will use data-driven direct marketing to identify and assist buildings to undertake efficiency upgrades and complete heavy heating oil conversions to cleaner fuels. The program will also provide streamlined information about available financing and incentives and connect local job-seekers and firms to increased demand for services.

Engage communities in creating energy efficient and resilient neighborhoods. To complement the Retrofit Accelerator, the City will launch a program to engage local communities to promote energy efficiency retrofits, with a focus on helping key neighborhoods that are facing pressures on housing affordability. The program will also complement new financing programs currently under development by the City's Department of Housing Preservation and Development (HPD), and will train and employ local residents in order to provide new opportunities for career advancement.

Expand access to information for mid-sized buildings. The Mayor's Office will work with City Council to expand the City's Greener, Greater Buildings Plan to include all buildings over 25,000 square feet in floor area, lowering the previous square footage cut off of over 50,000 square feet. This will bring an additional 11,400 properties (16,800 buildings) under the law, providing more building decision-makers with energy and water use information and creating new opportunities for savings. Expanding the laws will also increase the number of buildings that can be assisted by the Retrofit Accelerator.

Provide financing options for energy efficiency and clean energy. The New York City Energy Efficiency Corporation (NYCEEC) has developed innovative financing options for energy efficiency and resiliency measures that are ready to be scaled up, including green mortgages and direct lending products that underwrite energy savings into the loan. The City will also explore modifications to the J-51 housing tax credit and the use of Qualified Energy Conservation Bonds (QECBs) to encourage additional investments in efficiency measures.

Improve energy and water efficiency in affordable housing. HPD and the New York City Housing Development Corporation (HDC) will begin requiring a "green" capital needs assessment for all moderate rehabilitation projects that are financed by the City to ensure that opportunities to save energy and water are included within the scope of work. HPD will also develop a grant and loan program to assist owners of small- to mid-sized multifamily properties undertake efficiency improvements in exchange for entering into an affordable housing regulatory agreement. The program will be paired with a robust outreach and technical assistance effort that engages local partners, and will also create opportunities for workforce development and career advancement.

Bring solar power to new neighborhoods across New York City. Solar energy complements energy efficiency by providing a renewable energy source to power building systems and reducing grid dependency. The City's goal is to increase our solar power capacity by 250 MW over the next ten years on privately-owned properties. Expanding the NYC Solar Partnership provides a platform to coordinate solar programs and streamline processes to sustain the local solar industry's growth, expand equitable access to solar power, and promote community-shared and group purchasing of solar power.

Coordinate with the State to streamline financing and incentive programs. The City will work with the New York State Energy Research and Development Authority (NYSERDA) and the NY Green Bank to coordinate programs in order to provide a more streamlined customer experience, build the local workforce, and provide appropriate financing options for the affordable multifamily sector. The City will also continue advocating for an equitable allocation of State funding to the downstate region.

Collaborate with local utilities to promote energy efficiency. The City will work with local utilities including Consolidated Edison (Con Edison), National Grid, PSEG Long Island, and the New York Power Authority (NYPA) to improve the quality of and access to customer utility data, support the development of renewable sources of energy, manage citywide load growth, and reduce load growth in priority areas. This includes efforts to collaborate within the Northern Brooklyn-Southern Queens load pocket, where energy efficiency retrofits can help manage stresses on utility infrastructure and mitigate rising housing costs brought on by neighborhood growth.

Expand the goals and reach of the NYC Carbon Challenge. Expanding the City's voluntary carbon reduction program by adding new sectors and participants will allow the City to partner with private sector leaders and identify best practices for deep carbon reductions. The City will also challenge existing participants to increase their carbon reduction goals to lead the way on the citywide pathway towards 80 by 50.

Train the next generation of building operators. The City will help improve the efficiency and lifespan of building equipment and upgrade the skills of building staff by offering low- to no-cost training in energy efficiency best practices. These expanded trainings will reach new constituencies, including building superintendents and operators who speak English as a second language.

Expand NYC CoolRoofs. The City will continue the current mission of the Cool-Roofs program to coat one million square feet of rooftops white each year, which reduces building energy use and helps mitigate urban heat. The City will also expand the program's mission to focus on small- and mid-sized multifamily buildings and will enhance opportunities for green workforce training.

Help New Yorkers reduce energy use at home. New Yorkers can take simple steps in their own homes to reduce energy use that can lower their energy bills. GreeNYC is the City's public education program that engages New Yorkers to take actions to live more sustainably. Through GreeNYC, the City will empower New Yorkers to take simple energy-saving measures such as switching to more efficient light bulbs, adjusting thermostats, unplugging chargers, and using appliances more efficiently.

We will develop world class green building and energy codes.
Raise the standards for our building and energy codes. Working together with industry leaders and City Council, the City will continue to improve standards for energy performance and sustainable building practices in new construction. Implemented standards will raise the bar towards better construction practices, higher efficiency equipment, and improved operations and maintenance to improve the quality of our building stock and lower energy costs for residents.

Enhance Energy Code enforcement. Strong enforcement and education are necessary to ensure existing and new energy performance standards are met. Additional resources must be allocated to City agencies that will enforce these requirements in both the design phase and during construction.

We will become a global hub for clean energy technology and innovation.
Explore innovative technologies for New York City buildings. Reaching 80 by 50 will depend in part on identifying and scaling up new clean energy technologies and strategies for efficiency. The City will study promising new solutions to explore their adaptability to the New York City market and develop best practice guidelines for implementation.

Support emerging entrepreneurs in clean energy and energy efficiency. The City will expand clean technology incubator programs in the city to support entrepreneurs and promote local company growth, including "step-out" and prototyping space that will allow emerging companies to stay in New York City as they grow.

Analysis

On the day of the Peoples Climate March, which attracted more than 100,000 participants in New York City, Mayor Bill de Blasio's administration released *One City Built to Last: Transforming New York City's Buildings for a Low-Carbon Future*. It is a far reaching plan for the city's future which is highlighted by a commitment to reduce greenhouse emissions by 80 percent of 2005 levels by 2050. Although *One City Built to Last* is intended to mitigate the most severe consequences of climate change, it also promises to New York City's residents from all social classes and ethnic groups significant improvements in housing and employment opportunities.

Further Reading

Engel, Kirsten. 2006. "State and Local Climate Initiatives: What is Motivating State and Local Governments to Address a Global Problem and What Does This Say About Federalism and Environmental Law?" *The Urban Lawyer* 38: 1015–1029.

Lundquist, Lennart J., and Anders Biel, eds. 2016. *From Kyoto to the Town Hall: Making International and National Climate Policy Work at the Local Level.* New York: Routledge.

U.S.–China Joint Announcement on Climate Change

Date: November 12, 2014 (The date in Washington, D.C., was November 11, 2014)
Location: Beijing, China
Significance: President Barack Obama and China's President Xi Jinping announced that their respective countries had each set goals to cut CO2 emissions by 2030.
Source: The White House, Office of the Press Secretary. 2014. *U.S.–China Joint Announcement on Climate Change.* https://www.whitehouse.gov/the-press-office/2014/11/11/us-china-joint-announcement-climate-change (accessed September 26, 2015).

1. The United States of America and the People's Republic of China have a critical role to play in combating global climate change, one of the greatest threats facing humanity. The seriousness of the challenge calls upon the two sides to work constructively together for the common good.
2. To this end, President Barack Obama and President Xi Jinping reaffirmed the importance of strengthening bilateral cooperation on climate change and will work together, and with other countries, to adopt a protocol, another legal instrument or an agreed outcome with legal force under the Convention applicable to all Parties at the United Nations Climate Conference in Paris in 2015. They are committed to reaching an ambitious 2015 agreement that reflects the principle of common but differentiated responsibilities and respective capabilities, in light of different national circumstances.
3. Today, the Presidents of the United States and China announced their respective post–2020 actions on climate change, recognizing that these actions are part of the longer range effort to transition to low-carbon economies, mindful of the global temperature goal of 2°C. The United States intends to achieve an economy-wide target of reducing its emissions by 26%–28% below its 2005 level in 2025 and to make best efforts to reduce its emissions by 28%. China intends to achieve the peaking of CO2 emissions around 2030 and to make best efforts to peak early and intends to increase the share of non-fossil fuels in primary energy consumption to around 20% by 2030. Both sides intend to continue to work to increase ambition over time.
4. The United States and China hope that by announcing these targets now, they can inject momentum into the global climate negotiations and inspire other countries to join in coming forward with ambitious actions as soon as possible, preferably by the first quarter of 2015. The two Presidents resolved to work closely together over the next year to address major impediments to reaching a successful global climate agreement in Paris.

5. The global scientific community has made clear that human activity is already changing the world's climate system. Accelerating climate change has caused serious impacts. Higher temperatures and extreme weather events are damaging food production, rising sea levels and more damaging storms are putting our coastal cities increasingly at risk and the impacts of climate change are already harming economies around the world, including those of the United States and China. These developments urgently require enhanced actions to tackle the challenge.
6. At the same time, economic evidence makes increasingly clear that smart action on climate change now can drive innovation, strengthen economic growth and bring broad benefits—from sustainable development to increased energy security, improved public health and a better quality of life. Tackling climate change will also strengthen national and international security.
7. Technological innovation is essential for reducing the cost of current mitigation technologies, leading to the invention and dissemination of new zero and low-carbon technologies and enhancing the capacity of countries to reduce their emissions. The United States and China are two of the world's largest investors in clean energy and already have a robust program of energy technology cooperation. The two sides have, among other things:
    - established the U.S.-China Climate Change Working Group (CCWG), under which they have launched action initiatives on vehicles, smart grids, carbon capture, utilization and storage, energy efficiency, greenhouse gas data management, forests and industrial boilers;
    - agreed to work together towards the global phase down of hydrofluorocarbons (HFCs), very potent greenhouse gases;
    - created the U.S.-China Clean Energy Research Center, which facilitates collaborative work in carbon capture and storage technologies, energy efficiency in buildings, and clean vehicles; and
    - agreed on a joint peer review of inefficient fossil fuel subsidies under the G-20.
8. The two sides intend to continue strengthening their policy dialogue and practical cooperation, including cooperation on advanced coal technologies, nuclear energy, shale gas and renewable energy, which will help optimize the energy mix and reduce emissions, including from coal, in both countries. To further support achieving their ambitious climate goals, today the two sides announced additional measures to strengthen and expand their cooperation, using the existing vehicles, in particular the U.S.-China Climate Change Working Group, the U.S.-China Clean Energy Research Center and the U.S.-China Strategic and Economic Dialogue. These include:
    - **Expanding Joint Clean Energy Research and Development:** A renewed commitment to the U.S.-China Clean Energy Research Center, including continued funding for three existing tracks on building efficiency, clean vehicles and advanced coal technology and launching a new track on the energy-water nexus;
    - **Advancing Major Carbon Capture, Utilization and Storage Demonstrations:** Establishment of a major new carbon storage project based in China through an international public-private consortium led by the United States and China to intensively study and monitor carbon storage using industrial CO2 and also work together on a new Enhanced Water Recovery (EWR)

pilot project to produce fresh water from CO2 injection into deep saline aquifers;
- **Enhancing Cooperation on HFCs:** Building on the historic Sunnylands agreement between President Obama and President Xi regarding HFCs, highly potent greenhouse gases, the two sides will enhance bilateral cooperation to begin phasing-down the use of high global warming potential HFCs and work together in a multilateral context as agreed by the two Presidents at their meeting in St. Petersburg on 6 September 2013;
- **Launching a Climate-Smart/Low-Carbon Cities Initiative:** In response to growing urbanization and increasingly significant greenhouse gas emissions from cities and recognizing the potential for local leaders to undertake significant climate action, the United States and China will establish a new initiative on Climate-Smart/Low-Carbon Cities under the CCWG. As a first step, the United States and China will convene a Climate-Smart/ Low-Carbon Cities Summit where leading cities from both countries will share best practices, set new goals and celebrate city-level leadership in reducing carbon emissions and building resilience;
- **Promoting Trade in Green Goods:** Encouraging bilateral trade in sustainable environmental goods and clean energy technologies, including through a U.S. trade mission led by Secretaries Moniz and Pritzker in April 2015 that will focus on smart low-carbon cities and smart low-carbon growth technologies; and
- **Demonstrating Clean Energy on the Ground:** Additional pilot programs, feasibility studies and other collaborative projects in the areas of building efficiency, boiler efficiency, solar energy and smart grids.

Analysis

Through this agreement, President Barack Obama and China's President Xi Jinping agreed to limit their respective country's greenhouse gas emissions in an effort to curb the acceleration of global climate change. Although hailed as a landmark agreement, it should be noted that its terms were nonbinding. Since China had recently surpassed the United States as the leading producer of greenhouse gases in the world, President Obama needed the agreement if he was to have any hope of getting domestic climate change legislation approved by both houses of Congress. A commitment from China was also desired by the international community as a necessary step to ensure the success of the planned United Nations Climate Summit, to be held in Paris, France, from November 30 to December 11, 2015.

In this agreement, President Obama pledged that the United States would reduce its CO2 levels by 26 to 28 percent of the amount emitted in 2005 by 2025. China pledged to reach its peak emission of CO2 by 2030. By that year, China planned to have approximately twenty percent of its energy coming from non-fossil fuel sources. The Obama administration touted the agreement as a major step in protecting the environment, but the targets agreed to by both the United States and China pale in comparison to the forty percent reduction from 2005 levels that the European Union planned to achieve by 2030.

Although prominent Democrats, such as former Vice President Al Gore, lauded the agreement, it failed to help President Obama's administration advance climate change legislation in Congress because Republican critics, most notably Senator Mitch McConnell and Congressman John Boehner, observed that the agreement would damage the United States economy without really requiring China to do anything since its goals were fairly nebulous and nonbinding. Regardless of whether China ultimately meets its stated goals by 2030, this agreement proved an important first step towards the drafting of the September 2015 *Joint Presidential Statement on Climate Change with New Domestic Policy Commitments and a Common Vision for an Ambitious Global Climate Agreement in Paris.*

Further Reading

Jiang, Kenjun, et al. 2013. "China's Role in Attaining the Global 2°C Target." *Climate Policy* 13: 55–69.

Nadin, Rebecca, Sarah Opitz-Stapleton, and Xu Yinlong, eds. 2015. *Climate Change: Risk and Resilience in China.* New York: Routledge.

Rapkin, David P., and William R. Thompson. 2013. *Transition Scenarios: China and the United States in the Twenty-First Century.* Chicago: University of Chicago Press.

Sze, Julie. 2015. *Fantasy Islands: Chinese Dreams and Ecological Fears in an Age of Climate Crisis.* Berkeley: University of California Press.

Wang, Weguang, et al., eds. 2012. *China's Climate Change Policies.* New York: Routledge.

Senate Bill 66

To Prohibit Any Regulation Regarding Carbon Dioxide or Other Greenhouse Gas Emissions Reduction in the United States until China, India, and Russia Implement Similar Reductions

Date: January 7, 2015
Location: Washington, D.C.
Significance: In advance of the United Nations Climate Change Conference scheduled for Paris, France in 2015, Senator David Vitter, Republican from Louisiana, introduced this bill to signal to international negotiators that any agreement reached that did not apply the same requirements to the People's Republic of China, India, or the Russian Federation that it gave to the United States would ensure that the United States Congress would not ratify the document.
Source: 114th Congress, 1st Session, U.S. Senate. 2015. S.66. https://www.congress.gov/114/bills/s66/BILLS-114s66is.pdf (accessed July 21, 2015).

A Bill

To prohibit any regulation regarding carbon dioxide or other greenhouse gas emissions reduction in the United States until China, India, and Russia implement similar reductions.

Be it enacted by the Senate and House of Representatives of the United States of America in Congress assembled,

Section 1. Definition of Administrator.

In this Act, the term "Administrator" means the Administrator of the Environmental Protection Agency.

Sec. 2. Findings.

Congress finds that—
(1) in 1997, the Senate adopted Senate Resolution 98, 105th Congress, agreed to July 25, 1997, which expressed the sense of the Senate that the United States should

not accept any agreement that would mandate new commitments to limit or reduce greenhouse gas emissions by developed countries unless the agreement also mandated new specific scheduled commitments to limit or reduce greenhouse gas emissions by developing countries within the same compliance period; and
(2) the Administrator continues to move forward with the regulation of carbon dioxide emissions, however, the People's Republic of China, India, and the Russian Federation do not impose similar regulations on carbon dioxide emissions.

Sec. 3. Carbon Dioxide and Other Greenhouse Gas Emissions Reductions in China, India, and Russia.

Notwithstanding any other provision of law, the Administrator or the head of any other Federal agency or department shall not regulate or continue to implement or enforce any regulations, proposals, or actions establishing any carbon dioxide or greenhouse gas emissions reductions until the Administrator, the Administrator of the Energy Information Agency, and the Secretary of Commerce certify in writing that—
(1) the People's Republic of China, India, and the Russian Federation have proposed, implemented, and enforced measures requiring carbon dioxide and other greenhouse gas emissions reductions; and
(2) the reductions described in paragraph (1) are substantially similar to the carbon dioxide and other greenhouse gas emission reductions proposed by the Administrator or the head of any other Federal agency or department for the United States.

Sec. 4. Repeal.

Any regulation, proposal, or action in effect before, on, or after the date of enactment of this Act, but before the date on which the certification under section 3 is made, that requires any carbon dioxide or other greenhouse gas emissions reduction shall have no force or effect.

Analysis

In many international agreements regarding climate change, the United States had been expected to make much larger cuts in greenhouse gas (GHG) emissions than other major economies, most notably the People's Republic of China, India, and the Russian Federation. Through this document, the Senate called for measures to ensure that the Republic of China, India, and the Russian Federation all had commitments that were comparable to those of the United States in the agreement that emerged from the 2015 negotiations in Paris, France. If the wishes of the United States Senate were not honored, then it was improbable that the agreement would be ratified by that body.

Governor Brown Establishes Most Ambitious Greenhouse Gas Reduction Target in North America

Date: April 29, 2015
Location: Sacramento, California
Significance: California Governor Edmund G. Brown Jr.'s Executive Order set a goal of reducing the state's greenhouse gas emissions to forty percent of 1990 levels by 2030.
Source: Office of Governor Edmund G. Brown Jr. 2015. *Governor Brown Establishes Most Ambitious Greenhouse Gas Reduction Target in North America*. http://gov.ca.gov/news.php?id=18938 (accessed August 7, 2015).

Executive Order B-30-15

WHEREAS climate change poses an ever-growing threat to the well-being, public health, natural resources, economy, and the environment of California, including loss of snowpack, drought, sea level rise, more frequent and intense wildfires, heat waves, more severe smog, and harm to natural and working lands, and these effects are already being felt in the state; and

WHEREAS the Intergovernmental Panel on Climate Change concluded in its Fifth Assessment Report, issued in 2014, that "warming of the climate system is unequivocal, and since the 1950s, many of the observed changes are unprecedented over decades to millennia" and that "continued emission of greenhouse gases will cause further warming and long-lasting changes in all components of the climate system, increasing the likelihood of severe, pervasive and irreversible impacts for people and ecosystems"; and

WHEREAS projections of climate change show that, even under the best-case scenario for global emission reductions, additional climate change impacts are inevitable, and these impacts pose tremendous risks to the state's people, agriculture, economy, infrastructure and the environment; and

WHEREAS climate change will disproportionately affect the state's most vulnerable citizens; and

WHEREAS building on decades of successful actions to reduce pollution and increase energy efficiency the California Global Warming Solutions Act of 2006 placed California at the forefront of global and national efforts to reduce the threat of climate change; and

WHEREAS the Intergovernmental Panel on Climate Change has identified limiting global warming to 2 degrees Celsius or less by 2050 as necessary to avoid potentially catastrophic climate change impacts, and remaining below this threshold requires accelerated reductions of greenhouse gas emissions; and

WHEREAS California has established greenhouse gas emission reduction targets to reduce greenhouse gas emissions to 1990 levels by 2020 and further reduce such emissions to 80 percent below 1990 levels by 2050; and

WHEREAS setting an interim target of emission reductions for 2030 is necessary to guide regulatory policy and investments in California in the midterm, and put California on the most cost-effective path for long term emission reductions; and

WHEREAS all agencies with jurisdiction over sources of greenhouse gas emissions will need to continue to develop and implement emissions reduction programs to reach the state's 2050 target and attain a level of emissions necessary to avoid dangerous climate change; and

WHEREAS taking climate change into account in planning and decision making will help the state make more informed decisions and avoid high costs in the future.

NOW, THEREFORE, I, EDMUND G. BROWN JR., Governor of the State of California, in accordance with the authority vested in me by the Constitution and statutes of the State of California, in particular Government Code sections 8567 and 8571 of the California Government Code, do hereby issue this Executive Order, effective immediately

IT IS HEREBY ORDERED THAT:

1. A new interim statewide greenhouse gas emission reduction target to reduce greenhouse gas emissions to 40 percent below 1990 levels by 2030 is established in order to ensure California meets its target of reducing greenhouse gas emissions to 80 percent below 1990 levels by 2050.
2. All state agencies with jurisdiction over sources of greenhouse gas emissions shall implement measures, pursuant to statutory authority, to achieve reductions of greenhouse gas emissions to meet the 2030 and 2050 greenhouse gas emissions reductions targets.
3. The California Air Resources Board shall update the Climate Change Scoping Plan to express the 2030 target in terms of million metric tons of carbon dioxide equivalent.
4. The California Natural Resources Agency shall update every three years the state's climate adaptation strategy, Safeguarding California, and ensure that its provisions are fully implemented. The Safeguarding California plan will:
    - Identify vulnerabilities to climate change by sector and regions, including, at a minimum, the following sectors: water, energy, transportation, public health, agriculture, emergency services, forestry, biodiversity and habitat, and ocean and coastal resources;
    - Outline primary risks to residents, property, communities and natural systems from these vulnerabilities, and identify priority actions needed to reduce these risks; and
    - Identify a lead agency or group of agencies to lead adaptation efforts in each sector.
5. Each sector lead will be responsible to:
    - Prepare an implementation plan by September 2015 to outline the actions that will be taken as identified in Safeguarding California, and

- Report back to the California Natural Resources Agency by June 2016 on actions taken.
6. State agencies shall take climate change into account in their planning and investment decisions, and employ full life-cycle cost accounting to evaluate and compare infrastructure investments and alternatives.
7. State agencies' planning and investment shall be guided by the following principles
    - Priority should be given to actions that both build climate preparedness and reduce greenhouse gas emissions;
    - Where possible, flexible and adaptive approaches should be taken to prepare for uncertain climate impacts;
    - Actions should protect the state's most vulnerable populations; and
    - Natural infrastructure solutions should be prioritized.
8. The state's Five-Year Infrastructure Plan will take current and future climate change impacts into account in all infrastructure projects
9. The Governor's Office of Planning and Research will establish a technical, advisory group to help state agencies incorporate climate change impacts into planning and investment decisions.
10. The state will continue its rigorous climate change research program focused on understanding the impacts of climate change and how best to prepare and adapt to such impacts.

This Executive Order is not intended to create, and does not, create any rights or benefits, whether substantive or procedural, enforceable at law or in equity, against the State of California, its agencies, departments, entities, officers, employees, or any other person.

I FURTHER DIRECT that as soon as hereafter possible, this Order be filed in the Office of the Secretary of State and that widespread publicity and notice be given to this Order.

IN WITNESS WHEREOF I have hereunto set my hand and caused the Great Seal of the State of California to be affixed this 29th day of April 2015.

Analysis

In advance of the United Nations Climate Change Conference scheduled for Paris, France in 2015, California Governor Edmund G. Brown Jr.'s Executive Order set a goal of reducing the state's greenhouse gas emissions to forty percent of 1990 levels by 2030. While that goal set the standard in the United States, it mirrored the goal established by the European Union in October 2014.

Further Reading

Brown, Jerry. 2014. "Jerry Brown: Climate Change Policy in California—and Beyond." *Bulletin of the Atomic Scientists* 70: 1–7.

Engel, Kirsten. 2006. "State and Local Climate Initiatives: What is Motivating State and Local Governments to Address a Global Problem and What Does This Say About Federalism and Environmental Law?" *The Urban Lawyer* 38: 1015–1029.

National Security Implications of Climate-Related Risks and a Changing Climate

Date: July 23, 2015
Location: Washington, D.C.
Significance: This document provides insights into how the Department of Defense perceives the present and future impacts of climate change on national security. Since this report was produced in response to a request from the U.S. Senate's Committee on Appropriations, it also gives information on the effect of climate change on the federal budget.
Source: Department of Defense. 2015. *National Security Implications of Climate-Related Risks and a Changing Climate.* http://archive.defense.gov/pubs/150724-congressional-report-on-national-implications-of-climate-change.pdf (accessed October 4, 2015).

Response to Congressional Inquiry on National Security Implications of Climate-Related Risks and a Changing Climate, July 2015

Elements of Request for Report

This report responds to the request by the Senate Committee on Appropriations for information on the *National Security Implications of Climate Change* made in the report to accompany H.R. 4870, the Department of Defense (DoD) Appropriations Act for the fiscal year ending September 30, 2015. Specifically, the Committee requested that the Under Secretary of Defense for Policy provide a report to:

- Identify the most serious and likely climate related security risks for each Combatant Command.
- Identify ways Combatant Commands integrate risk mitigation in their planning processes, including in the areas of:
  - Humanitarian disaster relief;
  - Security cooperation;
  - Building partner capacity; and
  - Sharing best practices for mitigation of installation vulnerabilities.
- Describe resources required for an effective response and the timeline of resources needs.

This report is organized into three primary sections:
I. Common Conceptions of Risk and Response
II. Geographic Combatant Commands (GCCs)—Specific Aspects
III. Conclusion

I. Common Conceptions of Risk and Response

DoD recognizes the reality of climate change and the significant risk it poses to U.S. interests globally. The National Security Strategy, issued in February 2015, is clear that climate change is an urgent and growing threat to our national security, contributing to increased natural disasters, refugee flows, and conflicts over basic resources such as food and water. These impacts are already occurring, and the scope, scale, and intensity of these impacts are projected to increase over time.

The Department's defense strategy, as reflected in the 2014 Quadrennial Defense Review (QDR), emphasizes three pillars: protect the homeland, build security globally, and project power and win decisively. A changing climate increases the risk of instability and conflict overseas, and has implications for DoD on operations, personnel, installations, and the stability, development, and human security of other nations. This is why DoD released the Climate Change Adaptation Roadmap (CCAR) in October 2014. The CCAR identifies three overarching goals: to identify and assess the effects of a changing climate on the Department's infrastructure, mission, and activities; to identify, manage, and integrate climate change considerations across the full range of Department missions and activities; and to collaborate with internal and external entities on understanding and assessing the challenges of a changing climate and developing appropriate responses to those challenges.

Geographic Combatant Commands (GCCs) incorporate the risks posed by current and projected climate variations into their planning, resource requirements, and operational considerations. GCCs, often at the request of partner nations, cooperate with other nations on adaptation practices, resilience, environmental considerations, and risk reduction.

Climate-Related Security Risks

Global climate change will have wide-ranging implications for U.S. national security interests over the foreseeable future because it will aggravate existing problems—such as poverty, social tensions, environmental degradation, ineffectual leadership, and weak political institutions—that threaten domestic stability in a number of countries. Each GCC's assessment of risk reflects how this range of factors will affect security in its Area of Responsibility (AOR).

GCCs generally view climate change as a security risk because it impacts human security and, more indirectly, the ability of governments to meet the basic needs of their populations. Communities and states that are already fragile and have limited resources are significantly more vulnerable to disruption and far less likely to respond effectively and be resilient to new challenges. Case studies indicate that in addition to exacerbating existing risks from other factors (e.g. social, economic and political fault lines), climate-induced stress can generate new vulnerabilities (e.g.water scarcity) and thus contribute to instability and conflict even in situations not previously considered at risk. GCCs have identified four general areas of climate-related security risks:

- **Persistently recurring conditions such as flooding drought, and higher temperatures** increase the strain on fragile states and vulnerable populations by dampening economic activity and burdening public health through loss of agriculture and electricity production, the change in known infectious disease patterns and the rise of new ones, and increases in respiratory and cardiovascular diseases. This could result in increased intra-and interstate migration, and generate other negative effects on human security. For example, from 2006–2011, a severe multiyear drought affected Syria and contributed to massive agriculture failures and population displacements. Large movements of rural dwellers to city centers coincided with the presence of large numbers of Iraqi refugees in Syrian cities, effectively overwhelming institutional capacity to respond constructively to the changing service demands. These kinds of impacts in regions around the world could necessitate greater DoD involvement in the provision of humanitarian assistance and other aid.

- **More frequent and/or more severe extreme weather events** that may require substantial involvement of DoD units, personnel, and as sets in humanitarian assistance and disaster relief (HA/DR) abroad and in Defense Support of Civil Authorities (DSCA) at home. Massive flooding in Pakistan in 2010 was the country's worst in recorded history, killing more than 2,000 people and affecting 18 million; DoD delivered humanitarian relief to otherwise inaccessible areas. Super Storm Sandy in New York and New Jersey in 2012 resulted in over 14,000 DoD personnel mobilized to provide direct support, and at least an additional 10,000 who supported the operation in various capacities in the areas of power restoration, fuel resupply, transportation infrastructure repair, water and meal distribution, temporary housing and sheltering, and debris removal. The need for HADR and DSCA will likely rise as cities expand to encompass the majority of the global population and because flood risk threatens more people than any other natural hazard, especially in urban areas. Many growing cities are located in low- and middle-income countries with limited resources. Building partner nation capacity for HA/DR capabilities and civilian-military partnerships for DSCA are important parts of GCC security cooperation efforts. The Office of U.S. Foreign Disaster Assistance (OFDA) is responsible for leading and coordinating the U.S. Government's response to disasters overseas.
- **Sea level rise and temperature changes** lead to greater chance of flooding in coastal communities and increase adverse impacts to navigation safety, damages to port facilities and cooperative security locations, and displaced populations. Sea level rise may require more frequent or larger-scale DoD involvement in HADR and DSCA. Measures will also likely be required to protect military installations, both in the United States and abroad, and to work with partner nations that support DoD operations and activities. Sea level rise, increased ocean acidification, and increased ocean warming pose threats to fish stocks, coral, mangroves, recreation and tourism, and the control of disease affecting the economies, and ultimately stability, of DoD's partner nations. Some Pacific island nations face the risk of being entirely submerged by rising seas, and most island nations' freshwater supplies will be threatened by saltwater intrusion well before then. Loss of land, especially highly populated and agriculturally rich coastal land, also poses second-order effects on human displacement and economic and food stability, and may further exacerbate challenges associated with disease vectors.
- **Decreases in Arctic ice cover, type, and thickness** will lead to greater access for tourism, shipping, resource exploration and extraction, and military activities. Land access—which depends on frozen ground in the Arctic—will diminish as permafrost

thaws. These factors may increase the need for search and rescue (SAR) capabilities, monitoring of increased shipping and other human activity, and the capability to respond to crises or contingencies in the region. Difficult and unpredictable weather conditions, large distances, and scarce resources make emergency response in the Arctic difficult. Arctic operations are expensive and dangerous for military forces that are unprepared for the austere operating environment. DoD continues to evaluate the need for specific Arctic capabilities.

Integration of Climate-Related Security Risks into Planning Processes

GCCs use their Theater Campaign Plans, Operation Plans, Contingency Plans, and Theater Security Cooperation Plans as a means to identify or take into account climate risks. To assist with historical climatology and climate change near-term assessments, the Air Force's 14th Weather Squadron (14WS) provides authoritative data sets and tailored decision aids to GCCs. In addition, the National Oceanic and Atmospheric Administration (NOAA) provides long-term global climate projections, weather forecasts, and other services to all federal agencies within the United States—including DoD.

Cooperation and building partner capacity. Although activities vary, all GCCs are working with partner nations to increase partner abilities to reduce risks and implications from environmental impacts and climate change, including severe weather and other hazards. GCCs work with partner nations in three lines of activities:

- *Building infrastructure*: examples include emergency operations centers, disaster response warehouses, disaster shelters, and construction of dams for hydroelectric power.
- *Training*: provides disaster management and response personnel in partner nations with the knowledge and skills to meet the basic needs of the populace. These programs are often coordinated with the in-country security cooperation officers and the U.S. Agency for International Development (USAID).
- *Equipping*: GCCs support the delivery of Non-governmental Organization (NGO) emergency donations through the Defense Security Cooperation Agency's Funded Transportation Program and Denton Program. U.S. Southern Command (USSOUTHCOM) funds National Preparedness Baseline Assessments, which include gap analyses and five-year plans to build capability and capacity within partner nations, helping to provide a better picture of specific vulnerabilities to climatic events across a wide range of indicators. Other GCCs have similar disaster risk reduction cooperation programs or are evaluating such programs.

Sharing best practices for mitigation of installation vulnerabilities. Military Departments are responsible for assessing and addressing the vulnerabilities of DoD installations. DoD has directed a global screening level vulnerability assessment to determine DoD installations' vulnerability to climate changes, water and environmental requirements, and sea level rise. The Military Departments will use information from the screening assessments to identify serious vulnerabilities and to develop adaptation strategies as necessary for installations.

Resources and Timeline

Resources for assessing and responding to the impacts of climate change are provided within existing DoD missions, funds, and capabilities. Activities associated with climate resiliency planning in GCCs are subsumed under existing risk management processes. Training for GCC personnel in accessing and using climate and weather data related to climate change may incur costs. Attending coordination meetings with partner nation organizations focusing on climate impacts may also add to GCC personnel responsibilities. In the future, other adaptation and resilience costs may become clearer and will need to be evaluated.

Humanitarian assistance and disaster relief is costly, particularly for first response activities. Common requirements include strategic lift assets, water and water purification capability, vertical lift assets for dispersal of relief supplies, engineering equipment for removing debris from critical infrastructure, medical care, robust communications, repair of overloaded or nonfunctioning electricity generation, SAR, and port opening and traffic control capabilities. DoD has put many of these capabilities in use for humanitarian assistance following the earthquakes in Nepal in 2015. The main source of funding for the GCCs' HA/DR programs is the Overseas Humanitarian, Disaster, and Civic Aid (OHDACA) appropriation, which enables the Combatant Commands to provide immediate life-saving assistance to countries in their regions.

II. GCCs–Specific Aspects

Geographic Combatant Commands independently have assessed climatic risks in the context of their respective AORs. For instance, U.S. Africa Command (USAFRICOM) assesses humanitarian crisis as the most likely climate-related risk within its AOR, foremost due to the impact that devastating events like drought and disease could have on vulnerable populations and on state stability in places already struggling with fragility and conflict. North American Aerospace Defense Command / U.S. Northern Command (NORAD/USNORTHCOM) and USEUCOM are concerned with security risks arising from increased shipping, military operations, and resource exploration in the Arctic as the ice-cap melts. U.S. Pacific Command (USPACOM) considers rising sea-levels to be a particularly significant threat to people in geographically vulnerable locations. USSOUTHCOM similarly highlights the threat that sea level rise and ocean acidification and warming pose to fish stocks, coral, mangroves, recreation and tourism, and the control of disease. USSOUTHCOM also identifies coastal flooding to be a particular concern for parts of the Caribbean basin due to climate change-related sea level rise.

GCCs assess that, in line with the Intergovernmental Panel on Climate Change (IPCC) conclusion, climate change will have the greatest impact on areas and environments already prone to instability, which aligns with DoD's wider assessment of climate change as a threat multiplier. USPACOM already reflects this likely implication in its planning processes by addressing not only the direct effects of climate change but also the imperative this implication creates for environmental and resource management. U.S. Central Command (USCENTCOM) similarly monitors resource scarcity (e.g., water, food, energy) in its arid AOR, and accounts for this factor in its planning. Although the context differs, USAFRICOM assesses that climate change will exacerbate existing economic, social, and environmental vulnerabilities, while conditions of drought, disease, and economic stagnation may tip states toward systemic breakdowns.

GCCs recognize the risk climate change poses to existing resource allocation. USAFRICOM highlights how climate change will alter the distribution and quality of natural resources, such as fresh water, arable land, coastal territory, and marine resources. USPACOM assesses that climate change will affect populations already living in unstable environments and already experiencing urban or rural conflict driven, in particular, by seasonal water shortage. USCENTCOM identifies that climate changes heighten competition at the national or sub-national level in an already arid region, and this competition could be more dangerous as actors seek to protect limited resources. However, interstate conflict risk is attenuated by the context of international treaties and agreements. USNORTHCOM identifies increased resource exploration in the Arctic as driving an increase in the future demand for SAR and environmental disaster response missions in support of other agencies and civil authorities.

Both USSOUTHCOM and USAFRICOM recognize the economic risks associated with climate change, again as a stressor on vulnerable populations. USAFRICOM and USPACOM both identify how technological, infrastructure, skills, and information constraints heighten vulnerability to climate stresses.

GCCs highlight the impact that climate change may have on the frequency and severity of weather-related events. In addition to humanitarian assistance in its AOR, USPACOM anticipates this will increase the demand for DSCA as well as pose a challenge to U.S. critical defense infrastructure. These concerns are consistent with the 2014 DoD Climate Adaptation Roadmap. GCCs are accordingly integrating climate-related risk mitigation into planning processes, the details of which are provided below:

USAFRICOM

Integration of Climate-Related Risks Management into Planning Processes
- USAFRICOM will consider climate change-related factors from key strategic documents in its annual Theater Campaign Plan (TCP) reviews.
- USAFRICOM's Foreign HA/DR Country Plans will be expanded to promote engagement with partner nations to enhance planning, responses, and resilience to the effects of climate change. USAFRICOM also works closely with USAID in HA/DR, and USAFRICOM's intervention in HA/DR operations will align with USAID's strategy.
- USAFRICOM's building partner capacity efforts are nested within its security cooperation programs and will adapt to a variety of trends and projections, including climate change.

Resources and Timeline
- USAFRICOM's resource planning takes into account the probabilities of climate events driving resource requirements for HA/DR-related program budgets. Embassy country teams throughout Africa work closely with interagency partners to ensure DoD contributions are represented in embassies' Integrated Country Strategy documents.

USCENTCOM

Integration of Climate-Related Risks Management into Planning Processes
- Current and historic climatic conditions are factored into TCPs, including water scarcity, which is a recurring issue in the AOR. Warning indicators are part of the deliberate planning process.
- Climate change is not a stand-alone topic. Real-world, actual climate conditions are

National Security Implications of Climate-Related Risks

taken into consideration for planning missions. HA/DR and security cooperation/building partner capacity are specific lines of effort addressed in the TCP.
- Service components will address specific needs for their installations; for example, the Navy will address sea level rise and access to ports.

Resources and Timeline
- The Military Departments/Services provide most of the resources for on-the-ground activities. USCENTCOM focuses on nearer-term (five years) projected changes in climate.

USEUCOM

Integration of Climate-Related Risks Management into Planning Processes
- The Arctic Security Forces Roundtable is USEUCOM's flagship engagement effort for nations that have security forces within the Arctic region. It is a forum in which senior military leaders from Arctic nations, and other stakeholders, confer and agree upon actions that can support stability and peaceful commercial activity in the region. Lessons learned from our Arctic allies and partners are used to enhance operational safety. An Arctic assessment is included in the current TCP and informs the development of the GCC's activities.
- In response to melting ice and newly accessible areas of the Arctic, EUCOM sponsors the ARCTIC ZEPHYR series of table-top exercises focused on SAR operations in the Arctic.

Resources and Timeline
- The Northern Sea Route generally opens for four weeks each year—usually the month of September. Arctic-specific SAR resources will need to be trained and ready as Arctic activity, particularly tourism, increases. In 2016, the cruise liner CRYSTAL SERENITY will attempt to transit the North West Passage with 900 expected passengers, the first leisure cruise ship to make the crossing.
- Future Arctic offshore drilling will also create a resource demand and the need for emergency response, risk reduction measures, and environmental protections.

NORAD/USNORTHCOM

Integration of Climate-Related Risks Management into Planning Processes
- USNORTHCOM routinely includes extreme weather-driven scenarios in training and exercise events. USNORTHCOM has also developed planning tools to guide operational response efforts for potential catastrophic events, including severe weather events.
- NORAD/USNORTHCOM has participated in the ARCTIC ZEPHYR series of table-top exercises, and has conducted cooperative Arctic SAR exercises with Canada.
- NORAD/USNORTHCOM has hosted two Arctic collaborative workshops with the full range of governments, international partners, the private sector, and academia to focus on projected climate changes in the Arctic.

250   Climate Change and American Policy

Resources and Timeline
- Acquisition and supply chain requirements for the Arctic are considerably longer than for the rest of the AOR and are much more costly. DoD will continue to partner with federal departments and agencies; state, local, and tribal agencies; other nations; and the private sector on services as appropriate.

USPACOM

Integration of Climate-Related Risks Management into Planning Processes
- USPACOM efforts addressing climate change risks are primarily outlined in the "All Hazards" Line of Effort in the USPACOM TCP. The focus is two-fold: readiness to respond to and be resilient to disasters, and sustainable resource management toward critical resource security. USPACOM coordinates these efforts with a variety of DoD interagency partners—including the Department of State, USAID, Department of Homeland Security, Department of the Interior, and NOAA—to ensure its readiness to execute concepts of operations for DSCA, Pandemic and Emerging Infectious Disease, and Foreign HA/DR.
- USPACOM Country Security Cooperation plans, under the Theater Security Cooperation Plan and its logistics Theater Security Cooperation branch, provide opportunities to work in collaboration with host nations to address disaster and critical resource security needs through a variety of operations, actions, and activities in country.
- USPACOM is developing a visual display tool that seeks to overlay historic disaster event data, climate and weather data, population and geographic data, Country Book information, resource scarcity data, and all hazards-related activities in the region into a comprehensive tool that will not only provide planners with historic and needs-based data to inform plans, but will also enable USPACOM and those with access to the information to leverage and co-execute events for the greatest overall impact in the region. USPACOM is working to integrate Australia, Canada, New Zealand, and the United Kingdom into all hazards-related activities to leverage lessons learned, best practices, and limited resources to support disaster response readiness and capability development more effectively, as well as work toward critical resource security as a means of protecting peace in the region.
- In June 2015, USPACOM and Thailand co-hosted the fifth annual Pacific Environmental Security Forum in Bangkok. The forum seeks to develop foreign nation capacity in several environmental security areas through combined projects within the USPACOM AOR. Sessions on project concept development followed three days of discussions on the DoD Climate Change Adaptation Roadmap, the protection of the commons in a civilian-military context, and military environmental programs.
- USPACOM is also developing partnerships focused on building the overall resilience of the State of Hawaii and nesting the resilience of USPACOM installations within that framework.

Resources and Timeline
- USPACOM has established Pacific Augmentation Teams around its AOR to identify quickly immediate needs that can be met with military assets. These teams represent

an effort to shorten disaster response times by allowing USPACOM to begin mobilizing a response in anticipation of a Secretary of Defense approved request. Proximity to the disaster is also a factor that dramatically reduces the response time. U.S. forces training, exercising, or operating in the vicinity of a disaster can be re-tasked to support relief efforts.

USSOUTHCOM

Integration of Climate-Related Risks Management into Planning Processes
Humanitarian Assistance and Disaster Relief
- Although the USSOUTHCOM HA/DR program does not explicitly incorporate climate change planning, it maintains communication with the disaster response actors and first responders at the relevant agencies of USSOUTHCOM partner nations.
- USSOUTHCOM assists partner nations in making sure that they have the infrastructure, training, and equipment necessary to respond to any disaster, and also assists them with the preparation of their disaster response plans. In this way, USSOUTHCOM fulfills one of its military end states: "Partner nations have the capability and capacity to conduct HA/DR operations to mitigate the effects of natural and man-made disasters."
- USSOUTHCOM is not the lead entity when it comes to support of a partner nation response to a disaster; it supports OFDA.
- As an example of a weather-related disaster preparation, although not necessarily climate change planning per se, for the last three years (after the Haiti earthquake in 2010) USSOUTHCOM has requested authorities to use OHDACA funding to preposition assets under its operational control when a severe storm threatens Haiti, in order to be able to respond immediately to a potential disaster response requirement.
- The National Preparedness Baseline Assessments, funded by USSOUTHCOM, include a gap analysis as well as a five-year plan to build capability and capacity within the countries in the region. They include data collected at a national and sub-national scale utilizing both current and historical information, combined with environmental, social, policy, and economic indicators. This analysis can provide a better picture of specific vulnerabilities to climatic disasters across a wide range of environmental, social, and economic indicators. The collection of sub-regional data will provide a more nuanced depiction of each country's risks and vulnerabilities to disasters that may be influenced by climate change as well as their readiness to respond to them.

Security Cooperation / Building Partner Capacity: USSOUTHCOM has conducted some security cooperation activities related to military adaptation to climate change. These activities have been nested under its HA/DR and critical access lines of effort. Several partners in the region have expressed interest in discussing and taking steps to address environmental and energy challenges for military forces, including climate change. As an example, in 2014 USSOUTHCOM completed a collaborative report on this topic with military experts from Chile, Colombia, El Salvador, and Trinidad & Tobago, and presented the outcomes to the Inter-American Defense Board.

Resources and Timeline
- USSOUTHCOM has identified the additional resources required to "Identify and assess the effects of climate change on the Department," which is the first goal in the DoD 2014 Climate Change Adaptation Roadmap. USSOUTHCOM will also seek appropriate resources to fund assessments to determine the effects of its most serious and likely climate-related risks.

III. Conclusion

The Department of Defense sees climate change as a present security threat, not strictly a long-term risk. We are already observing the impacts of climate change in shocks and stressors to vulnerable nations and communities, including in the United States, and in the Arctic, Middle East, Africa, Asia, and South America. Case studies have demonstrated measurable impacts on areas vulnerable to the impacts of climate change and in specific cases significant interaction between conflict dynamics and sensitivity to climate changes. Although climate-related stress will disproportionately affect fragile and conflict-affected states, even resilient, well-developed countries are subject to the effects of climate change in significant and consequential ways.

For these reasons, Combatant Commands are integrating climate-related impacts into their planning cycles. Depending on the region, risks to Combatant Commands vary, but all GCCs share a common assessment of its significance. The ability of the United States and other countries to cope with the risks and implications of climate change requires monitoring, analysis, and integration of those risks into existing overall risk management measures, as appropriate for each Combatant Command.

Although DoD and the Combatant Commands cannot prepare for every risk and situation, the Department is beginning to include the implications of a changing climate in its frameworks for managing operational and strategic risks prudently. Moreover, the Department is working with other U.S. Government departments and agencies, partner nations, and many other entities on addressing climate security risks and implications.

Analysis

As the twentieth-century was nearing its conclusion, the Pentagon began taking note of the national security implications of global climate change. Some of the impacts were obvious, such as the threat to naval installations by the rise in sea levels resulting from the melting of glaciers in polar regions. The issue was of such concern that military planners began addressing the topic during President George W. Bush's presidency. This was surprising considering this was a period when global climate change was largely ignored by the executive branch of the federal government. The political climate changed when President Barack Obama came into office. Subsequently, the consequences of ongoing and future climate change began to be incorporated into Department of Defense planning, including its Quadrennial Defense Review (QDR). The 2014 QDR was especially noteworthy because it labeled climate change as a "threat multiplier." As such, it had the potential to exacerbate existing problems, such as poverty and the competition for scarce resources like food and water. The report proved prescient in 2015–2016, as

Europe dealt with a massive influx of migrants from the Middle East and North Africa. Although many of those migrants were viewed as economic migrants, a significant portion came from areas already being devastated by climate change. They were forced to leave their homelands due to a lack of the natural resources required to sustain life. Another warning resulting from the QDR was that such desperation was also a breeding ground for terrorism, which also seems to have accompanied the migrants from the Middle East and North Africa to countries such as Belgium and France.

The *National Security Implications of Climate-Related Risks and a Changing Climate* is essentially a summary of the findings of the 2014 QDR. It succinctly highlights the manner in which the Department of Defense's respective Geographic Combatant Commands view the threats within their areas of influence. It is particularly sobering to consider how the perceived threats differ in importance in the different parts of the world. For instance, in the Caribbean, problems posed by rising sea waters are of the utmost importance. In North Africa and the Middle East, the lack of water is one of the primary issues.

Further Reading

Committee on National Security Implications of Climate Change for U.S. Naval Forces, National Research Council of the National Academies. 2011. *National Security Implications of Climate Change for U.S. Naval Forces*. Washington, D.C.: National Academies.

Light, Sarah E. 2014. "Valuing National Security: Climate Change, the Military, and Society." *UCLA Law Review* 61: 1772–1812.

Matthew, Richard A. 2011. "Is Climate Change a National Security Issue?" *Issues in Science & Technology* 27: 49–60.

Nevitt, Mark P. 2015. "The Commander in Chief's Authority to Combat Climate Change." *Cardozo Law Review* 37: 437–502.

White House. 2015. *Findings from Select Federal Reports: The National Security Implications of a Changing Climate*. Washington, D.C.: White House. https://www.whitehouse.gov/sites/default/files/docs/ National_Security_Implications_of_Changing Climate_Final_051915.pdf.

Federal Plan Requirements for Greenhouse Gas Emissions from Electric Utility Generating Units Constructed on or Before January 8, 2014; Model Trading Rules; Amendments to Framework Regulations (Executive Summary and Organization and Approach for this Proposed Rule)

Date: August 3, 2015
Location: Washington, D.C.
Significance: Known as the "Clean Power Plan," this document provides a mechanism for the states to create individual plans that would lead to the reduction of carbon pollution emitted by power plants by 32% from 2005 levels by 2030.
Source: Environmental Protection Agency. 2015. *Federal Plan Requirements for Greenhouse Gas Emissions from Electric Utility Generating Units Constructed on or Before January 8, 2014; Model Trading Rules; Amendments to Framework Regulations.* http://www.epa.gov/airquality/cpp/cpp-proposed-federal-plan.pdf (accessed August 3, 2015).

A. Executive Summary

In the CAA, Congress created a partnership between the EPA and the states. Under section 111(d) of the CAA, the EPA establishes emission performance levels based on its determination of the BSER for existing sources of air pollution and provides guidelines for state plans to apply standards of performance to their sources that meet the BSER level of performance. The EPA promulgated EGs under CAA section 111(d) which set source-level CO_2 emission performance rates for the EGUs at certain large fossil fuel-fired power plants ("affected EGUs"). States then apply these EGs to their sources in developing state plans to achieve these emission performance levels for EPA approval, or initial submittals, by September 6, 2016. The amount of reductions in CO_2 that the EPA determined to be achievable for these sources is based on its determination of what constitutes the BSER. This determination is finalized in the EGs, which are designed to maximize the flexibility of both states and affected EGUs in meeting CO_2 emissions performance rates. While states may impose the emission rates directly on their affected EGUs, states also have the option of submitting more tailored plans that meet state-specific emissions goals. The EGs also provide flexibility by allowing for emissions trading and multi-state compliance options.

While it has been the EPA's longstanding view that the statute identifies states as the preferred implementers of CAA programs, the agency makes clear in the EGs that states cannot and will not be penalized for failing to participate in this program. However, if a state does not submit an approvable plan under section 111(d) of the CAA, the EPA will develop, implement, and enforce a federal plan to reduce CO_2 from the fossil fuel-fired power plants in that state. This is wholly consistent with the "cooperative federalism" structure of the CAA and many of our nation's other environmental laws. In addition, we have heard from states and other stakeholders that it would be helpful for the agency to present model designs for state plans, and a federal plan would be an appropriate means of doing that.

Accordingly, the EPA proposes a federal plan under section 111(d) of the CAA for the control of CO_2, a GHG pollutant, from certain emitting fossil fuel-fired power plants, in the event that some states do not adopt their own plans. Specifically, the EPA is proposing approaches in the form of mass- and rate-based trading options that provide flexibility in implementing emission standards for a state's affected EGUs. Both proposed approaches to the federal plan would require affected EGUs to meet emission standards set using the CO_2 emission performance rates in the EGs. The federal plan will achieve the same levels of emissions performance as required of state plans under the EGs. The EPA will promulgate a final federal plan for only the affected EGUs in states that the EPA determines did not submit an approvable plan.

At the same time, these two proposed options offer states model trading rules that the states can follow in developing their own plans in order to capitalize on the flexibility built into the final EGs. Thus, this document proposes four discrete actions: (1) A rate-based federal plan for each state with affected EGUs; (2) a mass-based federal plan for each state with affected EGUs; (3) a rate-based model trading rule for potential use by any state; and (4) a mass-based model trading rule for potential use by any state. The regulatory text of each federal plan and corresponding model trading rule is identical, except as indicated otherwise within the text of the model rule (for instance, the EPA is providing model rule text for states to use related to the crediting of a broader set of clean energy resources than is being proposed in the federal plan).

The EPA intends to finalize both the rate-based and mass-based model trading rules in summer 2016. The EPA will finalize a federal plan for only a given state in the event that the state does not submit an approvable plan by the deadlines specified in the final EGs and the EPA takes action finding that the state has failed to submit a plan, or disapproving a submitted plan because it does not meet the requirements of the EGs. Indeed, states may simply choose to accept a federal plan for their sources rather than undertake the development of a plan of their own by not submitting a state plan. Under this proposed rule, a federal plan promulgated for a particular state would take the form of either the mass-based model trading rule or the rate-based model trading rule. The EPA currently intends to finalize a single approach (i.e., either the mass-based or rate-based approach) for every state in which it promulgates a federal plan, given the benefits of a broad trading program, as discussed in the following section of this preamble. We invite comment on which approach, i.e., either mass-based or rate-based trading, should be selected if we opt to finalize a single approach.

It is the EPA's intention to give the states as much opportunity as possible to set their own course for carrying out the EGs. Even where a federal plan is put in place for a particular state, that state will still be able to submit a plan, which, upon approval, will allow the state and its sources to exit the federal plan. In addition, as discussed in section VI.A of this preamble, states may take delegation of administrative aspects of the federal plan in order to become the primary implementers. And as discussed in sections V.E and VII.A of this preamble, states may submit partial state plans in order to take over the implementation of a portion of a federal plan. For instance, in a mass-based trading program, the agency proposes to allow states to submit partial state plans to replace the federal plan allowance-distribution provisions with their own allowance-distribution provisions, similar to the approach we have taken in prior trading programs. Finally, even in states in which the affected EGUs are operating under a federal plan, the agency recognizes that states may adopt complementary measures outside of CAA programming to facilitate compliance and lower costs that could benefit power generators and consumers, directly or indirectly.

A state program that adheres to the model trading rule provisions specified in this rulemaking would be presumptively approvable. States may submit means of meeting the EGs' requirements that differ from the model trading rule provisions, so long as the state demonstrates to the EPA's satisfaction in the state plan submittal that such alternative means of addressing requirements are at least as stringent as the presumptively approvable approach described here. Additionally, there are stand-alone portions of the model trading rules, such as the evaluation, measurement, and verification (EM&V) procedures, that would be approvable even if a state adopted an approach that differs from the federal plan. The model trading rules serve as a mechanism to facilitate larger trading markets since consistency with the federal plan allows trading across both the state and federal programs. The EPA expects a larger trading region is likely to result in lower overall costs. These and other aspects of the model trading rules and federal plan provide additional support for this rule as proposed. Thus, the proposed rule would ensure that congressionally mandated emission standards under authority of section 111 of the CAA are implemented, either by the states in the first instance, or by the EPA where needed.

The agency is proposing a finding that it is necessary or appropriate to implement a CAA section 111(d) federal plan for the affected EGUs located in Indian country. CO_2 emission performance rates for these facilities were finalized in the EGs. Tribes generally may seek "treatment as a state" (TAS) and submit a tribal plan to implement CAA programs, including programs under CAA section 111(d), and this proposed finding does not preclude tribes from doing that. However, tribes are not subject to the deadlines applicable to state action under the EGs and in the absence of a federal plan, CO_2 emissions from these EGUs could go unregulated. Therefore, as discussed in section VI.D of this preamble, we are proposing a necessary or appropriate finding.

This document also proposes certain enhancements to the process and timing for state submittals and EPA action in the CAA section 111(d) framework regulations of 40 CFR part 60, subpart B (these proposals are not a part of the federal plan or model trading rules). These changes, if finalized, would be applicable under the Clean Power Plan and other CAA section 111(d) rules. These changes clarify the availability of certain procedural mechanisms similar to those available under CAA section 110 (such as calls for plan revisions and the availability of "conditional approvals," etc.). They also extend the deadlines for EPA action, in part to conform with the timelines in the EGs. These changes do not alter the timelines for state action under the EGs and do not alter the submission requirements established in the EGs. Finally, the agency proposes to clarify and request comment on an interpretive issue raised in the Clean Power Plan proposal regarding whether a reconstruction or modification that is subject to a CAA section 111(b) standard moves an existing source out of a CAA section 111(d) program. These proposed changes are discussed in section VII of this preamble. The agency intends to finalize these changes earlier than the finalization of the model trading rules.

In proposing a federal plan, the EPA considered a variety of potential impacts that its action might have on the environment, on businesses, particularly in the energy sector, and on the reliability of the electrical grid. The agency gave extensive consideration to impacts on vulnerable communities, particularly low-income communities, communities of color, and indigenous communities. These considerations are discussed in sections III, VIII, IX, and X of this preamble.

The agency convened a Small Business Advocacy Review Panel under the Regulatory Flexibility Act and has completed an Initial Regulatory Flexibility Analysis (IRFA). Various recommendations from the Panel are found reflected throughout this proposal. In section X of this preamble, the agency explains how it has conducted or intends to conduct all other statutory or executive order (EO) reviews that apply to this proposed action. The EPA also explains in this document how it proposes to take into consideration the "remaining useful lives" of affected EGUs in the design of the proposed federal plan, as discussed below in section III.G of this preamble.

The agency considered the impacts this action could have on the electricity grid and developed options for compliance that are cost-effective and that provide substantial flexibility for the affected EGUs that will accommodate the parties charged with maintaining the reliability of electrical power. A key feature of the proposed federal plan and model trading rule is that the flexibility inherent in both of the two approaches (i.e., rate-based or mass-based trading) enables the EPA and the states to create a level of flexibility for affected EGUs that allows owners and operators to determine the best way to achieve emissions reductions, at the EGU-, state-, multi-state-, regional-, or national level. As a result, compliance strategies can mirror, or be integrated with, the ongoing operations of the current electricity grid as it continues to serve its primary critical function of ensuring an uninterrupted supply of affordable and reliable electricity. This flexibility is especially valuable whenever the need to address specific reliability concerns arises. It allows owners and operators of reliability-critical EGUs to continue to meet their compliance obligations while operating to maintain electric reliability.

The EPA outlined and initiated the Clean Energy Incentive Program (CEIP) in the final EGs (see section VIII of the final EGs). The program is designed to incentivize investment in certain types of renewable energy (RE) projects, as well as demand-side energy efficiency (EE) projects implemented in low income communities, that generate MWh or reduce end-use energy demand during 2020 and/or 2021. The EPA proposes to apply the CEIP in all states subject to either a rate-based or mass-based federal plan.

We also reviewed impacts that this action could have on the environment and the need to ensure environmental integrity of the program as well as avoid unintended environmental impacts. We took measures to ensure that the reductions in carbon emissions this plan will achieve are real, and not just apparent. As in the EGs, in both the rate- and mass-based approaches, the EPA has incorporated components to address the concern that the dynamics of either a rate- or mass-based trading program could incentivize shifting generation from existing units in ways that would result in more CO2 emissions than would otherwise be expected, or that undermine the purpose of the CAA section 111(d) program.

We considered whether compliance choices under a federal plan could lead to an unintended concentration of other air pollutants in certain overburdened communities, particularly low-income communities and communities of color. As discussed below, our analysis shows why we do not expect this to occur at any significant level. In general, as in the EGs, we anticipate that the federal plan will result in overall reductions of copollutants, in addition to reductions in CO2, with corresponding co-benefits to public health. We also reviewed whether this action could trigger an obligation to consult with other agencies responsible for implementing the Endangered Species Act, and propose to conclude that it will not.

In the final emission guidelines, the EPA emphasized the importance of state actions to ensure that in developing their respective compliance plans the states addressed the concerns and priorities of vulnerable communities. In the process of developing a final federal plan, the EPA will also take actions to address those concerns as well. In addition to the public hearings that the EPA will be holding for all members of the American public on this proposed rulemaking, we will also be conducting a national webinar and outreach meeting(s) in all ten regions on this proposed rulemaking for communities. The goal of these outreach activities is to provide communities with the information they need to understand how the proposed rulemaking will potentially impact their respective communities. At the same time, this information will be useful in helping communities engage the EPA during our comment period, as well as with their states during the state plan development process. We will also be providing other outreach and support activities for vulnerable communities, which are outlined in the community and environmental justice (EJ) considerations in section IX.B of this preamble.

B. Organization and Approach for this Proposed Rule

In this action, the EPA is proposing a federal plan to implement the Clean Power Plan EGs for affected fossil fuel fired EGUs operating in states that do not have approved state plans. Specifically, the EPA is co-proposing two different approaches to a federal plan to implement the Clean Power Plan EGs—a rate-based trading approach and a mass-based trading approach. While establishing emission standards for affected EGUs that would be directly enforceable against the owners and operators of the source, both approaches would grant EGUs substantial flexibility in meeting their compliance obligations. For this reason, among others, these proposed approaches also serve as two proposed model trading rules that states may adopt or tailor in designing their own plans.

The EGs provide that states have until September 6, 2016 (or upon making an initial submittal, until September 6, 2018) to submit state plans, and the EPA does not intend to finalize and implement the federal plan for any states prior to the agency's action of determining a failure to submit a state plan or disapproving a state plan. At the same time, in order to support states' consideration of adoption of one of the model trading rules as an approvable state plan, the agency intends to finalize either or both model rule options presented in this proposed rule by summer 2016, prior to the deadline for state submittals.

The EPA currently intends to finalize a single approach—*i.e.*, either a rate-based or a mass-based approach—in all promulgated federal plans for particular states in order to enhance the consistency of the federal trading program, achieve economies of scale through a single, broad trading program, ensure efficient administration of the program, and simplify compliance options for affected EGUs. The EPA recognizes that the mass-based trading approach would be more straightforward to implement compared to the rate-based trading approach, both for industry and for the implementing agency. The EPA, industry, and many state agencies have extensive knowledge of and experience with mass-based trading programs. The EPA has more than two decades of experience implementing federally-administered mass-based emissions budget trading programs including the Acid Rain Program (ARP) sulfur dioxide (SO_2) trading program, the Nitrogen Oxides (NO_x) Budget Trading Program, CAIR, and CSAPR. The tracking system infrastructure exists and is proven effective for implementing such programs. The EPA requests comment on which approach—mass-based or rate-based trading—is preferred for the federal plan. Some stakeholders have suggested there could be utility in the availability of both approaches based on the unique circumstances of particular states. The EPA recognizes that it remains potentially possible to finalize a different approach to a federal plan in some circumstances, but believes that in general, and consistent with prior federal trading programs such as CSAPR, creating a single, broad program has the most advantages.

The stringency of the proposed federal plan is the same as the CO_2 emission performance rates established for affected EGUs in the EGs. As explained in the final EGs, the EPA determined the CO_2 emission performance rates through the application of the BSER. In the EGs, the EPA has taken final action on the BSER for CO_2 emissions from existing fossil fuel-fired EGUs. Any comments on this proposed rule relating to the BSER, its stringency, rationale, or legal basis, will not be considered as, by definition, they will be beyond the scope of this action.

1. **The Rate-based Approach**

In the first approach, the EPA would implement a rate-based emissions trading program. In a rate-based program, affected EGUs must meet an emission standard, derived from the EGs, expressed as a rate of pounds of CO_2 per megawatt hour (lbs/MWh). If sources emit above their assigned rate, they must acquire a sufficient number of emission rate credits (ERC), each representing a zero-emitting megawatt hour (MWh), to bring their rate of emissions into compliance. ERCs may be generated by affected EGUs or by other entities that supply zero- or low emitting electricity resources to the grid through an approval and recognition process that the EPA will administer. ERCs may be bought and sold, or banked for use in later years. The rate-based approach is explained in greater detail in section IV of this preamble.

2. **The Mass-based Approach**

The second approach to a federal plan that the EPA is proposing in this action is a mass-based trading program. In a mass-based program, the EPA would create a state emissions budget equal to the total tons of CO_2 allowed to be emitted by the affected EGUs in each state, consistent with the mass goals established in the EGs. The EPA would initially distribute the allowances within each state budget—less three proposed allowance set-asides—to the affected EGUs based on their historical generation. Allowances may then be transferred, bought, and sold on the open market, or banked for future use. The compliance obligation on each of the affected EGUs is to surrender the number of allowances sufficient to cover the EGU's respective emissions at the end of a given compliance period. The EPA is also proposing as a part of the mass-based approach three set-asides of allowances: (1) For a Clean Energy Incentive Program; (2) to support RE projects; and (3) to allocate allowances based on an updating measurement of affected-EGU generation. The EPA is also proposing that a jurisdiction may choose to replace the federal-plan allocation provisions with its own allowance allocation provisions. The mass-based approach is explained in greater detail in section V of this preamble.

3. Other Proposed Actions

The EPA is proposing in this action a finding that it is necessary or appropriate to regulate affected EGUs in certain parts of Indian country via a federal plan. This is discussed in section VI.D of this preamble.

In this action, the EPA is also proposing a number of changes to the framework CAA section 111(d) regulations of 40 CFR part 60, subpart B. These changes generally are intended to provide enhancements to the process for state plan submissions and the timing of EPA actions related to state plans and the federal plan. Specifically, the EPA proposes six changes, to include: (1) Partial approval/disapproval mechanisms similar to CAA section 110(k)(3); (2) a conditional approval mechanism similar to CAA section 110(k)(4); (3) a mechanism for the EPA to make calls for plan revisions similar to the "SIP-call" provisions of CAA section 110(k)(5); (4) an error correction mechanism similar to CAA section 110(k)(6); (5) completeness criteria and a process for determining completeness of state plans and submittals similar to CAA section 110(k)(1) and (2); and (6) updates to the deadlines for EPA action. These proposed changes are explained in greater detail in section VII of this preamble. They are not a component of the proposed federal plan, or changes in the EGs. If these changes are finalized, they will be applicable to other CAA section 111(d) rules. The EPA intends to finalize these changes earlier than the finalization of the model trading rules.

Analysis

The Clean Power Plan was a controversial effort that was envisioned by President Barack Obama as a key part of his presidential legacy. Crafted by the Environmental Protection Agency (EPA), the plan called for a reduction of carbon pollution from power plants of 870 million tons, which would represent a cut of 32% by 2030 of the amount of carbon emitted in 2005. The size of the cut was politically divisive as it represented a 2% increase over the goal that appeared in previous drafts of the proposal. Politically within the United States Congress, the increase mattered naught as Republicans and a number of Democrats from states like Kentucky and West Virginia were already on record as opposing even the early versions of the Clean Power Plan. In all likelihood, the increase to an already ambitious goal was meant to signal to the world community that the United States was once again committed to assuming its leadership of global climate change initiatives. This was a role that was significantly relinquished by the United States' following its failure to ratify the Kyoto Protocol.

Domestically, it was inevitable that the Clean Power Plan's mandates would be challenged in the nation's courts. In a handful of states, where the coal industry forms a major part of the local economy, the plan was viewed as a job killer. Lawsuits were already promised from Kentucky Senator and U.S. Senate Majority Mitch McConnell. On another front, the plan marked a serious challenge to the authority of the respective states. Dissatisfied with the efforts of the states to address climate change issues, President Obama, through the EPA, was requiring the states to create plans to cut carbon emissions from power plants. If the EPA was dissatisfied with the respective plans proffered, the EPA could substitute its own "national" plan. The threat to substitute a national plan for an inadequate state plan may have been intended to motivate the states to craft serious proposals, but could have easily been interpreted as a federal effort to seize powers reserved to the states by the United States Constitution.

A coalition of twenty seven states opted to challenge the legislation through the federal courts. On February 9, 2015, they won a key victory as the Supreme Court decided to block implementation of the Clean Power Plan until all of the legal challenges against it were litigated. The 5–4 decision by the justices meant that a majority of them believed that the states had a reasonable chance to win their case against the Obama administration. The partisan nature of the decision was evident as the justices viewed as politically liberal publicly acknowledged that they were against stopping implementation of the EPA's rules.

Further Reading

Capito, Shelley Moore, et al. 2016. "The Pros and Cons of the Administration's Clean Power Plan." *Congressional Digest* 95: 10–31.

Remarks by the President at the GLACIER Conference—Anchorage, AK

Date: August 31, 2015
Location: Anchorage, Alaska
Significance: This speech encapsulated President Barack Obama's view on climate change. It marked a significant effort in ensuring that his efforts to protect the environment would form a major part of his presidential legacy.
Source: White House, Office of the Press Secretary. 2015. *Remarks by the President at the GLACIER Conference—Anchorage, AK.* https://www.whitehouse.gov/the-press-office/2015/09/01/remarks-president-glacier-conference-anchorage-ak (accessed September 1, 2015).

THE PRESIDENT: Thank you so much. Thank you. (Applause.) It is wonderful to be here in the great state of Alaska. (Applause.)

I want to thank Secretary Kerry and members of my administration for your work here today. Thank you to the many Alaskans, Alaska Natives and other indigenous peoples of the Arctic who've traveled a long way, in many cases, to share your insights and your experiences. And to all the foreign ministers and delegations who've come here from around the world—welcome to the United States, and thank you all for attending this GLACIER Conference.

The actual name of the conference is much longer. It's a mouthful, but the acronym works because it underscores the incredible changes that are taking place here in the Arctic that impact not just the nations that surround the Arctic, but have an impact for the entire world, as well.

I want to thank the people of Alaska for hosting this conference. I look forward to visiting more of Alaska over the next couple of days. The United States is, of course, an Arctic nation. And even if this isn't an official gathering of the Arctic Council, the United States is proud to chair the Arctic Council for the next two years. And to all the foreign dignitaries who are here, I want to be very clear—we are eager to work with your nations on the unique opportunities that the Arctic presents and the unique challenges that it faces. We are not going to—any of us—be able to solve these challenges by ourselves. We can only solve them together.

Of course, we're here today to discuss a challenge that will define the contours of this century more dramatically than any other—and that's the urgent and growing threat of a changing climate.

Our understanding of climate change advances each day. Human activity is disrupting the climate, in many ways faster than we previously thought. The science is stark. It is sharpening. It proves that this once-distant threat is now very much in the present.

In fact, the Arctic is the leading edge of climate change—our leading indicator of what the entire planet faces. Arctic temperatures are rising about twice as fast as the global average. Over the past 60 years, Alaska has warmed about twice as fast as the rest of the United States. Last year was Alaska's warmest year on record—just as it was for the rest of the world. And the impacts here are very real.

Thawing permafrost destabilizes the earth on which 100,000 Alaskans live, threatening homes, damaging transportation and energy infrastructure, which could cost billions of dollars to fix.

Warmer, more acidic oceans and rivers, and the migration of entire species, threatens the livelihoods of indigenous peoples, and local economies dependent on fishing and tourism. Reduced sea levels leaves villages unprotected from floods and storm surges. Some are in imminent danger; some will have to relocate entirely. In fact, Alaska has some of the swiftest shoreline erosion rates in the world.

I recall what one Alaska Native told me at the White House a few years ago. He said, "Many of our villages are ready to slide off into the waters of Alaska, and in some cases, there will be absolutely no hope—we will need to move many villages."

Alaska's fire season is now more than a month longer than it was in 1950. At one point this summer, more than 300 wildfires were burning at once. Southeast of here, in our Pacific Northwest, even the rainforest is on fire. More than 5 million acres in Alaska have already been scorched by fire this year—that's an area about the size of Massachusetts. If you add the fires across Canada and Siberia, we're talking 300 [30] million acres—an area about the size of New York.

This is a threat to many communities—but it's also an immediate and ongoing threat to the men and women who put their lives on the line to protect ours. Less than two weeks ago, three highly trained firefighters lost their lives fighting a fire in Washington State. Another has been in critical condition. We are thankful to each and every firefighter for their heroism—including the Canadian firefighters who've helped fight the fires in this state.

But the point is that climate change is no longer some far-off problem. It is happening here. It is happening now. Climate change is already disrupting our agriculture and ecosystems, our water and food supplies, our energy, our infrastructure, human health, human safety—now. Today. And climate change is a trend that affects all trends—economic trends, security trends. Everything will be impacted. And it becomes more dramatic with each passing year.

Already it's changing the way Alaskans live. And considering the Arctic's unique role in influencing the global climate, it will accelerate changes to the way that we all live.

Since 1979, the summer sea ice in the Arctic has decreased by more than 40 percent—a decrease that has dramatically accelerated over the past two decades. One new study estimates that Alaska's glaciers alone lose about 75 gigatons—that's 75 billion tons—of ice each year.

To put that in perspective, one scientist described a gigaton of ice as a block the size of the National Mall in Washington—from Congress all the way to the Lincoln Memorial, four times as tall as the Washington Monument. Now imagine 75 of those ice blocks. That's what Alaska's glaciers alone lose each year. The pace of melting is only getting faster. It's now twice what it was between 1950 and 2000—twice as fast as it was just a little over a decade ago. And it's one of the reasons why sea levels rose by about eight inches over the last century, and why they're projected to rise another one to four feet this century.

Consider, as well, that many of the fires burning today are actually burning through the permafrost in the Arctic. So this permafrost stores massive amounts of carbon. When the permafrost is no longer permanent, when it thaws or burns, these gases are released into our atmosphere over time, and that could mean that the Arctic may become a new source of emissions that further accelerates global warming.

So if we do nothing, temperatures in Alaska are projected to rise between six and 12 degrees by the end of the century, triggering more melting, more fires, more thawing of the permafrost, a negative feedback loop, a cycle—warming leading to more warming—that we do not want to be a part of.

And the fact is that climate is changing faster than our efforts to address it. That, ladies and gentlemen, must change. We're not acting fast enough.

I've come here today, as the leader of the world's largest economy and its second largest emitter, to say that the United States recognizes our role in creating this problem, and we embrace our responsibility to help solve it. And I believe we can solve it. That's the good news. Even if we cannot reverse the damage that we've already caused, we have the means—the scientific imagination and technological innovation—to avoid irreparable harm.

We know this because last year, for the first time in our history, the global economy grew and global carbon emissions stayed flat. So we're making progress; we're just not making it fast enough.

Here in the United States, we're trying to do our part. Since I took office six and a half years ago, the United States has made ambitious investments in clean energy, and ambitious reductions in our carbon emissions. We now harness three times as much electricity from wind and 20 times as much from the sun. Alaskans now lead the world in the development of hybrid wind energy systems from remote grids, and it's expanding its solar and biomass resources.

We've invested in energy efficiency in every imaginable way—in our buildings, our cars, our trucks, our homes, even the appliances inside them. We're saving consumers billions of dollars along the way. Here in Alaska, more than 15,000 homeowners have cut their energy bills by 30 percent on average. That collectively saves Alaskans more than $50 million each year. We've helped communities build climate-resilient infrastructure to prepare for the impacts of climate change that we can no longer prevent.

Earlier this month, I announced the first set of nationwide standards to end the limitless dumping of carbon pollution from our power plants. It's the single most important step America has ever taken on climate change. And over the course of the coming days, I intend to speak more about the particular challenges facing Alaska and the United States as an Arctic power, and I intend to announce new measures to address them.

So we are working hard to do our part to meet this challenge. And in doing so, we're proving that there doesn't have to be a conflict between a sound environment and strong economic growth. But we're not moving fast enough. None of the nations represented here are moving fast enough.

And let's be honest—there's always been an argument against taking action. The notion is somehow this will curb our economic growth. And at a time when people are anxious about the economy, that's an argument oftentimes for inaction. We don't want our lifestyles disrupted. In countries where there remains significant poverty, including here in the United States, the notion is, can we really afford to prioritize this issue. The irony, of course, is, is that few things will disrupt our lives as profoundly as climate change. Few things can have as negative an impact on our economy as climate change.

On the other hand, technology has now advanced to the point where any economic disruption from transitioning to a cleaner, more efficient economy is shrinking by the day. Clean energy and energy efficiency aren't just proving cost-effective, but also cost-saving. The unit costs of things like solar are coming down rapidly. But we're still under-investing in it.

Many of America's biggest businesses recognize the opportunities and are seizing them. They're choosing a new route. And a growing number of American homeowners are choosing to go solar every day. It works. All told, America's economy has grown more than 60 percent over the last 20 years, but our carbon emissions are roughly back to where they were 20 years ago. So we know how to use less dirty fuel and grow our economy at the same time. But we're not moving fast enough.

More Americans every day are doing their part, though. Thanks to their efforts, America will reach the emission target that I set six years ago. We're going to reduce our carbon emissions in the range of 17 percent below 2005 levels by 2020. And that's why, last year, I set a new target: America is going to reduce our emissions 26 to 28 percent below 2005 levels by 10 years from now.

And that was part of a historic joint announcement we made last year in Beijing. The United States will double the pace at which we cut our emissions, and China committed, for the first time, to limiting its emissions. Because the world's two largest economies and two largest emitters came together, we're now seeing other nations stepping up aggressively as well. And I'm determined to make sure American leadership continues to drive international action—because we can't do this alone. Even America and China together cannot do this alone. Even all the countries represented around here cannot do this alone. We have to do it together.

This year, in Paris, has to be the year that the world finally reaches an agreement to protect the one planet that we've got while we still can.

So let me sum up. We know that human activity is changing the climate. That is beyond dispute. Everything else is politics if people are denying the facts of climate change. We can have a legitimate debate about how we are going to address this problem; we cannot deny the science. We also know the devastating consequences if the current trend lines continue. That is not deniable. And we are going to have to do some adaptation, and we are going to have to help communities be resilient, because of these trend lines we are not going to be able to stop on a dime. We're not going to be able to stop tomorrow.

But if those trend lines continue the way they are, there's not going to be a nation on this Earth that's not impacted negatively. People will suffer. Economies will suffer. Entire nations will find themselves under severe, severe problems. More drought; more floods; rising sea levels; greater migration; more refugees; more scarcity; more conflict.

That's one path we can take. The other path is to embrace the human ingenuity that can do something about it. This is within our power. This is a solvable problem if we start now.

And we're starting to see that enough consensus is being built internationally and within each of our own body politics that we may have the political will—finally—to get moving.

So the time to heed the critics and the cynics and the deniers is past. The time to plead ignorance is surely past. Those who want to ignore the science, they are increasingly alone. They're on their own shrinking island. (Applause.)

And let's remember, even beyond the climate benefits of pursuing cleaner energy sources and more resilient, energy-efficient ways of living, the byproduct of it is, is that we also make our air cleaner and safer for our children to breathe. We're also making our economies more resilient to energy shocks on global markets. We're also making our countries less reliant on unstable parts of the world. We are gradually powering a planet on its way to 9 billion humans in a more sustainable way.

These are good things. This is not simply a danger to be avoided; this is an opportunity to be seized. But we have to keep going. We're making a difference, but we have to keep going. We are not moving fast enough.

If we were to abandon our course of action, if we stop trying to build a clean-energy economy and reduce carbon pollution, if we do nothing to keep the glaciers from melting faster, and oceans from rising faster, and forests from burning faster, and storms from growing stronger, we will condemn our children to a planet beyond their capacity to repair: Submerged countries. Abandoned cities. Fields no longer growing. Indigenous peoples who can't carry out traditions that stretch back millennia. Entire industries of people who can't practice their livelihoods. Desperate refugees seeking the sanctuary of nations not their own. Political disruptions that could trigger multiple conflicts around the globe.

That's not a future of strong economic growth. That is not a future where freedom and human rights are on the move. Any leader willing to take a gamble on a future like that—any so-called leader who does not take this issue seriously or treats it like a joke— is not fit to lead.

On this issue, of all issues, there is such a thing as being too late. That moment is almost upon us. That's why we're here today. That's what we have to convey to our people—tomorrow, and the next day, and the day after that. And that's what we have to do when we meet in Paris later this year. It will not be easy. There are hard questions to answer. I am not trying to suggest that there are not going to be difficult transitions that we all have to make. But if we unite our highest aspirations, if we make our best efforts to protect this planet for future generations, we can solve this problem.

And when you leave this conference center, I hope you look around. I hope you have the chance to visit a glacier. Or just look out your airplane window as you depart, and take in the God-given majesty of this place. For those of you flying to other parts of the world, do it again when you're flying over your home countries. Remind yourself that there will come a time when your grandkids—and mine, if I'm lucky enough to have some—they'll want to see this. They'll want to experience it, just as we've gotten to do in our own lives. They deserve to live lives free from fear, and want, and peril. And ask yourself, are you doing everything you can to protect it. Are we doing everything we can to make their lives safer, and more secure, and more prosperous?

Let's prove that we care about them and their long-term futures, not just short-term political expediency.

I had a chance to meet with some Native peoples before I came in here, and they described for me villages that are slipping into the sea, and the changes that are taking place— changing migratory patterns; the changing fauna so that what used to feed the animals that they, in turn, would hunt or fish beginning to vanish. It's urgent for them today. But that is the future for all of us if we don't take care.

Your presence here today indicates your recognition of that. But it's not enough just to have conferences. It's not enough just to talk the talk. We've got to walk the walk. We've got work to do, and we've got to do it together.

So, thank you. And may God bless all of you, and your countries. And thank you, Alaska, for your wonderful hospitality. Thank you. (Applause.)

Analysis

To highlight his administration's campaign to mitigate the consequences of climate change, President Barack Obama visited Alaska from August 31 to September 2, 2015. Alaska was chosen for the effort because melting glaciers were already negatively impacting the lives of the state's citizens, especially its native populations. On the first day of his visit, Obama changed the name of Mount McKinley to Denali. This enraged Ohioans, as the former name honored a native of their state and the 25th President of the United States, William McKinney.

The first day of the trip was capped by the speech delivered by President Obama at the Conference on Global Leadership in the Arctic: Cooperation, Innovation, Engagement and Resilience (GLACIER). In it, he made a plea for the global community to quit denying the existence of climate change because it was occurring and accelerating. A failure to act had the potential to result in the world community losing its window of opportunity to mitigate the consequences of global warming. It is important to note that President Obama's speech was directed towards two different groups. Domestically, he was targeting United States politicians, especially in Congress, who opposed most of the climate legislation that had been proposed throughout his presidency. The second audience was the delegates around the world who at the time were working on the climate change agreements due to be proposed in Paris, France at the United Nations Climate Change Conference, held November 30 to December 11, 2015.

Much of the remainder of the trip was touring the state and meeting the locals. Among the staged events held during the president's visit was a hike on a glacier to bring attention to the rise in sea levels resulting from glacial melt.

Further Reading
Goodell, Jeff. 2015. "Obama Takes on Climate Change." *Rolling Stone* 1245: 36–45.
Obama, Barack. 2016. "We Met the Moment." *Vital Speeches of the Day* 82: 38–39.

Joint Meeting to Hear an Address by Pope Francis of the Holy See

Date: September 24, 2015
Location: Washington, D.C.
Significance: Pope Francis urged the members of the United States Congress to set aside their partisan differences and work together to help lead the global effort to address both the causes and consequences of climate change.
Sources: United States House of Representatives. 2015. *Joint Meeting to Hear an Address by Pope Francis of the Holy See (House of Representatives–September 24, 2015)*. https://www.congress.gov/congressional-record/2015/09/24/house-section/article/H6191-8?loclr=twtho (accessed September 25, 2015).

The SPEAKER. Members of Congress, I have the high privilege and the distinct honor of presenting to you Pope Francis of the Holy See. (Applause, the Members rising.)

POPE FRANCIS. Mr. Vice President, Mr. Speaker, Honorable Members of Congress, dear friends, I am most grateful for your invitation to address this joint session of Congress in "the land of the free and the home of the brave." I would like to think that the reason for this is that I, too, am a son of this great continent from which we have all received so much and toward which we share a common responsibility.

Each son or daughter of a given country has a mission, a personal and social responsibility. Your own responsibility as Members of Congress is to enable this country, by your legislative activity, to grow as a nation. You are the face of its people, their representatives. You are called to defend and preserve the dignity of your fellow citizens in the tireless and demanding pursuit of the common good, for this is the chief aim of all politics.

A political society endures when it seeks, as a vocation, to satisfy common needs by stimulating the growth of all its members, especially those in situations of greater vulnerability or risk. Legislative activity is always based on care for the people. To this you have been invited, called, and convened by those who elected you.

Yours is a work which makes me reflect in two ways on the figure of Moses. On the one hand, the patriarch and lawgiver of the people of Israel symbolizes the need of peoples to keep alive their sense of unity by means of just legislation. On the other, the figure of Moses leads us directly to God and thus to the transcendent dignity of the human being. Moses provides us with a good synthesis of your work: you are asked to protect, by means of the law, the image and likeness fashioned by God on every human life.

Today I would like not only to address you, but, through you, the entire people of the United States. Here, together with their representatives, I would like to take this opportunity to dialogue with the many thousands of men and women who strive each day to do an honest day's work, to bring home their daily bread, to save money, and—one step at a time—to build a better life for their families.

These are men and women who are not concerned simply with paying their taxes but, in their own quiet way, sustain the life of society. They generate solidarity by their actions, and they create organizations which offer a helping hand to those most in need.

I would also like to enter into a dialogue with the many elderly persons who are a storehouse of wisdom forged by experience and who seek in many ways, especially through volunteer work, to share their stories and their insights. I know that many of them are retired but still active; they keep working to build up this land.

I also want to dialogue with all those young people who are working to realize their great and noble aspirations, who are not led astray by facile proposals, and who face difficult situations, often as a result of immaturity on the part of many adults. I wish to dialogue with all of you, and I would like to do so through the historical memory of your people.

My visit takes place at a time when men and women of goodwill are marking the anniversaries of several great Americans. The complexities of history and the reality of human weakness notwithstanding, these men and women, for all their many differences and limitations, were able by hard work and self-sacrifice—some at the cost of their lives—to build a better future. They shaped fundamental values which will endure forever in the spirit of the American people.

...

This common good also includes the Earth, a central theme of the encyclical which I recently wrote in order to "enter into dialogue with all people about our common home." "We need a conversation which includes everyone, since the environmental challenge we are undergoing, and its human roots, concern and affect us all."

In Laudato Si', I call for a courageous and responsible effort to "redirect our steps" and to avert the most serious effects of the environmental deterioration caused by human activity. I am convinced that we can make a difference. I am sure and I have no doubt that the United States and this Congress have an important role to play.

Now is the time for courageous actions and strategies aimed at implementing a "culture of care" and "an integrated approach to combating poverty, restoring dignity to the excluded, and at the same time protecting nature." "We have the freedom needed to limit and direct technology, to devise intelligent ways of ... developing and limiting our power," and to put technology "at the service of another type of progress, one which is healthier, more human, more social, more integral." In this regard, I am confident that America's outstanding academic and research institutions can make a vital contribution in the years ahead.

...

Analysis

In April 2015, Pope Francis hosted scientists and religious leaders at the Vatican to discuss global climate change. The conference publicly allowed the pontiff to weigh whether climate change was due to natural causes or was man-made. On June 18, 2015, Pope Francis accepted that the scientific evidence he had been provided proved that man's activities had caused global climate change through his encyclical "*Laudato Si*," which was published in the United States as *Encyclical on Climate Change and Inequality:*

On Care for Our Common Home. One of the speakers at the press conference announcing the release of the encyclical was Naomi Klein, a well-known environmental activist. Her endorsement of the text was a signal that Pope Francis' views on climate change would not be embraced by climate change deniers or those that believed that the alterations in global climate patterns were due to natural causes.

Upon arriving in the United States for a visit on September 23, 2015, Pope Francis quickly joined the political discussions involving climate change in the United States by endorsing President Barack Obama's *Federal Plan Requirements for Greenhouse Gas Emissions from Electric Utility Generating Units Constructed on or Before January 8, 2014; Model Trading Rules; Amendments to Framework Regulations*, better known as the "Clean Power Plan." Pope Francis' willingness to join the domestic political fray on the climate led Representative Paul Gosar of Arizona, a Republican and a Catholic, to boycott the Pope's speech to a joint session of Congress the next day.

In both his joint appearance with President Obama and his speech to Congress, Pope Francis made the case that political leaders needed to abandon any belief structure that viewed climate change as a partisan political issue because they had a moral responsibility for protecting the Earth for future generations. This was especially true for the United States' leadership, as their nation had much more capacity and ability than most other countries to address the threat posed globally by climate change.

Further Reading

Klein, Naomi. 2014. *This Changes Everything: Capitalism vs. the Climate*. New York: Simon & Schuster.
Pope Francis. 2015. *Encyclical on Climate Change and Inequality: On Care for Our Common Home*. New York: Melville House.

Fact Sheet

The United States and China Issue Joint Presidential Statement on Climate Change with New Domestic Policy Commitments and a Common Vision for an Ambitious Global Climate Agreement in Paris

Date: September 25, 2015
Location: Washington, D.C.
Significance: This statement, produced by the leaders of the two highest greenhouse gas emitting nations in the world, was intended to signal to the negotiators meeting in Paris, France in December 2015 the terms desired by both China and the United States in the final agreement.
Source: The White House, Office of the Press Secretary. 2015. *Fact Sheet: The United States and China Issue Joint Presidential Statement on Climate Change with New Domestic Policy Commitments and a Common Vision for an Ambitious Global Climate Agreement in Paris.* https://www.whitehouse.gov/the-press-office/2015/09/25/fact-sheet-united-states-and-china-issue-joint-presidential-statement (accessed September 25, 2015).

On the occasion of President Xi's State Visit to Washington, D.C., the United States and China today marked another major milestone in their joint leadership in the fight against climate change with the release of a U.S.–China Joint Presidential Statement on Climate Change. The Statement, which builds on last November's historic announcement by President Obama and President Xi of ambitious, respective post–2020 climate targets, describes a common vision for a new global climate agreement to be concluded in Paris this December. The Statement also includes significant domestic policy announcements and commitments to global climate finance, demonstrating the determination of both countries to act decisively to achieve the goals set last year.

- **Common vision for the Paris climate agreement**—As part of their commitment to a successful and ambitious Paris outcome, the two countries articulated a set of shared understandings for the agreement, including on the importance of a successful agreement that ramps-up ambition over time, pointing toward a low-carbon transformation of the global economy this century. They agreed on the need for an enhanced transparency system to build mutual trust and confidence and promote effective implementation including through reporting and review of action and support in an appropriate manner, and made new progress on the issue of differentiation between developed and developing countries.
- **Ambitious domestic policy announcements**—China confirmed today that it plans

to launch in 2017 a national emission trading system covering power generation, steel, cement, and other key industrial sectors, as well as implement a "green dispatch" system to favor low-carbon sources in the electric grid. These announcements complement the recent finalization of the U.S. Clean Power Plan, which will reduce emissions in the U.S. power sector by 32% by 2030. Both countries are developing new heavy-duty vehicle fuel efficiency standards, to be finalized in 2016 and implemented in 2019. Both countries are also stepping up their work to phase down super-polluting hydrofluorocarbons (HFCs).

- **Breaking new ground on climate finance**—Looking beyond their shores, the two countries announced further steps to help accelerate the transition to low-carbon development internationally, including a new climate finance commitment by China of CNY 20 billion ($3.1 billion) to help developing countries combat climate change and new steps to control public support for high carbon activities. The two countries also re-affirmed their commitment to bilateral cooperation, both at the federal and sub-national levels.

Building a Common Vision for the Paris Agreement

Presidents Obama and Xi are committed to an ambitious outcome at the Paris climate conference and have articulated a concrete set of shared understandings for the Paris agreement. On mitigating the impact of climate change, the two leaders agreed on three elements of a package to strengthen the ambition of the Paris outcome. First, they recognized that the emissions targets and policies that nations have put forward are crucial steps in a longer-range effort to transition to low-carbon economies and agreed that those policies should ramp up over time in the direction of greater ambition. Second, they underscored the importance of countries developing and making available mid-century strategies for the transition to low-carbon economies, mindful of the below 2 degrees Celsius global temperature goal. Third, they emphasized the need for the low-carbon transformation of the global economy this century.

The leaders agreed on the importance of an enhanced transparency system to build mutual trust and confidence and promote effective implementation including through reporting and review of action and support in an appropriate manner, and agreed that such a system should provide flexibility to those developing countries that need it in light of their capacities.

The leaders also made new progress on the issue of differentiation, including by reaffirming their commitment to an ambitious agreement in 2015 that reflects the principle of common but differentiated responsibilities and respective capabilities, in light of different national circumstances, and embeds differentiation in the relevant elements of the agreement in a manner appropriate to each individual element. They also agreed that adaptation needs to be elevated in the international talks, and that it is a key component of the long-term global response to climate change.

On financial assistance for developing countries, in addition to specific new climate finance announcements, the two sides reiterated the 2020 climate finance mobilization goal that developed countries committed to in 2009 and they underscored the importance of continued financial support beyond 2020 to help developing countries build low-carbon and climate-resilient societies, urging continued support from developed countries and encouraging such support by other countries willing to do so.

Finally, the two sides recognized the crucial role of major technological advancement in the transition to low-carbon economies, and endorsed significant increases in basic research and development into clean energy technologies in the coming years.

This set of agreements on key issues in the Paris climate negotiations reflects the commitment of the U.S. and China to work together and with all other countries to reach an historic climate agreement in Paris.

Ambitious Domestic Climate Policies

Today, President Obama and President Xi announced substantial, domestic climate actions that are designed towards implementing and achieving their respective climate goals announced in November 2014.

Power Sector and Industry.
- **The U.S. is Taking Steps to Implement the Clean Power Plan in 2016.** The recently finalized Clean Power Plan will reduce carbon pollution from the power sector by 32 percent below 2005 levels by 2030. Today, the United States committed to finalize, in 2016, a federal plan to implement carbon emission standards for power plants in states that choose not to design their own implementation plans.
- **China is launching a national emissions trading system in 2017.** China's national cap-and-trade system will support emissions reductions in power generation, iron and steel, chemicals, building materials including cement, paper-making, and non-ferrous metals. These sectors together produce a substantial percentage of China's carbon pollution.
- **China is implementing a new green dispatch approach in its power sector.** China's "green dispatch" system will prioritize power generation from renewable sources, and establish guidelines to accept electricity first from the most efficient and lowest-polluting fossil fuel generators. This approach will accelerate the phase down of high-polluting, energy intensive power while supporting the deployment of renewable and non-fossil sources, and will better utilize China's rapidly growing solar and wind capacity while supporting its ambitious non-fossil energy targets of 15 percent by 2020 and around 20 percent by 2030.
- **Transportation.** Today, the U.S. and China are committing to finalize respective next-stage fuel efficiency standards for heavy-duty vehicles in 2016, and both countries are committing to implement them in 2019. This announcement marks a commitment from China to match the proposed timeline for the introduction of new heavy-duty vehicle standards in the United States.
- **Buildings and Cities.** Today, the United States is committing to finalize over 20 efficiency standards for appliances and equipment by the end of 2016—a commitment that will enable us to meet our goal of cutting 3 billion metric tons of carbon pollution from these measures by 2030. Consistent with the ambitious early peaking targets announced by Chinese cities last week at the U.S.-China Climate Leaders Summit in Los Angeles, China is also affirming that 50 percent of new buildings in urban areas will meet green building standards by 2020. Additionally, China has

affirmed that the share of public transport in motorized urban travel will reach 30 percent by the same year.
- **Methane and HFCs.** Building on the U.S. Strategy to Reduce Methane Emissions, in January 2015, the Administration announced a goal to cut methane emissions from the oil and gas sector by 40 to 45 percent from 2012 levels by 2025 and has committed to finalize recently proposed standards for methane emissions from the oil and gas sector in 2016. Today, the U.S. is committing to finalize two standards to reduce methane emissions from landfills in 2016. The U.S. EPA also recently finalized a rule to prohibit some of the most harmful HFCs in various end-uses under our Significant New Alternatives Policy, and the United States is committing to pursue new actions in 2016 to reduce HFC use and emissions, including announcing progress against private sector commitments to reduce HFCs equivalent to 700 million metric tons of carbon pollution, and a new round of additional private sector commitments to reduce emissions of HFCs. China is also planning to accelerate its efforts to control HFCs, including effectively controlling HFC-23 emissions by 2020.

Breaking New Ground on Climate Finance

President Obama and President Xi emphasized the importance of mobilizing climate finance to support low-carbon, climate-resilient development in developing countries. Consistent with the United States' $3 billion pledge to the Green Climate Fund (GCF)—the multilateral funding mechanism established to support climate action in developing countries—China announced that it will make available CNY 20 billion ($3.1 billion) through a bilateral fund designed to help developing countries combat climate change. This is by far China's most significant commitment to climate finance to date, and reinforces that the two countries are working together to ensure that developing countries have the tools they need to develop sustainably and prepare for the impacts of climate change.

The United States and China reached an important new understanding on the need to control financing for high-carbon projects internationally. Today, China—one of the largest providers of public financing for infrastructure worldwide—agreed to work towards strictly controlling public investment flowing into projects with high pollution and carbon emissions both domestically and internationally. This follows a commitment in 2013 by the United States to end public financing for new conventional coal-fired power plants except in the poorest countries, and a growing number of other countries and financing institutions moving in a similar direction.

Deepening Bilateral Cooperation on Climate Change

Over the last several years, the United States and China have deepened our bilateral engagement to tackle specific challenges each country faces in combating climate change. We have launched a number of sectoral initiatives, ramped up joint research and development, and built strong connections between regulators, scientists, engineers, and businesses working on shared challenges.

- **U.S.–China Climate Change Working Group.** Today, the two countries committed to further enhance and deepen our bilateral cooperation, including through the U.S.–China Climate Change Working Group. This cooperation includes technical and policy exchanges on the development of fuel efficiency standards for trucks and buses, continued progress towards implementation of CCUS projects with the selection of a project site in China as the large-scale CCUS demonstration project announced last November, cooperation on reducing HFCs, including private sector commitments to promote climate-friendly alternatives to HFCs, and continued support for phasing down HFCs through the Montreal Protocol, and additional cooperation on direct mitigation efforts across multiple other sectors.
- **Action by Sub-National Governments.** The United States and China attach special importance to the burgeoning cooperation between our sub-national governments to promote climate action. The Presidents welcomed the success of the **U.S.-China Climate Leaders Summit**, held September 15–16, 2015, in Los Angeles, California. The Summit featured the announcement of the new "Alliance of Peaking Pioneer Cities" in China—cities and provinces accounting for approximately 1.2 gigatons of annual CO_2 emissions (roughly equivalent to all emissions from Japan or Brazil) established, for the first time, peak years for carbon dioxide emissions that are earlier than the national goal to peak around 2030. U.S. cities, counties, and states also put forward ambitious, long-term emissions reduction targets, including a commitment by the State of California to reduce emissions by 80 percent-90 percent below 1990 levels by 2050, and a commitment from the City of Seattle to become carbon-neutral by 2050.
- **Clean Energy Research.** The United States and China also place great importance on exchanges and cooperation in the area of clean energy and recognize the collaboration and outcomes of the U.S.-China Clean Energy Research Center (CERC).
- **Transportation–**The two sides announced the launch of a new technical track under CERC to improve the energy efficiency of medium-duty to heavy-duty trucks. This initiative is expected to accelerate the development of high energy efficiency trucks and their introduction into the markets of both countries, leading to significant reductions in oil consumption and greenhouse gas emissions from the transportation sector. The United States and China additionally announced that they will develop collaborative Electric Vehicle (EV) Interoperability Centers with the goal of coordinating relative technical standards, promoting coordination, and providing technical support to the existing, successful electric vehicle work between our two countries.
- **Energy and water–**Under the new energy-water track of the CERC, the United States and China will work together to discover an array of innovative technologies to alleviate pressures on water resources and management related to energy production and use. The United States has also announced five projects to study the feasibility of using salty water—or brine—from CO_2 storage sites to produce fresh water.

Analysis

On November 12, 2014, Presidents Barack Obama and Xi Jinping announced the *U.S.-China Joint Announcement on Climate Change*, which set targets for both countries to reduce the amount of CO2 that was being emitted by the respective countries. During the press conference, it was revealed that the two countries were continuing to negotiate on the issue of climate change in hopes of drafting a more comprehensive agreement to be announced during President Xi's scheduled state visit to Washington, D.C., during September 2015.

This new agreement was not subtle about its primary purpose, which was to influence the negotiations being undertaken under the auspices of the United Nations on the subject of climate change in Paris, France from November 30 to December 11, 2015. Both China and the United States, as the two largest emitters of greenhouse gases in the world, wanted lower standards set for 2030 than were being proposed by the European Union (EU). The United States had offered a cut of 26 to 28 percent from 2005 levels by 2030, while the EU was advocating a 40 percent reduction. It was hoped by the leaders of China and the United States that their united front would outweigh the European Union during the negotiations United Nations Climate Summit.

One notable product of the statement was that it included the parameters for continued cooperation in researching how to address the changing climate and helping poorer countries to acquire technologies that would make it less necessary for them to burn fossil fuels. Unfortunately, the agreement was non-binding, which meant that neither country had any accountability for failing to even attempt, let alone achieve, the goals they stated in the text of the document.

Further Reading

Nadin, Rebecca, Sarah Opitz-Stapleton, and Xu Yinlong, eds. 2015. *Climate Change: Risk and Resilience in China*. New York: Routledge.
Rapkin, David P., and William R. Thompson. 2013. *Transition Scenarios: China and the United States in the Twenty-First Century*. Chicago: University of Chicago Press.
Sze, Julie. 2015. *Fantasy Islands: Chinese Dreams and Ecological Fears in an Age of Climate Crisis*. Berkeley: University of California Press.
Wang, Weguang, et al., eds. 2012. *China's Climate Change Policies*. New York: Routledge.

NASA Study
Mass Gains of Antarctic Ice Sheet Greater than Losses

Date: October 30, 2015
Location: Greenbelt, Maryland
Significance: Many of the studies produced by agencies of the United States government and the Intergovernmental Panel on Climate Change concluded that the amount of ice in Antarctica has been decreasing due to the effects of global warming. According to the authors of this study, Antarctica's ice is actually increasing in volume due to snowfall that began falling approximately 10,000 years ago.
Source: Garner, Rob, ed. 2015. *NASA Study: Mass Gains of Antarctic Ice Sheet Greater than Losses.* http://www.nasa.gov/feature/goddard/nasa-study-mass-gains-of-antarctic-ice-sheet-greater-than-losses (accessed November 4, 2015).

A new NASA study says that an increase in Antarctic snow accumulation that began 10,000 years ago is currently adding enough ice to the continent to outweigh the increased losses from its thinning glaciers.

The research challenges the conclusions of other studies, including the Intergovernmental Panel on Climate Change's (IPCC) 2013 report, which says that Antarctica is overall losing land ice.

According to the new analysis of satellite data, the Antarctic ice sheet showed a net gain of 112 billion tons of ice a year from 1992 to 2001. That net gain slowed to 82 billion tons of ice per year between 2003 and 2008.

"We're essentially in agreement with other studies that show an increase in ice discharge in the Antarctic Peninsula and the Thwaites and Pine Island region of West Antarctica," said Jay Zwally, a glaciologist with NASA Goddard Space Flight Center in Greenbelt, Maryland, and lead author of the study, which was published on Oct. 30 in the *Journal of Glaciology*. "Our main disagreement is for East Antarctica and the interior of West Antarctica—there, we see an ice gain that exceeds the losses in the other areas." Zwally added that his team "measured small height changes over large areas, as well as the large changes observed over smaller areas."

Scientists calculate how much the ice sheet is growing or shrinking from the changes in surface height that are measured by the satellite altimeters. In locations where the amount of new snowfall accumulating on an ice sheet is not equal to the ice flow downward and outward to the ocean, the surface height changes and the ice-sheet mass grows or shrinks.

But it might only take a few decades for Antarctica's growth to reverse, according to Zwally. "If the losses of the Antarctic Peninsula and parts of West Antarctica continue to increase at the same rate they've been increasing for the last two decades, the losses will catch up with the long-term gain in East Antarctica in 20 or 30 years—I don't think there will be enough snowfall increase to offset these losses."

The study analyzed changes in the surface height of the Antarctic ice sheet measured by radar altimeters on two European Space Agency European Remote Sensing (ERS) satellites, spanning from 1992 to 2001, and by the laser altimeter on NASA's Ice, Cloud, and land Elevation Satellite (ICESat) from 2003 to 2008.

Zwally said that while other scientists have assumed that the gains in elevation seen in East Antarctica are due to recent increases in snow accumulation, his team used meteorological data beginning in 1979 to show that the snowfall in East Antarctica actually decreased by 11 billion tons per year during both the ERS and ICESat periods. They also used information on snow accumulation for tens of thousands of years, derived by other scientists from ice cores, to conclude that East Antarctica has been thickening for a very long time.

"At the end of the last Ice Age, the air became warmer and carried more moisture across the continent, doubling the amount of snow dropped on the ice sheet," Zwally said.

The extra snowfall that began 10,000 years ago has been slowly accumulating on the ice sheet and compacting into solid ice over millennia, thickening the ice in East Antarctica and the interior of West Antarctica by an average of 0.7 inches (1.7 centimeters) per year. This small thickening, sustained over thousands of years and spread over the vast expanse of these sectors of Antarctica, corresponds to a very large gain of ice—enough to outweigh the losses from fast-flowing glaciers in other parts of the continent and reduce global sea level rise.

Zwally's team calculated that the mass gain from the thickening of East Antarctica remained steady from 1992 to 2008 at 200 billion tons per year, while the ice losses from the coastal regions of West Antarctica and the Antarctic Peninsula increased by 65 billion tons per year.

"The good news is that Antarctica is not currently contributing to sea level rise, but is taking 0.23 millimeters per year away," Zwally said. "But this is also bad news. If the 0.27 millimeters per year of sea level rise attributed to Antarctica in the IPCC report is not really coming from Antarctica, there must be some other contribution to sea level rise that is not accounted for."

"The new study highlights the difficulties of measuring the small changes in ice height happening in East Antarctica," said Ben Smith, a glaciologist with the University of Washington in Seattle who was not involved in Zwally's study.

"Doing altimetry accurately for very large areas is extraordinarily difficult, and there are measurements of snow accumulation that need to be done independently to understand what's happening in these places," Smith said.

To help accurately measure changes in Antarctica, NASA is developing the successor to the ICESat mission, ICESat-2, which is scheduled to launch in 2018. "ICESat-2 will measure changes in the ice sheet within the thickness of a No. 2 pencil," said Tom Neumann, a glaciologist at Goddard and deputy project scientist for ICESat-2. "It will contribute to solving the problem of Antarctica's mass balance by providing a long-term record of elevation changes."

Analysis

Scientists at NASA's Goddard Space Flight Center utilized satellites to measure Antarctica's ice sheet in order to gauge what was its mass and how it was being impacted by global climate change. After analyzing their data, they discovered that their findings contradicted much of the research conducted on the region over the course of the twenty-first century. Numerous studies had concluded that Antarctica has been losing ice due to global warming, thus contributing to the rise in global sea levels. The new study, published in the *Journal of Glaciology*, maintained that overall, Antarctica's ice sheet was actually growing. This was due to snow that began falling in the region roughly 10,000 years ago. Each year, some of that snow was converted to ice. Although, parts of Antarctica were, and are, melting at a significant pace, more ice was being created than was being lost. The authors of the study were quick to caution that although Antarctica was not at present contributing to the rise in global sea levels due to its continued growth, that trend could quickly reverse should global temperatures continue to rise.

Further Reading

Zwally, H. Jay, et al. 2015. "Mass Gains of the Antarctic Ice Sheet Exceed Losses." *Journal of Glaciology* 61: 1019–1036.

Sen. Cruz Confronts the Dogma of Climate Change Alarmism
"Public Policy Should Follow Actual Data, Not Political and Partisan Claims That Run Contrary to Evidence"
(Opening Statement)

Date: December 8, 2015
Location: Washington, D.C.
Significance: While negotiators from 196 countries were negotiating a major climate change agreement in Paris, France, Senator Ted Cruz convened a hearing before the Science Subcommittee of the Senate Commerce Committee to signal to both President Barack Obama's administration and the negotiators in Paris, France that many legislators and their constituents did not believe that the science undergirding many of the claims made by proponents of climate change were valid.
Source: Cruz, Ted. 2015. *Sen. Cruz Confronts the Dogma of Climate Change Alarmism: "Public Policy Should Follow Actual Data, Not Political and Partisan Claims That Run Contrary to Evidence."* http://www.cruz.senate.gov/?p=press_release&id=2548 (accessed December 10, 2015).

Good afternoon, everyone. Welcome to what I hope will be an important and informative hearing. This is a hearing on the science behind claims of global warming. Now, this is the Science Subcommittee of the Senate Commerce Committee, and we are hearing from distinguished scientists sharing their views, their interpretations, their analyses of the data and the evidence. Now, I am the son of two mathematicians, two computer programmers and scientists, and I believe that public policy should follow the actual science and the actual data and evidence, and not political and partisan claims that run contrary to the science and data and evidence.

On Nov 28, 2013, an intrepid band of explorers set off from New Zealand on a research expedition to the Antarctic. Among their goals was investigating the impact of global warming on the Antarctic continent and islands. On Christmas Eve, they became stuck in ice—ice that the climate-industrial complex had assured us was vanishing. This expedition was there to document how the ice was vanishing in the Antarctic, but the ship became stuck. It had run into an inconvenient truth, as Al Gore might put it. Three icebreakers tried and failed to reach the trapped ship because the ice was too thick. After a week of rescue attempts, the passengers were airlifted from the vessel.

Here are the inconvenient facts about the polar ice caps:
- The Artic is not ice-free. This year's minimum sea ice extent was well above the record low observed in 2011.
- In the Antarctic, a recent study from the Journal of Glaciology indicates that ice is

not only not decreasing, but is in fact increasing in mass, directly contrary to what the global warming alarmists had told us would be happening.
- This is not what the climate models projected.

Yet, these inconvenient facts never seem to get to the attention of people like John Kerry. And indeed I would note behind me—on August 31, 2009, then–Senator John Kerry said, 'Scientists project that the Arctic will be ice-free in the summer of 2013, not in 2050 but four years from now.' Well, the summer of 2013 has come and gone, and John Kerry was not just a little bit—he was wildly, extraordinarily, entirely wrong. Had the Antarctic expedition in the picture next to it not believed the global warming alarmists, had they actually looked to the science and the evidence, they wouldn't have gone down and been surprised when they got stuck in ice. Facts matter. Science matters. Data matter. That's what this hearing is about—data.

According to the satellite data, there has been no significant global warming for the past 18 years. Those are the data. The global warming alarmists don't like these data—they are inconvenient to their narrative. But facts and evidence matter. And I would note that many in the media reflexively take the side of the global warming alarmists, reflexively oppose anyone who actually says, "Well, was John Kerry's prediction accurate? No, it was stunningly and entirely false. Was the prediction of computer model after computer model that showed dramatic warming—were those predictions correct? No—the satellite data demonstrate no significant warming over 18 years."

Public policy should follow science and evidence and data. And I would note that I found it amusing that our friends on the Democratic side of the aisle, I discovered, held a press conference today as a prebuttal to this hearing. I suppose I should view that, in a sense, as a backhanded compliment. I'm reminded of the Bard, "Methinks she doth protest too much." What does it say when Members of the United States Senate are protesting, "How dare the Science Subcommittee in the United States Senate hear testimony from scientists about actual science? How dare we focus on such topics?' I think that is indeed exactly what we were elected to do."

Analysis

Senator Ted Cruz, a Republican from Texas and a candidate for his party's nomination for the presidency of the United States, convened a hearing before the Science Subcommittee of the Senate Commerce Committee to ostensibly examine whether man-made climate change was actually underway. Cruz, a long-time critic of climate change science, primarily invited known climate change deniers to testify during the hearings. Predictably, what emerged were criticisms of President Barack Obama's administration. Among the charges made were that they used false data from scientists who Cruz called "climate change alarmists" to justify actions that hurt the nation's economy and that it suppresses scientific data produced by government agencies that contradicts its ideology. Cruz specifically singled out U.S. Secretary of State John Kerry for criticism, which was not coincidental since Kerry at the time was one of the nation's leading negotiators in Paris, France. Many Democrats in Congress questioned the need for the hearing, claiming that most Americans believed that an exhaustive amount of scientific evidence had already confirmed that human activity is the leading cause of climate change.

Adoption of the Paris Agreement
Proposal by the President: Draft decision –/CP.21

Date: December 12, 2015
Location: Paris, France
Significance: The Paris Agreement is the most important climate change document presented to the international community for ratification since the Kyoto Protocol.
Source: United Nations Framework Convention on Climate Change. 2015. *Adoption of the Paris Agreement: Proposal by the President: Draft decision -/CP.21*. http://unfccc.int/resource/docs/2015/cop21/eng/l09.pdf (accessed December 12, 2015).

The *Conference of the Parties*,
Recalling decision 1/CP.17 on the establishment of the Ad Hoc Working Group on the Durban Platform for Enhanced Action,
Also recalling Articles 2, 3 and 4 of the Convention,
Further recalling relevant decisions of the Conference of the Parties, including decisions 1/CP.16, 2/CP.18, 1/CP.19 and 1/CP.20,
Welcoming the adoption of United Nations General Assembly resolution A/RES/70/1, "Transforming our world: the 2030 Agenda for Sustainable Development," in particular its goal 13, and the adoption of the Addis Ababa Action Agenda of the third International Conference on Financing for Development and the adoption of the Sendai Framework for Disaster Risk Reduction,
Recognizing that climate change represents an urgent and potentially irreversible threat to human societies and the planet and thus requires the widest possible cooperation by all countries, and their participation in an effective and appropriate international response, with a view to accelerating the reduction of global greenhouse gas emissions,
Also recognizing that deep reductions in global emissions will be required in order to achieve the ultimate objective of the Convention and emphasizing the need for urgency in addressing climate change,
Acknowledging that climate change is a common concern of humankind, Parties should, when taking action to address climate change, respect, promote and consider their respective obligations on human rights, the right to health, the rights of indigenous peoples, local communities, migrants, children, persons with disabilities and people in vulnerable situations and the right to development, as well as gender equality, empowerment of women and intergenerational equity,
Also acknowledging the specific needs and concerns of developing country Parties arising from the impact of the implementation of response measures and, in this regard, decisions 5/CP.7, 1/CP.10, 1/CP.16 and 8/CP.17,

Emphasizing with serious concern the urgent need to address the significant gap between the aggregate effect of Parties' mitigation pledges in terms of global annual emissions of greenhouse gases by 2020 and aggregate emission pathways consistent with holding the increase in the global average temperature to well below 2 °C above preindustrial levels and pursuing efforts to limit the temperature increase to 1.5 °C,

Also emphasizing that enhanced pre-2020 ambition can lay a solid foundation for enhanced post–2020 ambition,

Stressing the urgency of accelerating the implementation of the Convention and its Kyoto Protocol in order to enhance pre–2020 ambition,

Recognizing the urgent need to enhance the provision of finance, technology and capacity-building support by developed country Parties, in a predictable manner, to enable enhanced pre–2020 action by developing country Parties,

Emphasizing the enduring benefits of ambitious and early action, including major reductions in the cost of future mitigation and adaptation efforts,

Acknowledging the need to promote universal access to sustainable energy in developing countries, in particular in Africa, through the enhanced deployment of renewable energy,

Agreeing to uphold and promote regional and international cooperation in order to mobilize stronger and more ambitious climate action by all Parties and non–Party stakeholders, including civil society, the private sector, financial institutions, cities and other subnational authorities, local communities and indigenous peoples,

I. Adoption

1. *Decides* to adopt the Paris Agreement under the United Nations Framework Convention on Climate Change (hereinafter referred to as "the Agreement") as contained in the annex;
2. *Requests* the Secretary-General of the United Nations to be the Depositary of the Agreement and to have it open for signature in New York, United States of America, from 22 April 2016 to 21 April 2017;
3. *Invites* the Secretary-General to convene a high-level signature ceremony for the Agreement on 22 April 2016;
4. *Also invites* all Parties to the Convention to sign the Agreement at the ceremony to be convened by the Secretary-General, or at their earliest opportunity, and to deposit their respective instruments of ratification, acceptance, approval or accession, where appropriate, as soon as possible;
5. *Recognizes* that Parties to the Convention may provisionally apply all of the provisions of the Agreement pending its entry into force, and requests Parties to provide notification of any such provisional application to the Depositary;
6. *Notes* that the work of the Ad Hoc Working Group on the Durban Platform for Enhanced Action, in accordance with decision 1/CP.17, paragraph 4, has been completed;
7. *Decides* to establish the Ad Hoc Working Group on the Paris Agreement under the same arrangement, mutatis mutandis, as those concerning the election of officers to the Bureau of the Ad Hoc Working Group on the Durban Platform for Enhanced Action; 1

8. *Also decides* that the Ad Hoc Working Group on the Paris Agreement shall prepare for the entry into force of the Agreement and for the convening of the first session of the Conference of the Parties serving as the meeting of the Parties to the Paris Agreement;
9. *Further decides* to oversee the implementation of the work programme resulting from the relevant requests contained in this decision;
10. *Requests* the Ad Hoc Working Group on the Paris Agreement to report regularly to the Conference of the Parties on the progress of its work and to complete its work by the first session of the Conference of the Parties serving as the meeting of the Parties to the Paris Agreement;
11. *Decides* that the Ad Hoc Working Group on the Paris Agreement shall hold its sessions starting in 2016 in conjunction with the sessions of the Convention subsidiary bodies and shall prepare draft decisions to be recommended through the Conference of the Parties to the Conference of the Parties serving as the meeting of the Parties to the Paris Agreement for consideration and adoption at its first session;

II. Intended Nationally Determined Contributions

12. *Welcomes* the intended nationally determined contributions that have been communicated by Parties in accordance with decision 1/CP.19, paragraph 2(b);
13. *Reiterates* its invitation to all Parties that have not yet done so to communicate to the secretariat their intended nationally determined contributions towards achieving the objective of the Convention as set out in its Article 2 as soon as possible and well in advance of the twenty-second session of the Conference of the Parties (November 2016) and in a manner that facilitates the clarity, transparency and understanding of the intended nationally determined contributions;
14. *Requests* the secretariat to continue to publish the intended nationally determined contributions communicated by Parties on the UNFCCC website;
15. *Reiterates* its call to developed country Parties, the operating entities of the Financial Mechanism and any other organizations in a position to do so to provide support for the preparation and communication of the intended nationally determined contributions of Parties that may need such support;
16. *Takes note* of the synthesis report on the aggregate effect of intended nationally determined contributions communicated by Parties by 1 October 2015, contained in document FCCC/CP/2015/7;
17. *Notes* with concern that the estimated aggregate greenhouse gas emission levels in 2025 and 2030 resulting from the intended nationally determined contributions do not fall within least-cost 2°C scenarios but rather lead to a projected level of 55 gigatonnes in 2030, and also notes that much greater emission reduction efforts will be required than those associated with the intended nationally determined contributions in order to hold the increase in the global average temperature to below 2°C above pre-industrial levels by reducing emissions to 40 gigatonnes or to 1.5°C above pre-industrial levels by reducing to a level to be identified in the special report referred to in paragraph 21 below;
18. *Also notes, in this context,* the adaptation needs expressed by many developing country Parties in their intended nationally determined contributions;

19. *Requests* the secretariat to update the synthesis report referred to in paragraph 16 above so as to cover all the information in the intended nationally determined contributions communicated by Parties pursuant to decision 1/CP.20 by 4 April 2016 and to make it available by 2 May 2016;
20. *Decides* to convene a facilitative dialogue among Parties in 2018 to take stock of the collective efforts of Parties in relation to progress towards the long-term goal referred to in Article 4, paragraph 1, of the Agreement and to inform the preparation of nationally determined contributions pursuant to Article 4, paragraph 8, of the Agreement;
21. *Invites* the Intergovernmental Panel on Climate Change to provide a special report in 2018 on the impacts of global warming of 1.5 °C above pre-industrial levels and related global greenhouse gas emission pathways;

III. Decisions to Give Effect to the Agreement

22. *Invites* Parties to communicate their first nationally determined contribution no later than when the Party submits its respective instrument of ratification, accession, or approval of the Paris Agreement. If a Party has communicated an intended nationally determined contribution prior to joining the Agreement, that Party shall be considered to have satisfied this provision unless that Party decides otherwise;
23. *Urges* those Parties whose intended nationally determined contribution pursuant to decision 1/CP.20 contains a time frame up to 2025 to communicate by 2020 a new nationally determined contribution and to do so every five years thereafter pursuant to Article 4, paragraph 9, of the Agreement;
24. *Requests* those Parties whose intended nationally determined contribution pursuant to decision 1/CP.20 contains a time frame up to 2030 to communicate or update by 2020 these contributions and to do so every five years thereafter pursuant to Article 4, paragraph 9, of the Agreement;
25. *Decides* that Parties shall submit to the secretariat their nationally determined contributions referred to in Article 4 of the Agreement at least 9 to 12 months in advance of the relevant meeting of the Conference of the Parties serving as the meeting of the Parties to the Paris Agreement with a view to facilitating the clarity, transparency and understanding of these contributions, including through a synthesis report prepared by the secretariat;
26. *Requests* the Ad Hoc Working Group on the Paris Agreement to develop further guidance on features of the nationally determined contributions for consideration and adoption by the Conference of the Parties serving as the meeting of the Parties to the Paris Agreement at its first session;
27. *Agrees* that the information to be provided by Parties communicating their nationally determined contributions, in order to facilitate clarity, transparency and understanding, may include, as appropriate, inter alia, quantifiable information on the reference point (including, as appropriate, a base year), time frames and/or periods for implementation, scope and coverage, planning processes, assumptions and methodological approaches including those for estimating and accounting for anthropogenic greenhouse gas emissions and, as appropriate, removals, and how the Party considers that its nationally determined contribution is fair and ambitious,

in the light of its national circumstances, and how it contributes towards achieving the objective of the Convention as set out in its Article 2;

28. *Requests* the Ad Hoc Working Group on the Paris Agreement to develop further guidance for the information to be provided by Parties in order to facilitate clarity, transparency and understanding of nationally determined contributions for consideration and adoption by the Conference of the Parties serving as the meeting of the Parties to the Paris Agreement at its first session;

29. *Also requests* the Subsidiary Body for Implementation to develop modalities and procedures for the operation and use of the public registry referred to in Article 4, paragraph 12, of the Agreement, for consideration and adoption by the Conference of the Parties serving as the meeting of the Parties to the Paris Agreement at its first session;

30. *Further requests* the secretariat to make available an interim public registry in the first half of 2016 for the recording of nationally determined contributions submitted in accordance with Article 4 of the Agreement, pending the adoption by the Conference of the Parties serving as the meeting of the Parties to the Paris Agreement of the modalities and procedures referred to in paragraph 29 above;

31. *Requests* the Ad Hoc Working Group on the Paris Agreement to elaborate, drawing from approaches established under the Convention and its related legal instruments as appropriate, guidance for accounting for Parties' nationally determined contributions, as referred to in Article 4, paragraph 13, of the Agreement, for consideration and adoption by the Conference of the Parties serving as the meeting of the Parties to the Paris Agreement at its first session, which ensures that:

   (a) Parties account for anthropogenic emissions and removals in accordance with common methodologies and metrics assessed by the Intergovernmental Panel on Climate Change and adopted by the Conference of the Parties serving as the meeting of the Parties to the Paris Agreement;

   (b) Parties ensure methodological consistency, including on baselines, between the communication and implementation of nationally determined contributions;

   (c) Parties strive to include all categories of anthropogenic emissions or removals in their nationally determined contributions and, once a source, sink or activity is included, continue to include it;

   (d) Parties shall provide an explanation of why any categories of anthropogenic emissions or removals are excluded;

32. *Decides* that Parties shall apply the guidance mentioned in paragraph 31 above to the second and subsequent nationally determined contributions and that Parties may elect to apply such guidance to their first nationally determined contribution;

33. *Also decides* that the Forum on the Impact of the Implementation of response measures, under the subsidiary bodies, shall continue, and shall serve the Agreement;

34. *Further decides* that the Subsidiary Body for Scientific and Technological Advice and the Subsidiary Body for Implementation shall recommend, for consideration and adoption by the Conference of the Parties serving as the meeting of the Parties to the Paris Agreement at its first session, the modalities, work programme and functions of the Forum on the Impact of the Implementation of response measures to address the effects of the implementation of response measures under the Agreement by enhancing cooperation amongst Parties on understanding the impacts of mitigation

actions under the Agreement and the exchange of information, experiences, and best practices amongst Parties to raise their resilience to these impacts;

35. *Decides* that the guidance under paragraph 31 above shall ensure that double counting is avoided on the basis of a corresponding adjustment by both Parties for anthropogenic emissions by sources and/or removals by sinks covered by their nationally determined contributions under the Agreement;

36. *Invites* Parties to communicate, by 2020, to the secretariat mid-century, long-term low greenhouse gas emission development strategies in accordance with Article 4, paragraph 19, of the Agreement, and requests the secretariat to publish on the UNFCCC website Parties' low greenhouse gas emission development strategies as communicated;

37. *Requests* the Subsidiary Body for Scientific and Technological Advice to develop and recommend the guidance referred to under Article 6, paragraph 2, of the Agreement for adoption by the Conference of the Parties serving as the meeting of the Parties to the Paris Agreement at its first session, including guidance to ensure that double counting is avoided on the basis of a corresponding adjustment by Parties for both anthropogenic emissions by sources and removals by sinks covered by their nationally determined contributions under the Agreement;

38. *Recommends* that the Conference of the Parties serving as the meeting of the Parties to the Paris Agreement adopt rules, modalities and procedures for the mechanism established by Article 6, paragraph 4, of the Agreement on the basis of:
    (a) Voluntary participation authorized by each Party involved;
    (b) Real, measurable, and long-term benefits related to the mitigation of climate change;
    (c) Specific scopes of activities;
    (d) Reductions in emissions that are additional to any that would otherwise occur;
    (e) Verification and certification of emission reductions resulting from mitigation activities by designated operational entities;
    (f) Experience gained with and lessons learned from existing mechanisms and approaches adopted under the Convention and its related legal instruments;

39. *Requests* the Subsidiary Body for Scientific and Technological Advice to develop and recommend rules, modalities and procedures for the mechanism referred to in paragraph 38 above for consideration and adoption by the Conference of the Parties serving as the meeting of the Parties to the Paris Agreement at its first session;

40. *Also requests* the Subsidiary Body for Scientific and Technological Advice to undertake a work programme under the framework for non-market approaches to sustainable development referred to in Article 6, paragraph 8, of the Agreement, with the objective of considering how to enhance linkages and create synergy between, inter alia, mitigation, adaptation, finance, technology transfer and capacity-building, and how to facilitate the implementation and coordination of non-market approaches;

41. *Further requests* the Subsidiary Body for Scientific and Technological Advice to recommend a draft decision on the work programme referred to in paragraph 40 above, taking into account the views of Parties, for consideration and adoption by the Conference of the Parties serving as the meeting of the Parties to the Paris Agreement at its first session;

42. *Requests* the Adaptation Committee and the Least Developed Countries Expert

Group to jointly develop modalities to recognize the adaptation efforts of developing country Parties, as referred to in Article 7, paragraph 3, of the Agreement, and make recommendations for consideration and adoption by the Conference of the Parties serving as the meeting of the Parties to the Paris Agreement at its first session;

43. *Also requests* the Adaptation Committee, taking into account its mandate and its second three-year workplan, and with a view to preparing recommendations for consideration and adoption by the Conference of the Parties serving as the meeting of the Parties to the Paris Agreement at its first session:

    (a) To review, in 2017, the work of adaptation-related institutional arrangements under the Convention, with a view to identifying ways to enhance the coherence of their work, as appropriate, in order to respond adequately to the needs of Parties;

    (b) To consider methodologies for assessing adaptation needs with a view to assisting developing countries, without placing an undue burden on them;

44. *Invites* all relevant United Nations agencies and international, regional and national financial institutions to provide information to Parties through the secretariat on how their development assistance and climate finance programmes incorporate climate-proofing and climate resilience measures;

45. *Requests* Parties to strengthen regional cooperation on adaptation where appropriate and, where necessary, establish regional centres and networks, in particular in developing countries, taking into account decision 1/CP.16, paragraph 13;

46. *Also requests* the Adaptation Committee and the Least Developed Countries Expert Group, in collaboration with the Standing Committee on Finance and other relevant institutions, to develop methodologies, and make recommendations for consideration and adoption by the Conference of the Parties serving as the meeting of the Parties to the Paris Agreement at its first session on:

    (a) Taking the necessary steps to facilitate the mobilization of support for adaptation in developing countries in the context of the limit to global average temperature increase referred to in Article 2 of the Agreement;

    (b) Reviewing the adequacy and effectiveness of adaptation and support referred to in Article 7, paragraph 14(c), of the Agreement;

47. *Further requests* the Green Climate Fund to expedite support for the least developed countries and other developing country Parties for the formulation of national adaptation plans, consistent with decisions 1/CP.16 and 5/CP.17, and for the subsequent implementation of policies, projects and programmes identified by them;

48. *Decides* on the continuation of the Warsaw International Mechanism for Loss and Damage associated with Climate Change Impacts, following the review in 2016;

49. *Requests* the Executive Committee of the Warsaw International Mechanism to establish a clearinghouse for risk transfer that serves as a repository for information on insurance and risk transfer, in order to facilitate the efforts of Parties to develop and implement comprehensive risk management strategies;

50. *Also requests* the Executive Committee of the Warsaw International Mechanism to establish, according to its procedures and mandate, a task force to complement, draw upon the work of and involve, as appropriate, existing bodies and expert groups under the Convention including the Adaptation Committee and the Least Developed Countries Expert Group, as well as relevant organizations and expert

bodies outside the Convention, to develop recommendations for integrated approaches to avert, minimize and address displacement related to the adverse impacts of climate change;

51. *Further requests* the Executive Committee of the Warsaw International Mechanism to initiate its work, at its next meeting, to operationalize the provisions referred to in paragraphs 49 and 50 above, and to report on progress thereon in its annual report;

52. *Agrees* that Article 8 of the Agreement does not involve or provide a basis for any liability or compensation;

53. *Decides* that, in the implementation of the Agreement, financial resources provided to developing countries should enhance the implementation of their policies, strategies, regulations and action plans and their climate change actions with respect to both mitigation and adaptation to contribute to the achievement of the purpose of the Agreement as defined in Article 2;

54. *Further decides* that, in accordance with Article 9, paragraph 3, of the Agreement, developed countries intend to continue their existing collective mobilization goal through 2025 in the context of meaningful mitigation actions and transparency on implementation; prior to 2025 the Conference of the Parties serving as the meeting of the Parties to the Paris Agreement shall set a new collective quantified goal from a floor of USD 100 billion per year, taking into account the needs and priorities of developing countries;

55. *Recognizes* the importance of adequate and predictable financial resources, including for results-based payments, as appropriate, for the implementation of policy approaches and positive incentives for reducing emissions from deforestation and forest degradation, and the role of conservation, sustainable management of forests and enhancement of forest carbon stocks; as well as alternative policy approaches, such as joint mitigation and adaptation approaches for the integral and sustainable management of forests; while reaffirming the importance of non-carbon benefits associated with such approaches; encouraging the coordination of support from, inter alia, public and private, bilateral and multilateral sources, such as the Green Climate Fund, and alternative sources in accordance with relevant decisions by the Conference of the Parties;

56. *Decides* to initiate, at its twenty-second session, a process to identify the information to be provided by Parties, in accordance with Article 9, paragraph 5, of the Agreement with the view to providing a recommendation for consideration and adoption by the Conference of the Parties serving as the meeting of the Parties to the Paris Agreement at its first session;

57. *Also decides* to ensure that the provision of information in accordance with Article 9, paragraph 7 of the Agreement shall be undertaken in accordance with modalities, procedures and guidelines referred to in paragraph 96 below;

58. *Requests* Subsidiary Body for Scientific and Technological Advice to develop modalities for the accounting of financial resources provided and mobilized through public interventions in accordance with Article 9, paragraph 7, of the Agreement for consideration by the Conference of the Parties at its twenty-fourth session (November 2018), with the view to making a recommendation for consideration and adoption by the Conference of the Parties serving as the meeting of the Parties to the Paris Agreement at its first session;

59. *Decides* that the Green Climate Fund and the Global Environment Facility, the entities entrusted with the operation of the Financial Mechanism of the Convention, as well as the Least Developed Countries Fund and the Special Climate Change Fund, administered by the Global Environment Facility, shall serve the Agreement;
60. *Recognizes* that the Adaptation Fund may serve the Agreement, subject to relevant decisions by the Conference of the Parties serving as the meeting of the Parties to the Kyoto Protocol and the Conference of the Parties serving as the meeting of the Parties to the Paris Agreement;
61. *Invites* the Conference of the Parties serving as the meeting of the Parties to the Kyoto Protocol to consider the issue referred to in paragraph 60 above and make a recommendation to the Conference of the Parties serving as the meeting of the Parties to the Paris Agreement at its first session;
62. *Recommends* that the Conference of the Parties serving as the meeting of the Parties to the Paris Agreement shall provide guidance to the entities entrusted with the operation of the Financial Mechanism of the Convention on the policies, programme priorities and eligibility criteria related to the Agreement for transmission by the Conference of the Parties;
63. *Decides* that the guidance to the entities entrusted with the operations of the Financial Mechanism of the Convention in relevant decisions of the Conference of the Parties, including those agreed before adoption of the Agreement, shall apply mutatis mutandis;
64. *Also decides* that the Standing Committee on Finance shall serve the Agreement in line with its functions and responsibilities established under the Conference of the Parties;
65. *Urges* the institutions serving the Agreement to enhance the coordination and delivery of resources to support country-driven strategies through simplified and efficient application and approval procedures, and through continued readiness support to developing country Parties, including the least developed countries and small island developing States, as appropriate;
66. *Takes note of* the interim report of the Technology Executive Committee on guidance on enhanced implementation of the results of technology needs assessments as referred to in document FCCC/SB/2015/INF.3;
67. *Decides* to strengthen the Technology Mechanism and requests the Technology Executive Committee and the Climate Technology Centre and Network, in supporting the implementation of the Agreement, to undertake further work relating to, inter alia:
    (a) Technology research, development and demonstration;
    (b) The development and enhancement of endogenous capacities and technologies;
68. *Requests* the Subsidiary Body for Scientific and Technological Advice to initiate, at its forty-fourth session (May 2016), the elaboration of the technology framework established under Article 10, paragraph 4, of the Agreement and to report on its findings to the Conference of the Parties, with a view to the Conference of the Parties making a recommendation on the framework to the Conference of the Parties serving as the meeting of the Parties to the Paris Agreement for consideration and adoption at its first session, taking into consideration that the framework should facilitate, inter alia:

(a) The undertaking and updating of technology needs assessments, as well as the enhanced implementation of their results, particularly technology action plans and project ideas, through the preparation of bankable projects;
(b) The provision of enhanced financial and technical support for the implementation of the results of the technology needs assessments;
(c) The assessment of technologies that are ready for transfer;
(d) The enhancement of enabling environments for and the addressing of barriers to the development and transfer of socially and environmentally sound technologies;

69. *Decides* that the Technology Executive Committee and the Climate Technology Centre and Network shall report to the Conference of the Parties serving as the meeting of the Parties to the Paris Agreement, through the subsidiary bodies, on their activities to support the implementation of the Agreement;

70. *Also decides* to undertake a periodic assessment of the effectiveness of and the adequacy of the support provided to the Technology Mechanism in supporting the implementation of the Agreement on matters relating to technology development and transfer;

71. *Requests* the Subsidiary Body for Implementation to initiate, at its forty-fourth session, the elaboration of the scope of and modalities for the periodic assessment referred to in paragraph 70 above, taking into account the review of the Climate Technology Centre and Network as referred to in decision 2/CP.17, annex VII, paragraph 20 and the modalities for the global stocktake referred to in Article 14 of the Agreement, for consideration and adoption by the Conference of the Parties at its twenty-fifth session (November 2019);

72. *Decides* to establish the Paris Committee on Capacity-building whose aim will be to address gaps and needs, both current and emerging, in implementing capacity-building in developing country Parties and further enhancing capacity-building efforts, including with regard to coherence and coordination in capacity-building activities under the Convention;

73. *Also decides* that the Paris Committee on Capacity-building will manage and oversee the work plan mentioned in paragraph 74 below;

74. *Further decides* to launch a work plan for the period 2016–2020 with the following activities:
    (a) Assessing how to increase synergies through cooperation and avoid duplication among existing bodies established under the Convention that implement capacity-building activities, including through collaborating with institutions under and outside the Convention;
    (b) Identifying capacity gaps and needs and recommending ways to address them;
    (c) Promoting the development and dissemination of tools and methodologies for the implementation of capacity-building;
    (d) Fostering global, regional, national and subnational cooperation;
    (e) Identifying and collecting good practices, challenges, experiences, and lessons learned from work on capacity-building by bodies established under the Convention;
    (f) Exploring how developing country Parties can take ownership of building and maintaining capacity over time and space;

(g) Identifying opportunities to strengthen capacity at the national, regional, and subnational level;
(h) Fostering dialogue, coordination, collaboration and coherence among relevant processes and initiatives under the Convention, including through exchanging information on capacity-building activities and strategies of bodies established under the Convention;
(i) Providing guidance to the secretariat on the maintenance and further development of the web-based capacity-building portal;

75. *Decides* that the Paris Committee on Capacity-building will annually focus on an area or theme related to enhanced technical exchange on capacity-building, with the purpose of maintaining up-to-date knowledge on the successes and challenges in building capacity effectively in a particular area;

76. *Requests* the Subsidiary Body for Implementation to organize annual in-session meetings of the Paris Committee on Capacity-building;

77. *Also requests* the Subsidiary Body for Implementation to develop the terms of reference for the Paris Committee on Capacity-building, in the context of the third comprehensive review of the implementation of the capacity-building framework, also taking into account paragraphs 75, 76, 77 and 78 above and paragraphs 82 and 83 below, with a view to recommending a draft decision on this matter for consideration and adoption by the Conference of the Parties at its twenty-second session;

78. *Invites* Parties to submit their views on the membership of the Paris Committee on Capacity-building by 9 March 2016; 2

79. *Requests* the secretariat to compile the submissions referred to in paragraph 78 above into a miscellaneous document for consideration by the Subsidiary Body for Implementation at its forty-fourth session;

80. *Decides* that the inputs to the Paris Committee on Capacity-building will include, inter alia, submissions, the outcome of the third comprehensive review of the implementation of the capacity-building framework, the secretariat's annual synthesis report on the implementation of the framework for capacity-building in developing countries, the secretariat's compilation and synthesis report on capacity-building work of bodies established under the Convention and its Kyoto Protocol, and reports on the Durban Forum and the capacity-building portal;

81. *Requests* the Paris Committee on Capacity-building to prepare annual technical progress reports on its work, and to make these reports available at the sessions of the Subsidiary Body for Implementation coinciding with the sessions of the Conference of the Parties;

82. *Also requests* the Conference of the Parties at its twenty-fifth session (November 2019), to review the progress, need for extension, the effectiveness and enhancement of the Paris Committee on Capacity-building and to take any action it considers appropriate, with a view to making recommendations to the Conference of the Parties serving as the meeting of the Parties to the Paris Agreement at its first session on enhancing institutional arrangements for capacity-building consistent with Article 11, paragraph 5, of the Agreement;

83. *Calls upon* all Parties to ensure that education, training and public awareness, as reflected in Article 6 of the Convention and in Article 12 of the Agreement are adequately considered in their contribution to capacity-building;

84. *Invites* the Conference of the Parties serving as the meeting of the Parties to the Paris Agreement at its first session to explore ways of enhancing the implementation of training, public awareness, public participation and public access to information so as to enhance actions under the Agreement;
85. *Decides* to establish a Capacity-building Initiative for Transparency in order to build institutional and technical capacity, both pre- and post–2020. This initiative will support developing country Parties, upon request, in meeting enhanced transparency requirements as defined in Article 13 of the Agreement in a timely manner;
86. *Also decides* that the Capacity-building Initiative for Transparency will aim:
    (a) To strengthen national institutions for transparency-related activities in line with national priorities;
    (b) To provide relevant tools, training and assistance for meeting the provisions stipulated in Article 13 of the Agreement;
    (c) To assist in the improvement of transparency over time;
87. *Urges and requests* the Global Environment Facility to make arrangements to support the establishment and operation of the Capacity-building Initiative for Transparency as a priority reporting-related need, including through voluntary contributions to support developing countries in the sixth replenishment of the Global Environment Facility and future replenishment cycles, to complement existing support under the Global Environment Facility;
88. *Decides* to assess the implementation of the Capacity-building Initiative for Transparency in the context of the seventh review of the financial mechanism;
89. *Requests* that the Global Environment Facility, as an operating entity of the financial mechanism include in its annual report to the Conference of the Parties the progress of work in the design, development and implementation of the Capacity-building Initiative for Transparency referred to in paragraph 85 above starting in 2016;
90. *Decides* that, in accordance with Article 13, paragraph 2, of the Agreement, developing countries shall be provided flexibility in the implementation of the provisions of that Article, including in the scope, frequency and level of detail of reporting, and in the scope of review, and that the scope of review could provide for in-country reviews to be optional, while such flexibilities shall be reflected in the development of modalities, procedures and guidelines referred to in paragraph 92 below;
91. *Also decides* that all Parties, except for the least developed country Parties and small island developing States, shall submit the information referred to in Article 13, paragraphs 7, 8, 9 and 10, as appropriate, no less frequently than on a biennial basis, and that the least developed country Parties and small island developing States may submit this information at their discretion;
92. *Requests* the Ad Hoc Working Group on the Paris Agreement to develop recommendations for modalities, procedures and guidelines in accordance with Article 13, paragraph 13, of the Agreement, and to define the year of their first and subsequent review and update, as appropriate, at regular intervals, for consideration by the Conference of the Parties, at its twenty-fourth session, with a view to forwarding them to the Conference of the Parties serving as the meeting of the Parties to the Paris Agreement for adoption at its first session;
93. *Also requests* the Ad Hoc Working Group on the Paris Agreement in developing the recommendations for the modalities, procedures and guidelines referred to in paragraph 92 above to take into account, inter alia:

(a) The importance of facilitating improved reporting and transparency over time;
(b) The need to provide flexibility to those developing country Parties that need it in the light of their capacities;
(c) The need to promote transparency, accuracy, completeness, consistency, and comparability;
(d) The need to avoid duplication as well as undue burden on Parties and the secretariat;
(e) The need to ensure that Parties maintain at least the frequency and quality of reporting in accordance with their respective obligations under the Convention;
(f) The need to ensure that double counting is avoided;
(g) The need to ensure environmental integrity;

94. *Further requests* the Ad Hoc Working Group on the Paris Agreement, when developing the modalities, procedures and guidelines referred to in paragraph 92 above, to draw on the experiences from and take into account other on-going relevant processes under the Convention;

95. *Requests* the Ad Hoc Working Group on the Paris Agreement, when developing modalities, procedures and guidelines referred to in paragraph 92 above, to consider, inter alia:
    (a) The types of flexibility available to those developing countries that need it on the basis of their capacities;
    (b) The consistency between the methodology communicated in the nationally determined contribution and the methodology for reporting on progress made towards achieving individual Parties' respective nationally determined contribution;
    (c) That Parties report information on adaptation action and planning including, if appropriate, their national adaptation plans, with a view to collectively exchanging information and sharing lessons learned;
    (d) Support provided, enhancing delivery of support for both adaptation and mitigation through, inter alia, the common tabular formats for reporting support, and taking into account issues considered by the Subsidiary Body for Scientific and Technological Advice on methodologies for reporting on financial information, and enhancing the reporting by developing countries on support received, including the use, impact and estimated results thereof;
    (e) Information in the biennial assessments and other reports of the Standing Committee on Finance and other relevant bodies under the Convention;
    (f) Information on the social and economic impact of response measures;

96. *Also requests* the Ad Hoc Working Group on the Paris Agreement, when developing recommendations for modalities, procedures and guidelines referred to in paragraph 92 above, to enhance the transparency of support provided in accordance with Article 9 of the Agreement;

97. Further requests the Ad Hoc Working Group on the Paris Agreement to report on the progress of work on the modalities, procedures and guidelines referred to in paragraph 92 above to future sessions of the Conference of the Parties, and that this work be concluded no later than 2018;

98. *Decides* that the modalities, procedures and guidelines developed under paragraph 92 above, shall be applied upon the entry into force of the Paris Agreement;

99. *Also decides* that the modalities, procedures and guidelines of this transparency framework shall build upon and eventually supercede the measurement, reporting and verification system established by paragraphs 40 to 47 and 60 to 64 of decision 1/CP.16 and paragraph 12 to 62 of decision 2/CP.17 immediately following the submission of the final biennial reports and biennial update reports;
100. *Requests* the Ad Hoc Working Group on the Paris Agreement to identify the sources of input for the global stocktake referred to in Article 14 of the Agreement and to report to the Conference of the Parties, with a view to the Conference of the Parties making a recommendation to the Conference of the Parties serving as the meeting of the Parties to the Paris Agreement for consideration and adoption at its first session, including, but not limited to:
    (a) Information on:
        (i) The overall effect of the nationally determined contributions communicated by Parties;
        (ii) The state of adaptation efforts, support, experiences and priorities from the communications referred to in Article 7, paragraphs 10 and 11, of the Agreement, and reports referred to in Article 13, paragraph 7, of the Agreement;
        (iii) The mobilization and provision of support;
    (b) The latest reports of the Intergovernmental Panel on Climate Change;
    (c) Reports of the subsidiary bodies;
101. *Also requests* the Subsidiary Body for Scientific and Technological Advice to provide advice on how the assessments of the Intergovernmental Panel on Climate Change can inform the global stocktake of the implementation of the Agreement pursuant to its Article 14 of the Agreement and to report on this matter to the Ad Hoc Working Group on the Paris Agreement at its second session;
102. *Further requests* the Ad Hoc Working Group on the Paris Agreement to develop modalities for the global stocktake referred to in Article 14 of the Agreement and to report to the Conference of the Parties, with a view to making a recommendation to the Conference of the Parties serving as the meeting of the Parties to the Paris Agreement for consideration and adoption at its first session;
103. *Decides* that the committee referred to in Article 15, paragraph 2, of the Agreement shall consist of 12 members with recognized competence in relevant scientific, technical, socio-economic or legal fields, to be elected by the Conference of the Parties serving as the meeting of the Parties to the Paris Agreement on the basis of equitable geographical representation, with two members each from the five regional groups of the United Nations and one member each from the small island developing States and the least developed countries, while taking into account the goal of gender balance;
104. *Requests* the Ad Hoc Working Group on the Paris Agreement to develop the modalities and procedures for the effective operation of the committee referred to in Article 15, paragraph 2, of the Agreement, with a view to the Ad Hoc Working Group on the Paris Agreement completing its work on such modalities and procedures for consideration and adoption by the Conference of the Parties serving as the meeting of the Parties to the Paris Agreement at its first session;
105. *Also requests* the secretariat, solely for the purposes of Article 21 of the Agreement, to make available on its website on the date of adoption of the Agreement as well

as in the report of the Conference of the Parties at its twenty-first session, information on the most up-to-date total and per cent of greenhouse gas emissions communicated by Parties to the Convention in their national communications, greenhouse gas inventory reports, biennial reports or biennial update reports;

IV. Enhanced Action Prior to 2020

106. *Resolves* to ensure the highest possible mitigation efforts in the pre–2020 period, including by:
    (a) Urging all Parties to the Kyoto Protocol that have not already done so to ratify and implement the Doha Amendment to the Kyoto Protocol;
    (b) Urging all Parties that have not already done so to make and implement a mitigation pledge under the Cancun Agreements;
    (c) Reiterating its resolve, as set out in decision 1/CP.19, paragraphs 3 and 4, to accelerate the full implementation of the decisions constituting the agreed outcome pursuant to decision 1/CP.13 and enhance ambition in the pre–2020 period in order to ensure the highest possible mitigation efforts under the Convention by all Parties;
    (d) Inviting developing country Parties that have not submitted their first biennial update reports to do so as soon as possible;
    (e) Urging all Parties to participate in the existing measurement, reporting and verification processes under the Cancun Agreements, in a timely manner, with a view to demonstrating progress made in the implementation of their mitigation pledges;
107. *Encourages* Parties to promote the voluntary cancellation by Party and non–Party stakeholders, without double counting of units issued under the Kyoto Protocol, including certified emission reductions that are valid for the second commitment period;
108. *Urges* host and purchasing Parties to report transparently on internationally transferred mitigation outcomes, including outcomes used to meet international pledges, and emission units issued under the Kyoto Protocol with a view to promoting environmental integrity and avoiding double counting;
109. *Recognizes* the social, economic and environmental value of voluntary mitigation actions and their co-benefits for adaptation, health and sustainable development;
110. *Resolves* to strengthen, in the period 2016–2020, the existing technical examination process on mitigation as defined in decision 1/CP.19, paragraph 5(a), and decision 1/CP.20, paragraph 19, taking into account the latest scientific knowledge, including by:
    (a) Encouraging Parties, Convention bodies and international organizations to engage in this process, including, as appropriate, in cooperation with relevant non–Party stakeholders, to share their experiences and suggestions, including from regional events, and to cooperate in facilitating the implementation of policies, practices and actions identified during this process in accordance with national sustainable development priorities;
    (b) Striving to improve, in consultation with Parties, access to and participation in this process by developing country Party and non–Party experts;

(c) Requesting the Technology Executive Committee and the Climate Technology Centre and Network in accordance with their respective mandates:
   (i) To engage in the technical expert meetings and enhance their efforts to facilitate and support Parties in scaling up the implementation of policies, practices and actions identified during this process;
   (ii) To provide regular updates during the technical expert meetings on the progress made in facilitating the implementation of policies, practices and actions previously identified during this process;
   (iii) To include information on their activities under this process in their joint annual report to the Conference of the Parties;
(d) Encouraging Parties to make effective use of the Climate Technology Centre and Network to obtain assistance to develop economically, environmentally and socially viable project proposals in the high mitigation potential areas identified in this process;

111. *Encourages* the operating entities of the Financial Mechanism of the Convention to engage in the technical expert meetings and to inform participants of their contribution to facilitating progress in the implementation of policies, practices and actions identified during the technical examination process;

112. *Requests* the secretariat to organize the process referred to in paragraph 110 above and disseminate its results, including by:
   (a) Organizing, in consultation with the Technology Executive Committee and relevant expert organizations, regular technical expert meetings focusing on specific policies, practices and actions representing best practices and with the potential to be scalable and replicable;
   (b) Updating, on an annual basis, following the meetings referred to in paragraph 112(a) above and in time to serve as input to the summary for policymakers referred to in paragraph 112(c) below, a technical paper on the mitigation benefits and co-benefits of policies, practices and actions for enhancing mitigation ambition, as well as on options for supporting their implementation, information on which should be made available in a user friendly online format;
   (c) Preparing, in consultation with the champions referred to in paragraph 122 below, a summary for policymakers, with information on specific policies, practices and actions representing best practices and with the potential to be scalable and replicable, and on options to support their implementation, as well as on relevant collaborative initiatives, and publishing the summary at least two months in advance of each session of the Conference of the Parties as input for the high-level event referred to in paragraph 121 below;

113. *Decides* that the process referred to in paragraph 110 above should be organized jointly by the Subsidiary Body for Implementation and the Subsidiary Body for Scientific and Technological Advice and should take place on an ongoing basis until 2020;

114. *Also decides* to conduct in 2017 an assessment of the process referred to in paragraph 110 above so as to improve its effectiveness;

115. *Resolves* to enhance the provision of urgent and adequate finance, technology and capacity-building support by developed country Parties in order to enhance the level of ambition of pre–2020 action by Parties, and in this regard strongly urges developed

country Parties to scale up their level of financial support, with a concrete roadmap to achieve the goal of jointly providing USD 100 billion annually by 2020 for mitigation and adaptation while significantly increasing adaptation finance from current levels and to further provide appropriate technology and capacity-building support;

116. *Decides* to conduct a facilitative dialogue in conjunction with the twenty-second session of the Conference of the Parties to assess the progress in implementing decision 1/CP.19, paragraphs 3 and 4, and identify relevant opportunities to enhance the provision of financial resources, including for technology development and transfer and capacity building support, with a view to identifying ways to enhance the ambition of mitigation efforts by all Parties, including identifying relevant opportunities to enhance the provision and mobilization of support and enabling environments;

117. *Acknowledges* with appreciation the results of the Lima-Paris Action Agenda, which build on the climate summit convened on 23 September 2014 by the Secretary-General of the United Nations;

118. *Welcomes* the efforts of non–Party stakeholders to scale up their climate actions, and encourages the registration of those actions in the Non-State Actor Zone for Climate Action platform;3

119. *Encourages* Parties to work closely with non–Party stakeholders to catalyse efforts to strengthen mitigation and adaptation action;

120. Also encourages non–Party stakeholders to increase their engagement in the processes referred to in paragraph 110 above and paragraph 125 below;

121. Agrees to convene, pursuant to decision 1/CP.20, paragraph 21, building on the Lima-Paris Action Agenda and in conjunction with each session of the Conference of the Parties during the period 2016–2020, a high-level event that:
    (a) Further strengthens high-level engagement on the implementation of policy options and actions arising from the processes referred to in paragraph 110 above and paragraph below, drawing on the summary for policymakers referred to in paragraph 112(c) above;
    (b) Provides an opportunity for announcing new or strengthened voluntary efforts, initiatives and coalitions, including the implementation of policies, practices and actions arising from the processes referred to in paragraph 110 above and paragraph 125 below and presented in the summary for policymakers referred to in paragraph 112(c) above;
    (c) Takes stock of related progress and recognizes new or strengthened voluntary efforts, initiatives and coalitions;
    (d) Provides meaningful and regular opportunities for the effective high-level engagement of dignitaries of Parties, international organizations, international cooperative initiatives and non–Party stakeholders;

122. *Decides* that two high-level champions shall be appointed to act on behalf of the President of the Conference of the Parties to facilitate through strengthened high-level engagement in the period 2016–2020 the successful execution of existing efforts and the scaling-up and introduction of new or strengthened voluntary efforts, initiatives and coalitions, including by:
    (a) Working with the Executive Secretary and the current and incoming Presidents of the Conference of the Parties to coordinate the annual high-level event referred to in paragraph 121 above;

(b) Engaging with interested Parties and non–Party stakeholders, including to further the voluntary initiatives of the Lima-Paris Action Agenda;

(c) Providing guidance to the secretariat on the organization of technical expert meetings referred to in paragraph 112(a) above and paragraph 130(a) below;

123. *Also decides* that the high-level champions referred to in paragraph 122 above should normally serve for a term of two years, with their terms overlapping for a full year to ensure continuity, such that:

(a) The President of the Conference of the Parties of the twenty-first session should appoint one champion, who should serve for one year from the date of the appointment until the last day of the Conference of the Parties at its twenty-second session;

(b) The President of the Conference of the Parties of the twenty-second session should appoint one champion who should serve for two years from the date of the appointment until the last day of the Conference of the Parties at its twenty-third session (November 2017);

(c) Thereafter, each subsequent President of the Conference of the Parties should appoint one champion who should serve for two years and succeed the previously appointed champion whose term has ended;

124. *Invites* all interested Parties and relevant organizations to provide support for the work of the champions referred to in paragraph 122 above;

125. *Decides* to launch, in the period 2016–2020, a technical examination process on adaptation;

126. *Also decides* that the technical examination process on adaptation referred to in paragraph 125 above will endeavour to identify concrete opportunities for strengthening resilience, reducing vulnerabilities and increasing the understanding and implementation of adaptation actions;

127. *Further decides* that the technical examination process referred to in paragraph 125 above should be organized jointly by the Subsidiary Body for Implementation and the Subsidiary Body for Scientific and Technological Advice, and conducted by the Adaptation Committee;

128. *Decides* that the process referred to in paragraph 125 above will be pursued by:

(a) Facilitating the sharing of good practices, experiences and lessons learned;

(b) Identifying actions that could significantly enhance the implementation of adaptation actions, including actions that could enhance economic diversification and have mitigation co-benefits;

(c) Promoting cooperative action on adaptation;

(d) Identifying opportunities to strengthen enabling environments and enhance the provision of support for adaptation in the context of specific policies, practices and actions;

129. *Also decides* that the technical examination process on adaptation referred to in paragraph 125 above will take into account the process, modalities, outputs, outcomes and lessons learned from the technical examination process on mitigation referred to in paragraph 110 above;

130. *Requests* the secretariat to support the technical examination process referred to in paragraph 125 above by:

(a) Organizing regular technical expert meetings focusing on specific policies, strategies and actions;

(b) Preparing annually, on the basis of the meetings referred to in paragraph 130(a) above and in time to serve as an input to the summary for policymakers referred to in paragraph 112(c) above, a technical paper on opportunities to enhance adaptation action, as well as options to support their implementation, information on which should be made available in a user-friendly online format;

131. *Decides* that in conducting the process referred to in paragraph 125 above, the Adaptation Committee will engage with and explore ways to take into account, synergize with and build on the existing arrangements for adaptation-related work programmes, bodies and institutions under the Convention so as to ensure coherence and maximum value;

132. *Also decides* to conduct, in conjunction with the assessment referred to in paragraph 120 above, an assessment of the process referred to in paragraph 125 above, so as to improve its effectiveness;

133. *Invites Parties* and observer organizations to submit information on the opportunities referred to in paragraph 126 above by 3 February 2016;

V. Non-Party Stakeholders

134. *Welcomes* the efforts of all non–Party stakeholders to address and respond to climate change, including those of civil society, the private sector, financial institutions, cities and other subnational authorities;

135. *Invites* the non–Party stakeholders referred to in paragraph 134 above to scale up their efforts and support actions to reduce emissions and/or to build resilience and decrease vulnerability to the adverse effects of climate change and demonstrate these efforts via the Non-State Actor Zone for Climate Action platform4 referred to in paragraph 118 above;

136. *Recognizes* the need to strengthen knowledge, technologies, practices and efforts of local communities and indigenous peoples related to addressing and responding to climate change, and establishes a platform for the exchange of experiences and sharing of best practices on mitigation and adaptation in a holistic and integrated manner;

137. *Also recognizes* the important role of providing incentives for emission reduction activities, including tools such as domestic policies and carbon pricing;

VI. Administrative and Budgetary Matters

138. *Takes note* of the estimated budgetary implications of the activities to be undertaken by the secretariat referred to in this decision and requests that the actions of the secretariat called for in this decision be undertaken subject to the availability of financial resources;

139. *Emphasizes* the urgency of making additional resources available for the implementation of the relevant actions, including actions referred to in this decision, and the implementation of the work programme referred to in paragraph 9 above;

140. *Urges Parties* to make voluntary contributions for the timely implementation of this decision.

Annex

Paris Agreement

The Parties to this Agreement,

Being Parties to the United Nations Framework Convention on Climate Change, hereinafter referred to as "the Convention,"

Pursuant to the Durban Platform for Enhanced Action established by decision 1/CP.17 of the Conference of the Parties to the Convention at its seventeenth session,

In pursuit of the objective of the Convention, and being guided by its principles, including the principle of equity and common but differentiated responsibilities and respective capabilities, in the light of different national circumstances,

Recognizing the need for an effective and progressive response to the urgent threat of climate change on the basis of the best available scientific knowledge,

Also recognizing the specific needs and special circumstances of developing country Parties, especially those that are particularly vulnerable to the adverse effects of climate change, as provided for in the Convention,

Taking full account of the specific needs and special situations of the least developed countries with regard to funding and transfer of technology,

Recognizing that Parties may be affected not only by climate change, but also by the impacts of the measures taken in response to it,

Emphasizing the intrinsic relationship that climate change actions, responses and impacts have with equitable access to sustainable development and eradication of poverty,

Recognizing the fundamental priority of safeguarding food security and ending hunger, and the particular vulnerabilities of food production systems to the adverse impacts of climate change,

Taking into account the imperatives of a just transition of the workforce and the creation of decent work and quality jobs in accordance with nationally defined development priorities,

Acknowledging that climate change is a common concern of humankind, Parties should, when taking action to address climate change, respect, promote and consider their respective obligations on human rights, the right to health, the rights of indigenous peoples, local communities, migrants, children, persons with disabilities and people in vulnerable situations and the right to development, as well as gender equality, empowerment of women and intergenerational equity,

Recognizing the importance of the conservation and enhancement, as appropriate, of sinks and reservoirs of the greenhouse gases referred to in the Convention,

Noting the importance of ensuring the integrity of all ecosystems, including oceans, and the protection of biodiversity, recognized by some cultures as Mother Earth, and noting the importance for some of the concept of "climate justice," when taking action to address climate change,

Affirming the importance of education, training, public awareness, public participation, public access to information and cooperation at all levels on the matters addressed in this Agreement,

Recognizing the importance of the engagements of all levels of government and various actors, in accordance with respective national legislations of Parties, in addressing climate change,

Also recognizing that sustainable lifestyles and sustainable patterns of consumption and production, with developed country Parties taking the lead, play an important role in addressing climate change, Have agreed as follows:

Article 1

For the purpose of this Agreement, the definitions contained in Article 1 of the Convention shall apply. In addition:
1. "Convention" means the United Nations Framework Convention on Climate Change, adopted in New York on 9 May 1992.
2. "Conference of the Parties" means the Conference of the Parties to the Convention.
3. "Party" means a Party to this Agreement.

Article 2

1. This Agreement, in enhancing the implementation of the Convention, including its objective, aims to strengthen the global response to the threat of climate change, in the context of sustainable development and efforts to eradicate poverty, including by:
   (a) Holding the increase in the global average temperature to well below 2°C above pre-industrial levels and to pursue efforts to limit the temperature increase to 1.5°C above pre-industrial levels, recognizing that this would significantly reduce the risks and impacts of climate change;
   (b) Increasing the ability to adapt to the adverse impacts of climate change and foster climate resilience and low greenhouse gas emissions development, in a manner that does not threaten food production;
   (c) Making finance flows consistent with a pathway towards low greenhouse gas emissions and climate resilient development.
2. This Agreement will be implemented to reflect equity and the principle of common but differentiated responsibilities and respective capabilities, in the light of different national circumstances.

Article 3

As nationally determined contributions to the global response to climate change, all Parties are to undertake and communicate ambitious efforts as defined in Articles 4, 7, 9, 10, 11 and 13 with the view to achieving the purpose of this Agreement as set out in Article 2. The efforts of all Parties will represent a progression over time, while recognizing the need to support developing country Parties for the effective implementation of this Agreement.

Article 4

1. In order to achieve the long-term temperature goal set out in Article 2, Parties aim to reach global peaking of greenhouse gas emissions as soon as possible, recognizing that peaking will take longer for developing country Parties, and to undertake rapid reductions thereafter in accordance with best available science, so as to achieve a balance between anthropogenic emissions by sources and removals by sinks of greenhouse gases in the second half of this century, on the basis of equity, and in the context of sustainable development and efforts to eradicate poverty.
2. Each Party shall prepare, communicate and maintain successive nationally determined contributions that it intends to achieve. Parties shall pursue domestic mitigation measures with the aim of achieving the objectives of such contributions.

3. Each Party's successive nationally determined contribution will represent a progression beyond the Party's then current nationally determined contribution and reflect its highest possible ambition, reflecting its common but differentiated responsibilities and respective capabilities, in the light of different national circumstances.
4. Developed country Parties shall continue taking the lead by undertaking economy-wide absolute emission reduction targets. Developing country Parties should continue enhancing their mitigation efforts, and are encouraged to move over time towards economy-wide emission reduction or limitation targets in the light of different national circumstances.
5. Support shall be provided to developing country Parties for the implementation of this Article, in accordance with Articles 9, 10 and 11, recognizing that enhanced support for developing country Parties will allow for higher ambition in their actions.
6. The least developed countries and small island developing States may prepare and communicate strategies, plans and actions for low greenhouse gas emissions development reflecting their special circumstances.
7. Mitigation co-benefits resulting from Parties' adaptation actions and/or economic diversification plans can contribute to mitigation outcomes under this Article.
8. In communicating their nationally determined contributions, all Parties shall provide the information necessary for clarity, transparency and understanding in accordance with decision 1/CP.21 and any relevant decisions of the Conference of the Parties serving as the meeting of the Parties to the Paris Agreement.
9. Each Party shall communicate a nationally determined contribution every five years in accordance with decision 1/CP.21 and any relevant decisions of the Conference of the Parties serving as the meeting of the Parties to the Paris Agreement and be informed by the outcomes of the global stocktake referred to in Article 14.
10. The Conference of the Parties serving as the meeting of the Parties to the Paris Agreement shall consider common time frames for nationally determined contributions at its first session.
11. A Party may at any time adjust its existing nationally determined contribution with a view to enhancing its level of ambition, in accordance with guidance adopted by the Conference of the Parties serving as the meeting of the Parties to the Paris Agreement.
12. Nationally determined contributions communicated by Parties shall be recorded in a public registry maintained by the secretariat.
13. Parties shall account for their nationally determined contributions. In accounting for anthropogenic emissions and removals corresponding to their nationally determined contributions, Parties shall promote environmental integrity, transparency, accuracy, completeness, comparability and consistency, and ensure the avoidance of double counting, in accordance with guidance adopted by the Conference of the Parties serving as the meeting of the Parties to the Paris Agreement.
14. In the context of their nationally determined contributions, when recognizing and implementing mitigation actions with respect to anthropogenic emissions and removals, Parties should take into account, as appropriate, existing methods and guidance under the Convention, in the light of the provisions of paragraph 13 of this Article.
15. Parties shall take into consideration in the implementation of this Agreement the

concerns of Parties with economies most affected by the impacts of response measures, particularly developing country Parties.

16. Parties, including regional economic integration organizations and their member States, that have reached an agreement to act jointly under paragraph 2 of this Article shall notify the secretariat of the terms of that agreement, including the emission level allocated to each Party within the relevant time period, when they communicate their nationally determined contributions. The secretariat shall in turn inform the Parties and signatories to the Convention of the terms of that agreement.

17. Each party to such an agreement shall be responsible for its emission level as set out in the agreement referred to in paragraph 16 above in accordance with paragraphs 13 and 14 of this Article and Articles 13 and 15.

18. If Parties acting jointly do so in the framework of, and together with, a regional economic integration organization which is itself a Party to this Agreement, each member State of that regional economic integration organization individually, and together with the regional economic integration organization, shall be responsible for its emission level as set out in the agreement communicated under paragraph 16 of this Article in accordance with paragraphs 13 and 14 of this Article and Articles 13 and 15.

19. All Parties should strive to formulate and communicate long-term low greenhouse gas emission development strategies, mindful of Article 2 taking into account their common but differentiated responsibilities and respective capabilities, in the light of different national circumstances.

Article 5

1. Parties should take action to conserve and enhance, as appropriate, sinks and reservoirs of greenhouse gases as referred to in Article 4, paragraph 1(d), of the Convention, including forests.

2. Parties are encouraged to take action to implement and support, including through results-based payments, the existing framework as set out in related guidance and decisions already agreed under the Convention for: policy approaches and positive incentives for activities relating to reducing emissions from deforestation and forest degradation, and the role of conservation, sustainable management of forests and enhancement of forest carbon stocks in developing countries; and alternative policy approaches, such as joint mitigation and adaptation approaches for the integral and sustainable management of forests, while reaffirming the importance of incentivizing, as appropriate, non-carbon benefits associated with such approaches.

Article 6

1. Parties recognize that some Parties choose to pursue voluntary cooperation in the implementation of their nationally determined contributions to allow for higher ambition in their mitigation and adaptation actions and to promote sustainable development and environmental integrity.

2. Parties shall, where engaging on a voluntary basis in cooperative approaches that involve the use of internationally transferred mitigation outcomes towards nationally determined contributions, promote sustainable development and ensure environmental integrity and transparency, including in governance, and shall apply robust accounting

to ensure, inter alia, the avoidance of double counting, consistent with guidance adopted by the Conference of the Parties serving as the meeting of the Parties to the Paris Agreement.
3. The use of internationally transferred mitigation outcomes to achieve nationally determined contributions under this Agreement shall be voluntary and authorized by participating Parties.
4. A mechanism to contribute to the mitigation of greenhouse gas emissions and support sustainable development is hereby established under the authority and guidance of the Conference of the Parties serving as the meeting of the Parties to the Paris Agreement for use by Parties on a voluntary basis. It shall be supervised by a body designated by the Conference of the Parties serving as the meeting of the Parties to the Paris Agreement, and shall aim:
    (a) To promote the mitigation of greenhouse gas emissions while fostering sustainable development;
    (b) To incentivize and facilitate participation in the mitigation of greenhouse gas emissions by public and private entities authorized by a Party;
    (c) To contribute to the reduction of emission levels in the host Party, which will benefit from mitigation activities resulting in emission reductions that can also be used by another Party to fulfil its nationally determined contribution; and
    (d) To deliver an overall mitigation in global emissions.
5. Emission reductions resulting from the mechanism referred to in paragraph 4 of this Article shall not be used to demonstrate achievement of the host Party's nationally determined contribution if used by another Party to demonstrate achievement of its nationally determined contribution.
6. The Conference of the Parties serving as the meeting of the Parties to the Paris Agreement shall ensure that a share of the proceeds from activities under the mechanism referred to in paragraph 4 of this Article is used to cover administrative expenses as well as to assist developing country Parties that are particularly vulnerable to the adverse effects of climate change to meet the costs of adaptation.
7. The Conference of the Parties serving as the meeting of the Parties to the Paris Agreement shall adopt rules, modalities and procedures for the mechanism referred to in paragraph 4 of this Article at its first session.
8. Parties recognize the importance of integrated, holistic and balanced non-market approaches being available to Parties to assist in the implementation of their nationally determined contributions, in the context of sustainable development and poverty eradication, in a coordinated and effective manner, including through, inter alia, mitigation, adaptation, finance, technology transfer and capacity-building, as appropriate. These approaches shall aim to:
    (a) Promote mitigation and adaptation ambition;
    (b) Enhance public and private participation in the implementation of nationally determined contributions; and
    (c) Enable opportunities for coordination across instruments and relevant institutional arrangements.
9. A framework for non-market approaches to sustainable development is hereby defined to promote the nonmarket approaches referred to in paragraph 8 of this Article.

Article 7

1. Parties hereby establish the global goal on adaptation of enhancing adaptive capacity, strengthening resilience and reducing vulnerability to climate change, with a view to contributing to sustainable development and ensuring an adequate adaptation response in the context of the temperature goal referred to in Article 2.
2. Parties recognize that adaptation is a global challenge faced by all with local, subnational, national, regional and international dimensions, and that it is a key component of and makes a contribution to the long-term global response to climate change to protect people, livelihoods and ecosystems, taking into account the urgent and immediate needs of those developing country Parties that are particularly vulnerable to the adverse effects of climate change.
3. The adaptation efforts of developing country Parties shall be recognized, in accordance with the modalities to be adopted by the Conference of the Parties serving as the meeting of the Parties to the Paris Agreement at its first session.
4. Parties recognize that the current need for adaptation is significant and that greater levels of mitigation can reduce the need for additional adaptation efforts, and that greater adaptation needs can involve greater adaptation costs.
5. Parties acknowledge that adaptation action should follow a country-driven, gender-responsive, participatory and fully transparent approach, taking into consideration vulnerable groups, communities and ecosystems, and should be based on and guided by the best available science and, as appropriate, traditional knowledge, knowledge of indigenous peoples and local knowledge systems, with a view to integrating adaptation into relevant socioeconomic and environmental policies and actions, where appropriate.
6. Parties recognize the importance of support for and international cooperation on adaptation efforts and the importance of taking into account the needs of developing country Parties, especially those that are particularly vulnerable to the adverse effects of climate change.
7. Parties should strengthen their cooperation on enhancing action on adaptation, taking into account the Cancun Adaptation Framework, including with regard to:
    (a) Sharing information, good practices, experiences and lessons learned, including, as appropriate, as these relate to science, planning, policies and implementation in relation to adaptation actions;
    (b) Strengthening institutional arrangements, including those under the Convention that serve this Agreement, to support the synthesis of relevant information and knowledge, and the provision of technical support and guidance to Parties;
    (c) Strengthening scientific knowledge on climate, including research, systematic observation of the climate system and early warning systems, in a manner that informs climate services and supports decisionmaking;
    (d) Assisting developing country Parties in identifying effective adaptation practices, adaptation needs, priorities, support provided and received for adaptation actions and efforts, and challenges and gaps, in a manner consistent with encouraging good practices;
    (e) Improving the effectiveness and durability of adaptation actions.
8. United Nations specialized organizations and agencies are encouraged to support

the efforts of Parties to implement the actions referred to in paragraph 7 of this Article, taking into account the provisions of paragraph 5 of this Article.
9. Each Party shall, as appropriate, engage in adaptation planning processes and the implementation of actions, including the development or enhancement of relevant plans, policies and/or contributions, which may include:
    (a) The implementation of adaptation actions, undertakings and/or efforts;
    (b) The process to formulate and implement national adaptation plans;
    (c) The assessment of climate change impacts and vulnerability, with a view to formulating nationally determined prioritized actions, taking into account vulnerable people, places and ecosystems;
    (d) Monitoring and evaluating and learning from adaptation plans, policies, programmes and actions; and
    (e) Building the resilience of socioeconomic and ecological systems, including through economic diversification and sustainable management of natural resources.
10. Each Party should, as appropriate, submit and update periodically an adaptation communication, which may include its priorities, implementation and support needs, plans and actions, without creating any additional burden for developing country Parties.
11. The adaptation communication referred to in paragraph 10 of this Article shall be, as appropriate, submitted and updated periodically, as a component of or in conjunction with other communications or documents, including a national adaptation plan, a nationally determined contribution as referred to in Article 4, paragraph 2, and/or a national communication.
12. The adaptation communications referred to in paragraph 10 of this Article shall be recorded in a public registry maintained by the secretariat.
13. Continuous and enhanced international support shall be provided to developing country Parties for the implementation of paragraphs 7, 9, 10 and 11 of this Article, in accordance with the provisions of Articles 9, 10 and 11.
14. The global stocktake referred to in Article 14 shall, inter alia:
    (a) Recognize adaptation efforts of developing country Parties;
    (b) Enhance the implementation of adaptation action taking into account the adaptation communication referred to in paragraph 10 of this Article;
    (c) Review the adequacy and effectiveness of adaptation and support provided for adaptation; and
    (d) Review the overall progress made in achieving the global goal on adaptation referred to in paragraph 1 of this Article.

Article 8

1. Parties recognize the importance of averting, minimizing and addressing loss and damage associated with the adverse effects of climate change, including extreme weather events and slow onset events, and the role of sustainable development in reducing the risk of loss and damage.
2. The Warsaw International Mechanism for Loss and Damage associated with Climate Change Impacts shall be subject to the authority and guidance of the Conference of the Parties serving as the meeting of the Parties to the Paris Agreement and may be enhanced and strengthened, as determined by the Conference of the Parties serving as the meeting of the Parties to the Paris Agreement.

3. Parties should enhance understanding, action and support, including through the Warsaw International Mechanism, as appropriate, on a cooperative and facilitative basis with respect to loss and damage associated with the adverse effects of climate change.
4. Accordingly, areas of cooperation and facilitation to enhance understanding, action and support may include:
    (a) Early warning systems;
    (b) Emergency preparedness;
    (c) Slow onset events;
    (d) Events that may involve irreversible and permanent loss and damage;
    (e) Comprehensive risk assessment and management;
    (f) Risk insurance facilities, climate risk pooling and other insurance solutions;
    (g) Non-economic losses;
    (h) Resilience of communities, livelihoods and ecosystems.
5. The Warsaw International Mechanism shall collaborate with existing bodies and expert groups under the Agreement, as well as relevant organizations and expert bodies outside the Agreement.

Article 9
1. Developed country Parties shall provide financial resources to assist developing country Parties with respect to both mitigation and adaptation in continuation of their existing obligations under the Convention.
2. Other Parties are encouraged to provide or continue to provide such support voluntarily.
3. As part of a global effort, developed country Parties should continue to take the lead in mobilizing climate finance from a wide variety of sources, instruments and channels, noting the significant role of public funds, through a variety of actions, including supporting country-driven strategies, and taking into account the needs and priorities of developing country Parties. Such mobilization of climate finance should represent a progression beyond previous efforts.
4. The provision of scaled-up financial resources should aim to achieve a balance between adaptation and mitigation, taking into account country-driven strategies, and the priorities and needs of developing country Parties, especially those that are particularly vulnerable to the adverse effects of climate change and have significant capacity constraints, such as the least developed countries and small island developing States, considering the need for public and grant-based resources for adaptation.
5. Developed country Parties shall biennially communicate indicative quantitative and qualitative information related to paragraphs 1 and 3 of this Article, as applicable, including, as available, projected levels of public financial resources to be provided to developing country Parties. Other Parties providing resources are encouraged to communicate biennially such information on a voluntary basis.
6. The global stocktake referred to in Article 14 shall take into account the relevant information provided by developed country Parties and/or Agreement bodies on efforts related to climate finance.
7. Developed country Parties shall provide transparent and consistent information on support for developing country Parties provided and mobilized through public interventions biennially in accordance with the modalities, procedures and guidelines

to be adopted by the Conference of the Parties serving as the meeting of the Parties to the Paris Agreement, at its first session, as stipulated in Article 13, paragraph 13. Other Parties are encouraged to do so.
8. The Financial Mechanism of the Convention, including its operating entities, shall serve as the financial mechanism of this Agreement.
9. The institutions serving this Agreement, including the operating entities of the Financial Mechanism of the Convention, shall aim to ensure efficient access to financial resources through simplified approval procedures and enhanced readiness support for developing country Parties, in particular for the least developed countries and small island developing States, in the context of their national climate strategies and plans.

Article 10

1. Parties share a long-term vision on the importance of fully realizing technology development and transfer in order to improve resilience to climate change and to reduce greenhouse gas emissions.
2. Parties, noting the importance of technology for the implementation of mitigation and adaptation actions under this Agreement and recognizing existing technology deployment and dissemination efforts, shall strengthen cooperative action on technology development and transfer.
3. The Technology Mechanism established under the Convention shall serve this Agreement.
4. A technology framework is hereby established to provide overarching guidance for the work of the Technology Mechanism in promoting and facilitating enhanced action on technology development and transfer in order to support the implementation of this Agreement, in pursuit of the long-term vision referred to in paragraph 1 of this Article.
5. Accelerating, encouraging and enabling innovation is critical for an effective, long-term global response to climate change and promoting economic growth and sustainable development. Such effort shall be, as appropriate, supported, including by the Technology Mechanism and, through financial means, by the Financial Mechanism of the Convention, for collaborative approaches to research and development, and facilitating access to technology, in particular for early stages of the technology cycle, to developing country Parties.
6. Support, including financial support, shall be provided to developing country Parties for the implementation of this Article, including for strengthening cooperative action on technology development and transfer at different stages of the technology cycle, with a view to achieving a balance between support for mitigation and adaptation. The global stocktake referred to in Article 14 shall take into account available information on efforts related to support on technology development and transfer for developing country Parties.

Article 11

1. Capacity-building under this Agreement should enhance the capacity and ability of developing country Parties, in particular countries with the least capacity, such as the least developed countries, and those that are particularly vulnerable to the adverse effects of climate change, such as small island developing States, to take

effective climate change action, including, inter alia, to implement adaptation and mitigation actions, and should facilitate technology development, dissemination and deployment, access to climate finance, relevant aspects of education, training and public awareness, and the transparent, timely and accurate communication of information.
2. Capacity-building should be country-driven, based on and responsive to national needs, and foster country ownership of Parties, in particular, for developing country Parties, including at the national, subnational and local levels. Capacity-building should be guided by lessons learned, including those from capacity-building activities under the Convention, and should be an effective, iterative process that is participatory, cross-cutting and gender-responsive.
3. All Parties should cooperate to enhance the capacity of developing country Parties to implement this Agreement. Developed country Parties should enhance support for capacity-building actions in developing country Parties.
4. All Parties enhancing the capacity of developing country Parties to implement this Agreement, including through regional, bilateral and multilateral approaches, shall regularly communicate on these actions or measures on capacity-building. Developing country Parties should regularly communicate progress made on implementing capacity-building plans, policies, actions or measures to implement this Agreement.
5. Capacity-building activities shall be enhanced through appropriate institutional arrangements to support the implementation of this Agreement, including the appropriate institutional arrangements established under the Convention that serve this Agreement. The Conference of the Parties serving as the meeting of the Parties to the Paris Agreement shall, at its first session, consider and adopt a decision on the initial institutional arrangements for capacity-building.

Article 12

Parties shall cooperate in taking measures, as appropriate, to enhance climate change education, training, public awareness, public participation and public access to information, recognizing the importance of these steps with respect to enhancing actions under this Agreement.

Article 13

1. In order to build mutual trust and confidence and to promote effective implementation, an enhanced transparency framework for action and support, with built-in flexibility which takes into account Parties' different capacities and builds upon collective experience is hereby established.
2. The transparency framework shall provide flexibility in the implementation of the provisions of this Article to those developing country Parties that need it in the light of their capacities. The modalities, procedures and guidelines referred to in paragraph 13 of this Article shall reflect such flexibility.
3. The transparency framework shall build on and enhance the transparency arrangements under the Convention, recognizing the special circumstances of the least developed countries and small island developing States, and be implemented in a facilitative, non-intrusive, non-punitive manner, respectful of national sovereignty, and avoid placing undue burden on Parties.
4. The transparency arrangements under the Convention, including national

communications, biennial reports and biennial update reports, international assessment and review and international consultation and analysis, shall form part of the experience drawn upon for the development of the modalities, procedures and guidelines under paragraph 13 of this Article.

5. The purpose of the framework for transparency of action is to provide a clear understanding of climate change action in the light of the objective of the Convention as set out in its Article 2, including clarity and tracking of progress towards achieving Parties' individual nationally determined contributions under Article 4, and Parties' adaptation actions under Article 7, including good practices, priorities, needs and gaps, to inform the global stocktake under Article 14.
6. The purpose of the framework for transparency of support is to provide clarity on support provided and received by relevant individual Parties in the context of climate change actions under Articles 4, 7, 9, 10 and 11, and, to the extent possible, to provide a full overview of aggregate financial support provided, to inform the global stocktake under Article 14.
7. Each Party shall regularly provide the following information:
    (a) A national inventory report of anthropogenic emissions by sources and removals by sinks of greenhouse gases, prepared using good practice methodologies accepted by the Intergovernmental Panel on Climate Change and agreed upon by the Conference of the Parties serving as the meeting of the Parties to the Paris Agreement;
    (b) Information necessary to track progress made in implementing and achieving its nationally determined contribution under Article 4.
8. Each Party should also provide information related to climate change impacts and adaptation under Article 7, as appropriate.
9. Developed country Parties shall, and other Parties that provide support should, provide information on financial, technology transfer and capacity-building support provided to developing country Parties under Article 9, 10 and 11.
10. Developing country Parties should provide information on financial, technology transfer and capacity-building support needed and received under Articles 9, 10 and 11.
11. Information submitted by each Party under paragraphs 7 and 9 of this Article shall undergo a technical expert review, in accordance with decision 1/CP.21. For those developing country Parties that need it in the light of their capacities, the review process shall include assistance in identifying capacity-building needs. In addition, each Party shall participate in a facilitative, multilateral consideration of progress with respect to efforts under Article 9, and its respective implementation and achievement of its nationally determined contribution.
12. The technical expert review under this paragraph shall consist of a consideration of the Party's support provided, as relevant, and its implementation and achievement of its nationally determined contribution. The review shall also identify areas of improvement for the Party, and include a review of the consistency of the information with the modalities, procedures and guidelines referred to in paragraph 13 of this Article, taking into account the flexibility accorded to the Party under paragraph 2 of this Article. The review shall pay particular attention to the respective national capabilities and circumstances of developing country Parties.
13. The Conference of the Parties serving as the meeting of the Parties to the Paris

Agreement shall, at its first session, building on experience from the arrangements related to transparency under the Convention, and elaborating on the provisions in this Article, adopt common modalities, procedures and guidelines, as appropriate, for the transparency of action and support.
14. Support shall be provided to developing countries for the implementation of this Article.
15. Support shall also be provided for the building of transparency-related capacity of developing country Parties on a continuous basis.

Article 14

1. The Conference of the Parties serving as the meeting of the Parties to the Paris Agreement shall periodically take stock of the implementation of this Agreement to assess the collective progress towards achieving the purpose of this Agreement and its long-term goals (referred to as the "global stocktake"). It shall do so in a comprehensive and facilitative manner, considering mitigation, adaptation and the means of implementation and support, and in the light of equity and the best available science.
2. The Conference of the Parties serving as the meeting of the Parties to the Paris Agreement shall undertake its first global stocktake in 2023 and every five years thereafter unless otherwise decided by the Conference of the Parties serving as the meeting of the Parties to the Paris Agreement.
3. The outcome of the global stocktake shall inform Parties in updating and enhancing, in a nationally determined manner, their actions and support in accordance with the relevant provisions of this Agreement, as well as in enhancing international cooperation for climate action.

Article 15

1. A mechanism to facilitate implementation of and promote compliance with the provisions of this Agreement is hereby established.
2. The mechanism referred to in paragraph 1 of this Article shall consist of a committee that shall be expert-based and facilitative in nature and function in a manner that is transparent, non-adversarial and non-punitive. The committee shall pay particular attention to the respective national capabilities and circumstances of Parties.
3. The committee shall operate under the modalities and procedures adopted by the Conference of the Parties serving as the meeting of the Parties to the Paris Agreement at its first session and report annually to the Conference of the Parties serving as the meeting of the Parties to the Paris Agreement.

Article 16

1. The Conference of the Parties, the supreme body of the Convention, shall serve as the meeting of the Parties to this Agreement.
2. Parties to the Convention that are not Parties to this Agreement may participate as observers in the proceedings of any session of the Conference of the Parties serving as the meeting of the Parties to this Agreement. When the Conference of the Parties serves as the meeting of the Parties to this Agreement, decisions under this Agreement shall be taken only by those that are Parties to this Agreement.
3. When the Conference of the Parties serves as the meeting of the Parties to this

Agreement, any member of the Bureau of the Conference of the Parties representing a Party to the Convention but, at that time, not a Party to this Agreement, shall be replaced by an additional member to be elected by and from amongst the Parties to this Agreement.
4. The Conference of the Parties serving as the meeting of the Parties to the Paris Agreement shall keep under regular review the implementation of this Agreement and shall make, within its mandate, the decisions necessary to promote its effective implementation. It shall perform the functions assigned to it by this Agreement and shall:
   (a) Establish such subsidiary bodies as deemed necessary for the implementation of this Agreement; and
   (b) Exercise such other functions as may be required for the implementation of this Agreement.
5. The rules of procedure of the Conference of the Parties and the financial procedures applied under the Convention shall be applied mutatis mutandis under this Agreement, except as may be otherwise decided by consensus by the Conference of the Parties serving as the meeting of the Parties to the Paris Agreement.
6. The first session of the Conference of the Parties serving as the meeting of the Parties to the Paris Agreement shall be convened by the secretariat in conjunction with the first session of the Conference of the Parties that is scheduled after the date of entry into force of this Agreement. Subsequent ordinary sessions of the Conference of the Parties serving as the meeting of the Parties to the Paris Agreement shall be held in conjunction with ordinary sessions of the Conference of the Parties, unless otherwise decided by the Conference of the Parties serving as the meeting of the Parties to the Paris Agreement.
7. Extraordinary sessions of the Conference of the Parties serving as the meeting of the Parties to the Paris Agreement shall be held at such other times as may be deemed necessary by the Conference of the Parties serving as the meeting of the Parties to the Paris Agreement or at the written request of any Party, provided that, within six months of the request being communicated to the Parties by the secretariat, it is supported by at least one third of the Parties.
8. The United Nations and its specialized agencies and the International Atomic Energy Agency, as well as any State member thereof or observers thereto not party to the Convention, may be represented at sessions of the Conference of the Parties serving as the meeting of the Parties to the Paris Agreement as observers. Any body or agency, whether national or international, governmental or non-governmental, which is qualified in matters covered by this Agreement and which has informed the secretariat of its wish to be represented at a session of the Conference of the Parties serving as the meeting of the Parties to the Paris Agreement as an observer, may be so admitted unless at least one third of the Parties present object. The admission and participation of observers shall be subject to the rules of procedure referred to in paragraph 5 of this Article.

Article 17
1. The secretariat established by Article 8 of the Convention shall serve as the secretariat of this Agreement.
2. Article 8, paragraph 2, of the Convention on the functions of the secretariat, and

Article 8, paragraph 3, of the Convention, on the arrangements made for the functioning of the secretariat, shall apply mutatis mutandis to this Agreement. The secretariat shall, in addition, exercise the functions assigned to it under this Agreement and by the Conference of the Parties serving as the meeting of the Parties to the Paris Agreement.

Article 18
1. The Subsidiary Body for Scientific and Technological Advice and the Subsidiary Body for Implementation established by Articles 9 and 10 of the Convention shall serve, respectively, as the Subsidiary Body for Scientific and Technological Advice and the Subsidiary Body for Implementation of this Agreement. The provisions of the Convention relating to the functioning of these two bodies shall apply mutatis mutandis to this Agreement. Sessions of the meetings of the Subsidiary Body for Scientific and Technological Advice and the Subsidiary Body for Implementation of this Agreement shall be held in conjunction with the meetings of, respectively, the Subsidiary Body for Scientific and Technological Advice and the Subsidiary Body for Implementation of the Convention.
2. Parties to the Convention that are not Parties to this Agreement may participate as observers in the proceedings of any session of the subsidiary bodies. When the subsidiary bodies serve as the subsidiary bodies of this Agreement, decisions under this Agreement shall be taken only by those that are Parties to this Agreement.
3. When the subsidiary bodies established by Articles 9 and 10 of the Convention exercise their functions with regard to matters concerning this Agreement, any member of the bureau of those subsidiary bodies representing a Party to the Convention but, at that time, not a Party to this Agreement, shall be replaced by an additional member to be elected by and from amongst the Parties to this Agreement.

Article 19
1. Subsidiary bodies or other institutional arrangements established by or under the Convention, other than those referred to in this Agreement, shall serve this Agreement upon a decision of the Conference of the Parties serving as the meeting of the Parties to the Paris Agreement. The Conference of the Parties serving as the meeting of the Parties to the Paris Agreement shall specify the functions to be exercised by such subsidiary bodies or arrangements.
2. The Conference of the Parties serving as the meeting of the Parties to the Paris Agreement may provide further guidance to such subsidiary bodies and institutional arrangements.

Article 20
1. This Agreement shall be open for signature and subject to ratification, acceptance or approval by States and regional economic integration organizations that are Parties to the Convention. It shall be open for signature at the United Nations Headquarters in New York from 22 April 2016 to 21 April 2017. Thereafter, this Agreement shall be open for accession from the day following the date on which it is closed for signature. Instruments of ratification, acceptance, approval or accession shall be deposited with the Depositary.
2. Any regional economic integration organization that becomes a Party to this

Agreement without any of its member States being a Party shall be bound by all the obligations under this Agreement. In the case of regional economic integration organizations with one or more member States that are Parties to this Agreement, the organization and its member States shall decide on their respective responsibilities for the performance of their obligations under this Agreement. In such cases, the organization and the member States shall not be entitled to exercise rights under this Agreement concurrently.
3. In their instruments of ratification, acceptance, approval or accession, regional economic integration organizations shall declare the extent of their competence with respect to the matters governed by this Agreement. These organizations shall also inform the Depositary, who shall in turn inform the Parties, of any substantial modification in the extent of their competence.

Article 21

1. This Agreement shall enter into force on the thirtieth day after the date on which at least 55 Parties to the Convention accounting in total for at least an estimated 55 percent of the total global greenhouse gas emissions have deposited their instruments of ratification, acceptance, approval or accession.
2. Solely for the limited purpose of paragraph 1 of this Article, "total global greenhouse gas emissions" means the most up-to-date amount communicated on or before the date of adoption of this Agreement by the Parties to the Convention.
3. For each State or regional economic integration organization that ratifies, accepts or approves this Agreement or accedes thereto after the conditions set out in paragraph 1 of this Article for entry into force have been fulfilled, this Agreement shall enter into force on the thirtieth day after the date of deposit by such State or regional economic integration organization of its instrument of ratification, acceptance, approval or accession.
4. For the purposes of paragraph 1 of this Article, any instrument deposited by a regional economic integration organization shall not be counted as additional to those deposited by its member States.

Article 22

The provisions of Article 15 of the Convention on the adoption of amendments to the Convention shall apply mutatis mutandis to this Agreement.

Article 23

1. The provisions of Article 16 of the Convention on the adoption and amendment of annexes to the Convention shall apply mutatis mutandis to this Agreement.
2. Annexes to this Agreement shall form an integral part thereof and, unless otherwise expressly provided for, a reference to this Agreement constitutes at the same time a reference to any annexes thereto. Such annexes shall be restricted to lists, forms and any other material of a descriptive nature that is of a scientific, technical, procedural or administrative character.

Article 24

The provisions of Article 14 of the Convention on settlement of disputes shall apply mutatis mutandis to this Agreement.

Article 25
1. Each Party shall have one vote, except as provided for paragraph 2 of this Article.
2. Regional economic integration organizations, in matters within their competence, shall exercise their right to vote with a number of votes equal to the number of their member States that are Parties to this Agreement. Such an organization shall not exercise its right to vote if any of its member States exercises its right, and vice versa.

Article 26
The Secretary-General of the United Nations shall be the Depositary of this Agreement. Article 27 No reservations may be made to this Agreement.

Article 27
No reservations may be made to this Agreement.

Article 28
1. At any time after three years from the date on which this Agreement has entered into force for a Party, that Party may withdraw from this Agreement by giving written notification to the Depositary.
2. Any such withdrawal shall take effect upon expiry of one year from the date of receipt by the Depositary of the notification of withdrawal, or on such later date as may be specified in the notification of withdrawal.
3. Any Party that withdraws from the Convention shall be considered as also having withdrawn from this Agreement.

Article 29
The original of this Agreement, of which the Arabic, Chinese, English, French, Russian and Spanish texts are equally authentic, shall be deposited with the Secretary-General of the United Nations.

DONE at Paris this twelfth day of December two thousand and fifteen.

IN WITNESS WHEREOF, the undersigned, being duly authorized to that effect, have signed this Agreement.

Analysis

On September 21, 2014, approximately 400,000 people participated in the Peoples Climate March in New York City, New York. The demonstration was the single largest climate change related protest ever held. The group demanded that world leaders craft a document at the Conference of the Parties (COP) 21, scheduled for December 2015 in Paris, France, that would truly prevent the escalation of global climate change by creating a legally binding agreement to curtail the emission of greenhouse gases (GHGs). Similar events were held the same day in more than 160 countries.

In the months preceding the Paris meeting, countries around the world began trying to shape the accord through the media. Industrialized countries such as China and the United States began floating the idea that the goal of the meeting should be to hold the increase in global temperatures to 2° Celsius above the level that existed during pre-industrial times. For island nations, such as the Marshall Islands, that goal was unacceptable. If the global temperature were allowed to reach the 2° Celsius goal, their countries would be underwater. For nations whose territory was barely above sea level, the maximum acceptable level of temperature growth was 1.5° Celsius above the temperature prior to the Industrial Revolution. This fundamental disagreement colored much of the negotiations in Paris. Ultimately, in the agreement, both sides claimed to be satisfied with the compromise. The final goal was to hold the increase in global temperatures significantly below 2° Celsius. The hope was that the ultimate increase would be below the 1.5° Celsius mark.

Whether the 2° Celsius goal was even attainable at the meeting is open for debate. In advance of COP21, countries were required to set preliminary goals for their individual countries as a demonstration that they were serious about addressing the amount of GHGs emitted within their borders. Although the respective countries were lauded for their supposed commitment to the cause, their initial pledges signaled that much work needed to be done. Even if every country that made an initial commitment actually did what they pledged, the global temperature increase would have been significantly above the 2° Celsius above preindustrial level.

COP21 opened with great fanfare as it was billed as the starting-point of a global effort to eradicate GHGs. To accomplish that goal, just about every country would have to abandon their use of carbon-based fuels, such as coal, by 2050. Participants were justifiably optimistic because the two largest emitters of GHGs, namely China and the United States, had committed to helping lead the global effort. This was significant because since the Kyoto, Japan negotiations, both countries had been reticent to participate in climate change agreements negotiated under the auspices of the United Nations.

The Paris Agreement created a framework for the respective countries to be accountable to each other to ensure that all did their part to protect the environment. Initially, every country had to make an inventory of the sources of GHG emissions within their borders. Each then had to declare their "Nationally Determined Contribution" every five years, meaning how much their country would reduce GHG emissions during that period. To ensure compliance, every country was subject to review by representatives from other countries. Unfortunately, the framers of the agreement did not include penalties for countries that failed to live up to their obligations. It was believed that international peer pressure would suffice to encourage countries to do their part to fix a global problem.

Such a large global effort requires funding to succeed. Although for the first time in an international climate agreement every country was officially expected to help pay expenses, the onus to pay the vast majority of the $100 billion-a-year cost was assigned to industrialized countries such as the United States. "Developing countries," namely most of the world, were expected to contribute as much as they could, but in truth, they were intended to be the recipients, not source, of the money.

The document opened for signatures on April 22, 2016. Not coincidentally, that was Earth Day. A ceremony was held in New York City, New York at the United Nations where dignitaries from approximately 175 countries signed the Paris Agreement. The signing did not make the Paris Agreement international law, but was instead a symbolic commitment to have the agreement ratified within the respective countries.

The ratification process has been the undoing of many similar initiatives, most notably the Kyoto Protocol to the United Nations Framework Convention on Climate Change. During the Kyoto negotiations, the will of the United States Senate had been ignored. In a unanimous vote held months before negotiations began in Japan, the United States Senate had clearly stated that it would not ratify the Kyoto Protocol if certain conditions were not met. Their warning was ignored and the unacceptable conditions were included in the final document. The consequence was that the United States never ratified the Kyoto Protocol. That single act doomed the Kyoto Protocol's effectiveness since, at the time, the United States was the world's largest producer of GHGs. To avoid that fate, since Senators had already voiced similar threats for the Paris Agreement, negotiators from the United States attempted to bypass the Senate by carefully parsing words so that they could claim that the agreement was not a "treaty." Treaties, by law, have to be ratified by the Senate. If it was not a treaty, then President Barack Obama could make it official through Executive Action. Whether the Paris Agreement is a treaty or not will likely be determined by the federal courts sometime in the future. Despite the public assurances of the Obama administration that the Paris Agreement is binding to all parties, the selective nature of the wording almost certainly ensures that it is not. If it is binding, then it is a treaty. If the agreement is ultimately determined to be a treaty, it is doubtful that the Senate, as it is comprised in 2016, would ratify the document.

Further Reading

Afionis, Stavros. 2016. *The European Union in International Climate Change Negotiations.* New York: Routledge.

Cañete, Miguel Arias. 2016. "Paris Is More Than the Deal." *Vital Speeches of the Day* 82: 40–42.

Clémençon, Raymond. 2016. "The Two Sides of the Paris Climate Agreement." *Journal of Environment & Development* 25: 3–24.

Eastwood, Lauren. 2016. *Negotiating the Environment: Civil Society, Globalization, and the UN.* New York: Routledge.

Bibliography

Adger, W. Neil. 2006. *Fairness in Adaptation to Climate Change.* Cambridge, MA: MIT Press.

Afionis, Stavros. 2016. *The European Union in International Climate Change Negotiations.* New York: Routledge.

Agarwal, Anil, and Sunita Narain. 1991. *Global Warming in an Unequal World: A Case of Environmental Colonialism.* New Delhi: Centre for Science and Development.

Ager, Derek. 1993. *The New Catastrophism: The Importance of the Rare Event in Geological History.* New York: Cambridge University Press.

Alley, Richard B. 2000. *The Two-Mile Time Machine: Ice Cores, Climate Change, and Our Future.* Princeton, NJ: Princeton University Press.

Anderegg, William R. L., et al. 2010. "Expert Credibility in Climate Change." *Proceedings of the National Academy of Sciences* 107. http://www.pnas.org/content/107/27/12107.full.

Andrews, Richard N. L. 1999. *Managing the Environment, Managing Ourselves.* New Haven, CT: Yale University Press.

Ansolabehere, Stephen, and David M. Konisky. 2014. *Cheap and Clean: How Americans Think About Energy in the Age of Global Warming.* Cambridge, MA: MIT Press.

Antholis, William, and Strobe Talbot. 2010. *Fast Forward: Ethics and Politics in the Age of Global Warming.* Washington, D.C.: Brookings Institution Press.

Archer, David. 2008. *The Long Thaw: How Humans Are Changing the Next 100,000 Years of Earth's Climate.* Princeton, NJ: Princeton University Press.

Archer, David. 2011. *Global Warming: Understanding the Forecast.* 2d ed. New York: John Wiley & Sons.

Archer, David, and Raymond Pierrehumbert, eds. 2011. *The Warming Papers: The Scientific Foundation for the Climate Change Forecast.* New York: Wiley-Blackwell.

Arctic Climate Impact Assessment—Scientific Report. 2005. New York: Cambridge University Press.

Athanasiou, Tom. 1998. *Divided Planet: The Ecology of Rich and Poor.* Athens: University of Georgia Press.

Austin, Elizabeth. 2016. *Treading on Thin Air: Atmospheric Physics, Forensic Meteorology, and Climate Change: How Weather Shapes Our Everyday Lives.* New York: Pegasus.

Babiker, Mustafa, et al. 2002. "The Evolution of a Climate Regime: Kyoto to Marrakech and Beyond." *Environmental Science & Policy* 5: 195–206.

Barnosky, Anthony D. 2009. *Heatstroke: Nature in an Age of Global Warming.* Washington, D.C.: Island.

Bartsch, Ulrich, and Benito Müller. 2000. *Fossil Fuels in a Changing Climate: Impacts of the Kyoto Protocol and Developing Country Participation.* New York: Oxford University Press.

Benedick, Richard Elliott. 1998. *Ozone Diplomacy: New Directions in Safeguarding the Planet.* Cambridge, MA: Harvard University Press.

Bernard, Lucas, and Willi Semmler, eds. 2015. *The Oxford Handbook of the Macroeconomics of Global Warming.* New York: Oxford University Press.

Besel, Richard D. 2013. "Accommodating Climate Change Science: James Hansen and the Rhetorical/Political Emergence of Global Warming." *Science in Context* 26: 137–152.

Bierly, Eugene W. 1988. "The World Climate Program: Collaboration and Communication on a Global Scale." *Annals of the American Academy of Political and Social Science* 495: 106–116.

Black, Brian. 2012. *Crude Reality: Petroleum in World History.* New York: Rowman and Littlefield.

Black, Brian C., ed. 2013. *Climate Change: An Encyclopedia of Science and History.* Santa Barbara, CA: ABC-Clio.

Bodansky, Daniel. 1993. "The United Nations Framework Convention on Climate Change: A Commentary." *Yale Journal of International Law* 18: 451–558.

Bodansky, Daniel. 1995. "The Emerging Climate Change Regime." *Annual Review of Energy and Environment* 20:425–461.

Bodansky, Daniel, Jutta Brunnee, and Lavanya Rajamani. 2016. *International Climate Change Law*. New York: Oxford University Press.

Boehmer-Christiansen, Sonja, and Aynsley J. Kellow. 2002. *International Environmental Policy: Interests and the Failure of the Kyoto Process*. Northampton, MA: Edward Elgar.

Bolin, Bert. 2007. *A History of the Science and Politics of Climate Change: The Role of the Intergovernmental Panel on Climate Change*. New York: Cambridge University Press.

Bord, Richard J., et al. 1998. "Public Perceptions of Global Warming: United States and International Perspectives." *Climate Research* 11: 75–84.

Bostrom, Nick, and Milan M. Cirkovic, eds. 2008. *Global Catastrophic Risks*. New York: Oxford University Press.

Botkin, Daniel B. 1990. *Discordant Harmonies: A New Ecology for the Twenty-First Century*. New York: Oxford University Press.

Bowen, Mark. 2005. *Thin Ice: Unlocking the Secrets of Climate in the World's Highest Mountains*. New York: Henry Holt.

Bowen, Mark. 2008. *Censoring Science: Inside the Political Attack on Dr. James Hansen and the Truth of Global Warming*. New York: Dutton.

Boykoff, Maxwell T. 2011. *Who Speaks for the Climate? Making Sense of Media Reporting on Climate Change*. New York: Cambridge University Press.

Braasch, Gary. 2007. *Earth Under Fire: How Global Warming Is Changing the World*. Berkeley: University of California Press.

Brenton, Tony. 1994. *The Greening of Machiavelli: The Evolution of International Environmental Politics*. London: Earthscan.

Brewer, Thomas L. 2014. *The United States in a Warming World: The Political Economy of Government, Business, and Public Responses to Climate Change*. New York: Cambridge University Press.

Bristow, Tom, and Thomas H. Ford, eds. 2016. *A Cultural History of Climate Change*. New York: Routledge.

Broecker, Wallace S. 2010. *The Great Ocean Conveyer: Discovering the Trigger for Abrupt Climate Change*. Princeton, NJ: Princeton University Press.

Broecker, Wallace S., and Robert Kunzig. 2008. *Fixing Climate: What Past Climate Changes Reveal About the Current Threat—And How to Counter It*. New York: Hill and Wang.

Brooke, John L. 2014. *Climate Change and the Course of Global History: A Rough Journey*. New York: Cambridge University Press.

Brown, Lester R. 2009. *Plan B 4.0: Mobilizing to Save Civilization*. New York: W.W. Norton.

Bruce, James P., et al., eds. 1996. *Climate Change 1995: Economic and Social Dimensions of Climate Change*. New York: Cambridge University Press.

Bulkeley, Harriet. 2015. *Accomplishing Climate Governance*. New York: Cambridge University Press.

Bulkeley, Harriet A., and Peter Newell. 2015. *Governing Climate Change*. 2d ed. New York: Routledge.

Burch, John R., Jr. 2015. *Water Rights and the Environment in the United States: A Documentary and Reference Guide*. Santa Barbara, CA: Greenwood.

Burch, Sarah L., and Sara E. Harris. 2014. *Understanding Climate Change: Science, Policy, and Practice*. Toronto: University of Toronto Press.

Burkett, Virginia, and Margaret Davidson, eds. 2013. *Coastal Impacts, Adaptation, and Vulnerabilities: A Technical Input to the 2013 National Climate Assessment*. Washington, D.C.: Island.

Burroughs, William James. 2007. *Climate Change: A Multidisciplinary Approach*. 2d ed. New York: Cambridge University Press.

Calvin, William H. 2008. *Global Fever: How to Treat Climate Change*. Chicago: University of Chicago Press.

Campbell, Kurt M., ed. 2008. *Climatic Cataclysm: The Foreign Policy and National Security Implications of Climate Change*. Washington, D.C.: Brookings Institution.

Cañete, Miguel Arias. 2016. "Paris Is More than the Deal." *Vital Speeches of the Day* 82: 40–42.

Cannon, Jonathan Z. 2015. *Environment in the Balance: The Green Movement and the Supreme Court*. Cambridge, MA: Harvard University Press.

Carlarne, Cinnamon P., Kevin R. Gray, and Richard Tarasofsky, eds. 2016. *Oxford Handbook of International Climate Change Law*. New York: Oxford University Press.

Chambers, Frank, and Michael Ogle, eds. 2002. *Climate Change: Critical Concepts in the Environment and Physical Geography*. New York: Routledge.

Chan, Gabriel, et al. 2012. "The So2 Allowance-

Trading System and the Clean Air Act Amendments of 1990: Reflections on 20 Years of Policy Innovation." *National Tax Journal* 65: 419–452.

Chichilnisky, Graciela, and Kristin A. Sheeran. 2009. *Saving Kyoto: An Insider's Guide to How It Works, Why It Matters and What It Means for the Future.* London: New Holland.

Choi, Inho. 2005. "Global Climate Change and the Use of Economic Approaches: The Ideal Design Features of Domestic Greenhouse Gas Emissions Trading with an Analysis of the European Union's CO_2 Emissions Directive and the Climate Stewardship Act." *Natural Resources Journal* 45: 865–952.

Christian-Smith, Juliet, and Peter H. Gleick. 2012. *A Twenty-First Century US Water Policy.* New York: Oxford University Press.

Christie, Maureen. 2000. *The Ozone Layer: A Philosophy of Science Perspective.* New York: Cambridge University Press.

Ciplet, David, J. Timmons Roberts, and Mizan R. Khan. 2015. *Power in a Warming World: The New Global Politics of Climate Change and the Remaking of Environmental Inequality.* Cambridge, MA: MIT Press.

Clémençon, Raymond. 2016. "The Two Sides of the Paris Climate Agreement." *Journal of Environment & Development* 25: 3–24.

Committee on National Security Implications of Climate Change for U.S. Naval Forces, National Research Council of the National Academies. 2011. *National Security Implications of Climate Change for U.S. Naval Forces.* Washington, D.C.: National Academies Press.

Cooper, Gail. 2002. *Air-Conditioning America.* Baltimore: Johns Hopkins University Press.

Corfee-Morlot, Jan, Mark Maslin, and Jacqueline Burgess. 2007. "Climate Science in the Public Sphere." *Philosophical Transactions of the Royal Society* 365: 2741–2746.

Cotton, William R., and Roger A. Pielke Sr. 2007. *Human Impacts on Weather and Climate.* 2d ed. New York: Cambridge University Press.

Council on Environmental Quality and the Department of State. 1980. *The Global 2000 Report to the President: Entering the Twenty-First Century.* 3 vols. Washington, D.C.: Government Printing Office.

Cowie, Jonathan. 2007. *Climate Change: Biological and Human Aspects.* New York: Cambridge University Press.

Cox, John D. 2005. *Climate Crash: Abrupt Climate Change and What It Means for Our Future.* Washington, D.C.: National Academies.

Crosby, Alfred. 2006. *Children of the Sun: A History of Humanity's Unappeasable Appetite for Energy.* New York: Norton.

Cuff, David J., and Andrew S. Goudie. 2009. *Oxford Companion to Global Change,* New York: Oxford University Press.

Dalton, Meghan M., Philip Mote, and Amy K. Snover, eds. 2013. *Climate Change in the Northwest: Implications for Our Landscapes, Waters, and Communities.* Washington, D.C.: Island.

Danver, Steven L., and John R. Burch Jr., eds. 2011. *Encyclopedia of Water Politics and Policy in the United States.* Washington, D.C.: Congressional Quarterly Press.

Darst, Robert G. 2001. *Smokestack Diplomacy: Cooperation and Conflict in East—West Environmental Politics.* Cambridge, MA: MIT Press.

Davis, Neil. 2001. *Permafrost: A Guide to Frozen Ground in Transition.* Fairbanks: University of Alaska Press.

DeBuys, William Eno. 2011. *A Great Aridness: Climate Change and the Future of the American Southwest.* New York: Oxford University Press.

Dessler, Andrew E. 2015. *Introduction to Modern Climate Change.* 2d ed. New York: Cambridge University Press.

Dessler, Andrew E., and Edward A. Parson. 2006. *The Science and Politics of Global Climate Change: A Guide to the Debate.* New York: Cambridge University Press.

Diamond, Jared. 2004. *Collapse: How Societies Choose to Fail or Succeed.* New York: Viking.

DiMento, Joseph F. C., and Pamela Doughman, eds. 2007. *Climate Change: What It Means for Us, Our Children, and Our Grandchildren.* Cambridge, MA: MIT Press.

Dimitrov, Radoslav S. 2010. "Inside UN Climate Change Negotiations: The Copenhagen Conference." *Review of Policy Research* 27: 795–821.

Dong, Wenjie, et al. 2015. *Atlas of Climate Change: Responsibility and Obligation of Human Society.* New York: Springer.

Dornbusch, Rudiger, and James M. Poterba, eds. 1991. *Global Warming: Economic Policy Responses.* Cambridge, MA: MIT Press.

Dow, Kirstin, and Thomas E. Downing. 2011. *The Atlas of Climate Change: Mapping the World's Greatest Challenge.* 3d ed. Berkeley: University of California Press.

Dryzek, John S., Richard B. Norgaard, and David Schlosberg. 2013. *Climate-Challenged Society.* New York: Oxford University Press.

Dunlap, Riley E., and Aaron M. McCright. 2008. "A Widening Gap: Republican and Democratic Views on Climate Change." *Environment.* http://www.environmentmagazine.org/archives/back%20issues/september-october%202008/dunlap-full.html.

Dunlap, Riley E., and Peter J. Jacques. 2013. "Climate Change Denial Books and Conservative Think Tanks: Exploring the Connection." *American Behavioral Scientist* 57: 699–731.

Eastwood, Lauren. 2016. *Negotiating the Environment: Civil Society, Globalization, and the UN*. New York: Routledge.

Edwards, Paul N. 2010. *A Vast Machine: Computer Models, Climate Data, and the Politics of Global Warming*. Cambridge, MA: MIT Press.

Emanuel, Kerry A. 2007. *What We Know About Climate Change*. Cambridge, MA: MIT Press.

Energy Information Administration, Office of Integrated Analysis and Forecasting, U.S. Department of Energy. 2000. *Analysis of Strategies for Reducing Multiple Emissions from Power Plants: Sulfur Dioxide, Nitrogen Oxides, and Carbon Dioxide*. Washington, D.C.: Energy Information Administration. http://www.eia.gov/oiaf/servicerpt/powerplants/pdf/sroiaf%282000%2905.pdf.

Engel, Kirsten. 2006. "State and Local Climate Initiatives: What Is Motivating State and Local Governments to Address a Global Problem and What Does This Say About Federalism and Environmental Law?" *The Urban Lawyer* 38: 1015–1029.

Ferrey, Steven. 2012. "Changing Venue of International Governance and Finance: Exercising Legal Control Over the $100 Billion Per Year Climate Fund?" *Wisconsin International Law Journal* 30: 26–111.

Fisher, Dana R. 2006. "Bringing the Material Back In: Understanding the U.S. Position on Climate Change." *Sociological Forum* 21: 467–494.

Fisher, Elizabeth. 2013. "Climate Change Litigation, Obsession and Expertise: Reflecting on the Scholarly Response to Massachusetts v. EPA." *Law & Policy* 35: 236–260.

Flannery, Tim. 2005. *The Weather Makers: How Man Is Changing Climate and What It Means for Life on Earth*. New York: Atlantic Monthly.

Flannery, Tim. 2015. *Atmosphere of Hope: Searching for Solutions to the Climate Crisis*. New York: Atlantic Monthly.

Fleming, James R. 2007. *The Callendar Effect: The Life and Work of Guy Stewart Callendar (1898–1964), the Scientist Who Established the Carbon Dioxide Theory of Climate Change*. Boston: American Meteorological Society.

Fleming, James R. 2010. *Fixing the Sky: The Checkered History of Weather and Climate Control*. New York: Columbia University Press.

Fleming, James Rodger. 1998. *Historical Perspectives on Climate Change*. New York: Oxford University Press.

Freeman, Jody, and Adrian Vermeule. 2007. "Massachusetts v EPA: From Politics to Expertise." *Supreme Court Review* 2007: 51–110.

Friel, Howard. 2010. *The Lomborg Deception: Setting the Record Straight About Global Warming*. New Haven, CT: Yale University Press.

Gareau, Brian J. 2013. *From Precaution to Profit: Contemporary Challenges to Environmental Protection in the Montreal Protocol*. New Haven, CT: Yale University Press.

Garfin, Gregg, et al. 2013. *Assessment of Climate Change in Southwest United States: A Report Prepared for the National Climate Assessment*. Washington, D.C.: Island.

Gelbspan, Ross. 2004. *Boiling Point: How Politicians, Big Oil and Coal, Journalists and Activists Are Fueling the Climate Crisis—And What We Can Do to Avert Disaster*. New York: Basic.

Goodell, Jeff. 2006. *Big Coal: The Dirty Secret Behind America's Energy Future*. New York: Houghton-Mifflin.

Goodstein, Eban. 2007. *Fighting for Love in the Century of Extinction: How Passion and Politics Can Stop Global Warming*. Burlington: University of Vermont Press.

Gore, Al. 1992. *Earth in the Balance: Ecology and the Human Spirit*. New York: Houghton Mifflin.

Gore, Al. 2006. *An Inconvenient Truth: The Planetary Emergency of Global Warming and What We Can Do About It*. New York: Rodale.

Gore, Albert. 2009. *Our Choice: A Plan to Solve the Climate Crisis*. New York: Rodale.

Grenthe, Ingmar. 1995. *The Nobel Prize in Chemistry 1995*. http://www.nobelprize.org/nobel_prizes/chemistry/laureates/1995/presentation-speech.html (accessed March 30, 2016).

Griffis, Roger, and Jennifer Howard, eds. 2013. *Ocean and Marine Resources in a Changing Climate: A Technical Input to the 2013 National Climate Assessment*. Washington, D.C.: Island.

Grover, Velma, ed. 2008. *Global Warming and Climate Change: Ten Years After Kyoto and Still Counting*. Enfield, NH: Science.

Guptu, Joyeeta. 2014. *The History of Global Climate Governance*. New York: Cambridge University Press.

Guzman, Andrew. 2013. *Overheated: The Human Cost of Climate Change*. New York: Oxford University Press.

Hadden, Jennifer. 2015. *Networks in Contention: The Divisive Politics of Climate Change*. New York: Cambridge University Press.

Hansen, James E. 2009. *Storms of My Grandchildren: The Truth About the Coming Climate Catastrophe and Our Last Chance to Save Humanity*. New York: Bloomsbury.

Harris, Paul G., ed. 2000. *Climate Change and American Foreign Policy*. New York: St. Martin's.

Harris, Paul G., ed. 2001. *The Environment, International Relations, and U.S. Foreign Policy*. Washington, D.C.: Georgetown University Press.

Harris, Paul G. 2013. *What's Wrong with Climate Politics and How to Fix It*. New York: Polity.

Harris, Paul G. 2016. *Global Ethics and Climate Change*. 2d ed. Edinburgh: Edinburgh University Press.

Hatfield, Jerry L, et al., eds. 2014. *Climate Change in the Midwest: A Synthesis Report for the National Climate Assessment*. Washington, D.C.: Island.

Hays, Samuel P. 1993. *Beauty, Health, and Permanence: Environmental Politics in the United States*. New York: Cambridge University Press.

Hays, Samuel P. 1998. *Explorations in Environmental History: Essays by Samuel P. Hays*. Pittsburgh: University of Pittsburgh Press.

Helm, Dieter. 2012. *The Carbon Crunch: How We're Getting Climate Change Wrong—And How to Fix It*. New Haven, CT: Yale University Press.

Henson, Robert. 2014. *The Thinking Person's Guide to Climate Change*. Boston: American Meteorological Society.

Hewitt, William F. 2013. *A Newer World: Politics, Money, Technology, and What's Really Being Done to Solve the Climate Crisis*. Durham: New Hampshire University Press.

Hillstrom, Kevin, ed. 2010. *U.S. Environmental Policy and Politics: A Documentary History*. Washington, D.C.: Congressional Quarterly.

Hoffman, Andrew J. 2015. *How Culture Shapes the Climate Change Debate*. Stanford, CA: Stanford Briefs.

Hoffman, Matthew J. 2011. *Climate Governance at the Crossroads: A Global Response After Kyoto*. New York: Oxford University Press.

Hopgood, Stephen. 1998. *American Foreign Environmental Policy and the Power of the State*. New York: Oxford University Press.

Houghton, John T. 2015. *Global Warming: The Complete Briefing*. 5th ed. New York: Cambridge University Press.

Hovi, John, Detlef F. Sprinz, and Guri Bang. 2012. "Why the United States Did Not Become a Party to the Kyoto Protocol: German, Norwegian, and US Perspectives." *European Journal of International Relations* 18: 129–150.

Howe, Joshua P. 2014. *Behind the Curve: Science and the Politics of Global Warming*. Seattle: University of Washington Press.

Hoyt, Douglas V., and Kenneth Schatten. 1997. *The Role of the Sun in Climate Change*. New York: Oxford University Press.

Hulme, Mike. 2009. *Why We Disagree About Climate Change: Understanding Controversy, Inaction, and Opportunity*. New York: Cambridge University Press.

Incropera, Frank P. 2015. *Climate Change: A Wicked Problem: Complexity and Uncertainty at the Intersection of Science, Economics, Politics, and Human Behavior*. New York: Cambridge University Press.

Ingram, B. Lynn, and Frances Malamud-Roam. 2013. *The West Without Water: What Past Floods, Droughts, and Other Climate Clues Tell Us About Tomorrow*. Berkeley: University of California Press.

Ingram, Keith, et al., eds. 2013. *Climate of the Southeast United States: Variability, Change, Impacts, and Vulnerability*. Washington, D.C.: Island.

Inhofe, James. 2012. *The Greatest Hoax: How the Global Warming Conspiracy Threatens Your Future*. Washington, D.C.: WND.

Intergovernmental Panel on Climate Change. 2005. *Safeguarding the Ozone Layer and the Global Climate System*. New York: Cambridge University Press.

Intergovernmental Panel on Climate Change. 2007. *Climate Change 2007: The Physical Science Basis*. New York: Cambridge University Press.

Intergovernmental Panel on Climate Change. 2008. *Climate Change 2007: Impacts, Adaptation and Vulnerability*. New York: Cambridge University Press.

Intergovernmental Panel on Climate Change. 2008. *Climate Change 2007: Mitigation of Climate Change*. New York: Cambridge University Press.

Intergovernmental Panel on Climate Change. 2014. *Climate Change 2013: The Physical Science Basis*. New York: Cambridge University Press.

Intergovernmental Panel on Climate Change. 2014. *Climate Change 2014: Impacts, Adaptation, and Vulnerability: Part A: Global Sectoral Aspects: Vol. 1*. New York: Cambridge University Press.

Intergovernmental Panel on Climate Change. 2014. *Climate Change 2014: Impacts, Adaptation, and Vulnerability: Part B: Regional Aspects: Vol. 2*. New York: Cambridge University Press.

Intergovernmental Panel on Climate Change. 2015. *Climate Change 2014: Mitigation of Climate Change*. New York: Cambridge University Press.

Ionesco, Dina, Dana Mokhnacheva, and François Gemenne. 2016. *The Atlas of Environmental Migration*. New York: Routledge.

Jamieson, Dale. 2014. *Reason in a Dark Time: Why the Struggle Against Climate Change Failed—And What It Means for Our Future*. New York: Oxford University Press.

Jenkins, Jerry. 2010. *Climate Change in the Adirondacks: The Path to Sustainability*. Ithaca, NY: Comstock.

Jouzel, Jean, Claude Lorius, and Dominique Raynaud. 2013. *The White Planet: The Evolution and Future of Our Frozen World*. Princeton, NJ: Princeton University Press.

Kahn, Greg. 2003. "The Fate of the Kyoto Protocol Under the Bush Administration." *Berkeley Journal of International Law* 21: 548–571.

Kamieniecki, Sheldon, and Michael E. Kraft, eds. 2012. *Oxford Handbook of U.S. Environmental Politics*. New York: Oxford University Press.

Keener, Victoria, et al., eds. 2013. *Climate Change and Pacific Islands: Indicators and Impacts: Report for the 2012 Pacific Islands Regional Climate Assessment*. Washington, D.C.: Island.

Kiehl, Jeffrey T. 2016. *Facing Climate Change: An Integrated Path to the Future*. New York: Columbia University Press.

Klein, Naomi. 2014. *This Changes Everything: Capitalism Vs. the Climate*. New York: Simon & Schuster.

Kolbert, Elizabeth. 2006. *Field Notes from a Catastrophe: Man, Nature, and Climate Change*. New York: Bloomsbury.

Kolbert, Elizabeth. 2014. *The Sixth Extinction: An Unnatural History*. New York: Henry Holt.

Kollmuss, Anja, et al. 2010. *Handbook of Carbon Offset Programs: Trading Systems, Funds, Protocols and Standards*. Washington, D.C.: Earthscan.

Krupp, Fred, and Miriam Horn. 2008. *Earth: The Sequel: The Race to Reinvent Energy and Stop Global Warming*. New York: W.W. Norton.

Kutney, Gerald. 2014. *Carbon Politics and the Failure of the Kyoto Protocol*. New York: Routledge.

Labatt, Sonia, and Rodney R. White. 2007. *Carbon Finance: The Financial Implications of Climate Change*. New York: Wiley.

Launder, Brian, and J. Michael T. Thompson, eds. *Geo-Engineering Climate Change: Environmental Necessity or Pandora's Box*. New York: Cambridge University Press.

Lee, Peter. 2015. *Truth Wars: The Politics of Climate Change, Military Intervention and Financial Crisis*. New York: Palgrave Macmillan.

Leggett, Jeremy. 2006. *Half-Gone: Oil, Gas, Hot Air and the Global Energy Crisis*. London: Portobello.

Lever-Tracy, Constance, ed. 2010. *Routledge Handbook of Climate Change and Society*. New York: Routledge.

Llavador, Humberto, John E. Roemer, and Joaquim Silvestre. 2015. *Sustainability for a Warming Planet*. Cambridge, MA: Harvard University Press.

Lomborg, Bjørn. 2010. *Smart Solutions to Climate Change: Comparing Costs and Benefits*. New York: Cambridge University Press.

Lovejoy, Thomas E., and Lee Hannah, eds. 2006. *Climate Change and Biodiversity*. New Haven, CT: Yale University Press.

Lundquist, Lennart J., and Anders Biel, eds. 2016. *From Kyoto to the Town Hall: Making International and National Climate Policy Work at the Local Level*. New York: Routledge.

Mann, Michael E. 2012. *The Hockey Stick and the Climate Wars: Dispatches from the Front Lines*. New York: Columbia University Press.

Marino, Elizabeth. 2015. *Fierce Climate, Sacred Ground: An Ethnography of Climate Change in Shishmaref, Alaska*. Fairbanks: University of Alaska Press.

Marshall, George. 2014. *Don't Even Think About It: Why Our Brains Are Wired to Ignore Climate Change*. New York: Bloomsbury.

Maslin, Mark. 2014. *Climate Change: A Very Short Introduction*. 3d ed. New York: Oxford University Press.

Maslin, Mark, and Samuel Randalls, eds. 2012. *Future Climate Change*. 4 vols. New York: Routledge.

Mathez, Edmond A. 2009. *Climate Change: The Science of Global Warming and Our Energy Future*. New York: Columbia University Press.

Matthew, Richard A. 2011. "Is Climate Change a National Security Issue?" *Issues in Science & Technology* 27: 49–60.

McCright, Aaron M., and Riley E. Dunlap. "Defeating Kyoto: The Conservative Movement's Impact on U.S. Climate Change Policy." *Social Problems* 50: 348–373.

McGraw, Seamus. 2015. *Betting the Farm on a Drought: Stories from the Front Lines of Climate Change*. Austin: University of Texas Press.

McNeill, J. R., and Erin Stewart Mauldin, eds. 2012. *A Companion to Global Environmental History*. New York: Wiley-Blackwell.

Melillo, Jerry, et al., eds. 2009. *Global Climate Change Impacts in the United States*. New York: Cambridge University Press.

Metz, Bert, et al., eds. 2001. *Climate Change 2001: Mitigation*. New York: Cambridge University Press.

Michaels, Patrick J. 2004. *Meltdown: The Predictable Distortion of Global Warming by Scientists, Politicians, and the Media.* Washington, D.C.: Cato Institute.

Mintzer, Irving M., and J. A. Leonard, eds. 1994. *Negotiating Climate Change: The Inside Story of the Rio Convention.* New York: Cambridge University Press.

Mooney, Chris. 2005. *The Republican War on Science.* New York: Basic.

Mooney, Chris. 2007. *Storm World: Hurricanes, Politics, and the Battle Over Global Warming.* New York: Houghton Mifflin Harcourt.

Moser, Susanne C., and Lisa Dilling, eds. 2007. *Creating a Climate for Change: Communicating Climate Change and Facilitating Social Change.* New York: Cambridge University Press.

Musil, Robert K. 2008. *Hope for a Heated Planet: How Americans Are Fighting Global Warming and Building a Better Future.* New Brunswick, NJ: Rutgers University Press.

Nash, Steven. 2015. *Virginia Climate Fever: How Global Warming Will Transform Our Cities, Shorelines, and Forests.* Charlottesville: University of Virginia Press.

National Academy of Sciences and the Royal Society. 2014. *Climate Change: Evidence and Causes.* Washington, D.C.: National Academies.

National Assessment Synthesis Team. 2000. *Climate Change Impacts on the United States: The Potential Consequences of Climate Variability and Change: Overview.* New York: Cambridge University Press.

National Research Council of the National Academies. 2006. *Surface Temperature Reconstructions for the Last 2,000 Years.* Washington, D.C.: National Academies.

National Research Council of the National Academies. 2010. *Adapting to the Impacts of Climate Change.* Washington, D.C.: National Academies.

National Research Council of the National Academies. 2010. *Advancing the Science of Climate Change.* Washington, D.C.: National Academies.

National Research Council of the National Academies. 2010. *Informing an Effective Response to Climate Change.* Washington, D.C.: National Academies.

National Research Council of the National Academies. 2011. *America's Climate Choices.* Washington, D.C.: National Academies.

National Research Council of the National Academies. 2011. *Global Change and Extreme Hydrology: Testing Conventional Wisdom.* Washington, D.C.: National Academies.

National Research Council of the National Academies. 2011. *Understanding Earth's Deep Past: Lessons for Our Climate Future.* Washington, D.C.: National Academies.

National Research Council of the National Academies. 2012. *Sea-Level Rise for the Coasts of California, Oregon, and Washington: Past, Present, and Future.* Washington, D.C.: National Academies.

National Research Council of the National Academies. 2013. *Abrupt Impacts of Climate Change: Anticipating Surprises.* Washington, D.C.: National Academies.

National Research Council of the National Academies. 2014. *Linkages Between Arctic Warming and Mid-Latitude Weather Patterns: Summary of a Workshop.* Washington, D.C.: National Academies.

National Research Council of the National Academies. 2015. *Climate Intervention: Carbon Dioxide Removal and Reliable Sequestration.* Washington, D.C.: National Academies.

National Research Council of the National Academies. 2015. *Climate Intervention: Reflecting Sunlight to Cool Earth.* Washington, D.C.: National Academies.

Newell, Peter. 2000. *Climate for Change: Non-State Actors and the Global Politics of the Greenhouse.* New York: Cambridge University Press.

Nordhaus, William. 2013. *The Climate Casino: Risk, Uncertainty, and Economics for a Warming World.* New Haven, CT: Yale University Press.

Nuccitelli, Dana. 2015. *Climatology Versus Pseudoscience: Exposing the Failed Predictions of Global Warming Skeptics.* Santa Barbara, CA: Praeger.

Oberthür, Sebastian, Hermann E. Ott, with Richard T. Tarasofsky. 1999. *The Kyoto Protocol: International Climate Policy for the 21st Century.* New York: Springer.

Ojima, Dennis, et al., eds. 2015. *Great Plains Regional Technical Input Report.* Washington, D.C.: Island.

Oreskes, Naomi, and Erik Conway. 2011. *Merchants of Doubt: How a Handful of Scientists Obscured the Truth on Issues from Tobacco Smoke to Global Warming.* New York: Bloomsbury.

Orr, David W. 2009. *Down to the Wire: Confronting Climate Change.* New York: Oxford University Press.

Parker, Charles F., et al. 2012. "Fragmented Climate Change Leadership: Making Sense of the Ambiguous Outcome of COP-15." *Environmental Politics* 21: 268–286.

Paterson, Matthew. 2009. "Post-Hegemonic Climate Politics?" *British Journal of Politics & International Relations* 11: 140–158.

Patt, Anthony. 2015. *Transforming Energy: Solving Climate Change with Technology Policy.* New York: Cambridge University Press.

Pearce, Fred. 2007. *With Speed and Violence: Why Scientists Fear Tipping Points in Climate Change.* Boston: Beacon.

Peloso, Chris. 2010. "Crafting an International Climate Change Protocol: Applying the Lessons Learned from the Success of the Montreal Protocol and the Ozone Depletion Problem." *Journal of Land Use & Environmental Law* 25: 305–329.

Percival, Robert V. 2007. "Massachusetts v EPA: Escaping the Common Law's Growing Shadow." *Supreme Court Review* 2007: 111–160.

Peterson, David L., James M. Vose, and Toral Patel-Weynand, eds. 2014. *Climate Change and United States Forests.* New York: Springer.

Philander, S. George. 1998. *Is the Temperature Rising? The Uncertain Science of Climate Change.* Princeton, NJ: Princeton University Press.

Philander, S. George, ed. 2012. *Encyclopedia of Global Warming & Climate Change.* 2d ed. Los Angeles: Sage Reference.

Pielke, Roger, Jr. 2010. *The Climate Fix: What Scientists and Politicians Won't Tell You About Global Warming.* New York: Basic.

Pilkey, Orrin H., and Keith C. Pilkey. 2011. *Global Climate Change: A Primer.* Durham, NC: Duke University Press.

Pooley, Eric. 2010. *The Climate War: True Believers, Power Brokers, and the Fight to Save the Earth.* New York: Hyperion.

Pope Francis. 2015. *Encyclical on Climate Change and Inequality: On Care for Our Common Home.* New York: Melville House.

Powell, James Lawrence. 2011. *The Inquisition of Climate Science.* New York: Columbia University Press.

Press, Daniel. 2015. *American Environmental Policy: The Failures of Compliance, Abatement, and Mitigation.* Northampton, MA: Edward Elgar.

Primack, Richard B. 2014. *Walden Warming: Climate Change Comes to Thoreau's Woods.* Chicago: University of Chicago Press.

Princen, Thomas, Jack P. Manno, and Pamela L. Martin, eds. 2015. *Ending the Fossil Fuel Era.* Cambridge, MA: MIT Press.

Rabe, Barry G. 2004. *Statehouse and Greenhouse: The Emerging Politics of American Climate Change Policy.* Washington, D.C.: Brookings Institution.

Rabe, Barry G. 2016. "The Durability of Carbon Cap-And-Trade Policy." *Governance* 29: 103–119.

Rahm, Dianne. 2010. *Climate Change Policy in the United States: The Science, the Politics, and the Prospects for Change.* Jefferson, NC: McFarland.

Rapkin, David P., and William R. Thompson. 2013. *Transition Scenarios: China and the United States in the Twenty-First Century.* Chicago: University of Chicago Press.

Rapp, Donald. 2014. *Assessing Climate Change: Temperatures, Solar Radiation and Heat Balances.* 3d ed. New York: Springer.

Rappaport, Ann, and Sarah Hammond Creighton. 2007. *Degrees That Matter: Climate Change and the University.* Cambridge, MA: MIT Press.

Reed, Alan. 2006. *Precious Air: The Kyoto Protocol and Profit in the Global Warming Game.* Overland Park, KS: Leathers.

Rich, Frederic C. 2016. *Getting to Green: Saving Nature: A Bipartisan Solution.* New York: W.W. Norton.

Richardson, Katherine, Will Steffen, and Diana Liverman. 2011. *Climate Change: Global Risks, Challenges and Decisions.* New York: Cambridge University Press.

Richter, Burton. 2010. *Beyond Smoke and Mirrors: Climate Change and Energy in the 21st Century.* New York: Cambridge University Press.

Rigby, Catherine E. 2015. *Dancing with Disaster: Environmental Histories, Narratives, and Ethics for Perilous Times.* Charlottesville: University of Virginia Press.

Robin, Libby, Sverker Sörlin, and Paul Warde, eds. 2013. *The Future of Nature: Documents of Global Change.* New Haven, CT: Yale University Press.

Rosenzweig, Cynthia, and Daniel Hillel. 2008. *Climate Variability and the Global Harvest: Impacts of El Niño and Other Oscillations on Agro-Ecosystems.* New York: Oxford University Press.

Roston, Eric. 2008. *The Carbon Age: How Life's Core Element Has Become Civilization's Greatest Threat.* New York: Walker.

Ruddiman, William F. 2005. *Plows, Plagues, and Petroleum: How Humans Took Control of Climate.* Princeton, NJ: Princeton University Press.

Ruddiman, William F. 2013. *Earth's Climate: Past and Future.* 3d ed. New York: W.H. Freeman.

Sachs, Jeffrey D. 2005. *The End of Poverty: Economic Possibilities for Our Time.* New York: Penguin.

Sachs, Jeffrey D. 2015. *The Age of Sustainable Development.* New York: Columbia University Press.

Sackman, Douglas Cazaux, ed. 2010. *A Companion to American Environmental History*. New York: Wiley-Blackwell.

Safeguarding the Ozone Layer and the Global Climate System: Special Report of the Intergovernmental Panel on Climate Change. 2005. New York: Cambridge University Press.

Schapiro, Mark. 2014. *Carbon Shock: A Tale of Risk and Calculus on the Front Lines of the Disrupted Global Economy: How Carbon Is Changing the Cost of Everything*. White River Junction, VT: Chelsea Green.

Schmalensee, Richard, and Robert N. Stavins. 2013. "The SO_2 Allowance Trading System: The Ironic History of a Grand Policy Experiment." *Journal of Economic Perspectives* 27: 103–121.

Schmidt, Gavin, and Joshua Wolfe. 2009. *Climate Change: Picturing the Science*. New York: W.W. Norton.

Schneider, Stephen H., and Terry Root, eds. 2002. *Wildlife Responses to Climate Change: North American Case Studies*. Washington, D.C.: Island.

Seaman, Camille. 2014. *Melting Away: A Ten-Year Journey Through Our Endangered Polar Regions*. New York: Princeton Architectural.

Sejersen, Frank. 2015. *Rethinking Greenland and the Arctic in the Era of Climate Change: New Northern Horizons*. New York: Routledge.

Sim, Stuart. 2009. *The Carbon Footprint Wars: What Might Happen If We Retreat from Globalization*. Edinburgh: Edinburgh University Press.

Sinnott-Armstrong, Walter, and Richard B. Howarth, eds. 2006. *Perspectives on Climate Change: Science, Economics, Politics, Ethics*. San Diego, CA: Elsevier Science & Technology.

Sobel, Adam. 2014. *Storm Surge: Hurricane Sandy, Our Changing Climate, and Extreme Weather of the Past and Future*. New York: Harper Wave.

Soden, Dennis L., ed. 1999. *The Environmental Presidency*. Albany: State University of New York Press.

Spencer, Roy W. 2008. *Climate Confusion: How Global Warming Hysteria Leads to Bad Science, Pandering Politicians, and Misguided Policies That Hurt the Poor*. New York: Encounter.

Speth, James Gustave. 2004. *Red Sky at Morning: America and the Crisis of the Global Environment*. New Haven, CT: Yale University Press.

Steinberg, Paul F. 2015. *Who Rules the Earth? How Social Rules Shape Our Planet and Our Lives*. New York: Oxford University Press.

Stern, Nicholas. 2007. *The Economics of Climate Change: The Stern Review*. New York: Cambridge University Press.

Stern, Nicholas Herbert. 2015. *Why Are We Waiting? The Logic, Urgency, and Promise of Tackling Climate Change*. Cambridge, MA: MIT Press.

Stevenson, Hayley, and John S. Dryzek. 2014. *Democratizing Global Climate Governance*. New York: Cambridge University Press.

Stoknes, Per Espen. 2015. *What We Think About When We Try Not to Think About Global Warming: Toward a New Psychology of Climate Action*. White River Junction, VT: Chelsea Green.

Stone, Brian, Jr. 2012. *The City and the Coming Climate: Climate Change in the Places We Live*. New York: Cambridge University Press.

Stradling, David. 1999. *Smokestacks and Progressives: Environmentalists, Engineers, and Air Quality in America, 1881–1951*. Baltimore: Johns Hopkins University Press.

Surampalli, Rao Y., et al., eds. 2015. *Carbon Capture and Storage: Physical, Chemical, and Biological Methods*. Reston, VA: American Society of Civil Engineers.

Susskind, Lawrence E. 1994. *Environmental Diplomacy: Negotiating More Effective Global Agreements*. New York: Oxford University Press.

Sweet, William. 2006. *Kicking the Carbon Habit: Global Warming and the Case for Renewable and Nuclear Energy*. New York: Columbia University Press.

Szarka, Joseph. 2012. "The EU, the USA and the Climate Divide: Reappraising Strategic Choices." *European Political Science* 11: 31–40.

255 Members of the National Academy of Sciences. 2010. *Climate Change and the Integrity of Science: Lead Letter Published in Science Magazine, May 7, 2010*. http://www.pacinst.org/wp-content/uploads/sites/21/2013/02/climate_statement3.pdf.

United States Department of Energy. 2015. *Climate and Energy-Water-Land System Interactions: Technical Report to the U.S. Department of Energy in Support of the National Climate Assessment*. Alexandria, VA: National Technical Information Service.

U.S. Global Change Research Program. 2009. *Global Climate Change Impacts in the United States: A State of Knowledge Report*. New York: Cambridge University Press.

U.S. Global Change Research Program. "Globalchange.Gov." http://www.globalchange.gov.

Van Kooten, G. Cornelis. 2004. *Climate Change Economics: Why International Accords Fail*. Northampton, MA: Edward Elgar.

Van Kooten, G. Cornelis. 2013. *Climate Change,*

Climate Science and Economics: Prospects for an Alternative Energy Future. New York: Springer.

Victor, David G. 2001. *The Collapse of the Kyoto Protocol and the Struggle to Slow Global Warming.* Princeton, NJ: Princeton University Press.

Vig, Norman J., and Michael E. Kraft, eds. 2015. *Environmental Policy: New Directions for the Twenty-First Century.* 9th ed. Washington, D.C.: Congressional Quarterly.

Vogler, John. 2015. *Climate Change in World Politics.* New York: Palgrave Macmillan.

Volk, Tyler. 2008. *CO_2 Rising: The World's Greatest Environmental Challenge.* Cambridge, MA: MIT Press.

Wagner, Gernot, and Martin L. Weitzman. 2015. *Climate Shock: The Economic Consequences of a Hotter Planet.* Princeton, NJ: Princeton University Press.

Walker, Gabrielle, and Sir David King. 2008. *The Hot Topic: What We Can Do About Global Warming.* New York: Harcourt.

Ward, Peter Douglas. 2007. *Under a Green Sky: Global Warming, the Mass Extinctions of the Past, and What They Mean for Our Future.* Washington, D.C.: Smithsonian Institution.

Ward, Peter Douglas. 2010. *The Flooded Earth: Our Future in a World Without Ice Caps.* New York: Basic.

Watson, Robert T., ed. 2002. *Climate Change 2001: Synthesis Report.* New York: Cambridge University Press.

Weart, Spencer R. 2008. *The Discovery of Global Warming.* Rev. and expanded ed. Cambridge, MA: Harvard University Press.

Weiss, Charles, and William B. Bonvillian. 2009. *Structuring an Energy Technology Revolution.* Cambridge, MA: MIT Press.

Westmoreland, Joshua K. 2010. "Global Warming and Originalism: The Role of the EPA in the Obama Administration." *Boston College Environmental Law Review* 37: 225–256.

Wilbanks, Thomas J., ed. 2014. *Climate Change and Energy Supply and Use: Technical Report for the U.S. Department of Energy in Support of the National Climate Assessment.* Washington, D.C.: Island.

Wilbanks, Thomas J., and Steven J. Fernandez, eds. 2014. *Climate Change and Infrastructure, Urban Systems, and Vulnerabilities: Technical Report for the U.S. Department of Energy in Support of the National Climate Assessment.* Washington, D.C.: Island.

Winston, Andrew S. 2014. *The Big Pivot: Radically Practical Strategies for a Hotter, Scarcer, and More Open World.* Cambridge, MA: Harvard Business Review.

Wolinsky-Nahmias, Yael. 2014. *Changing Climate Politics: U.S. Policies and Civic Action.* Washington, D.C.: Congressional Quarterly.

Woodworth, Paddy. 2013. *Our Once and Future Planet: Restoring the World in the Climate Change Century.* Chicago: University of Chicago Press.

Worster, Donald. 2016. *Shrinking the Earth: The Rise and Decline of American Abundance.* New York: Oxford University Press.

Wright, Christopher, and Daniel Nyberg. 2015. *Climate Change, Capitalism, and Corporations: Processes of Creative Self-Destruction.* New York: Cambridge University Press.

Wu, Yutian, Lorenzo M. Polvani, and Richard Seager. 2013. "The Importance of the Montreal Protocol in Protecting Earth's Hydroclimate." *Journal of Climate* 26: 4049–4068.

Yamin, Farhana, and Joanna Depledge. 2004. *The International Climate Change Regime: A Guide to Rules, Institutions and Procedures.* New York: Cambridge University Press.

Zwally, H. Jay, et al. 2015. "Mass Gains of the Antarctic Ice Sheet Exceed Losses." *Journal of Glaciology* 61: 1019–1036.

Index

ABC News 137
acid rain 2, 13, 39–44
Adoption of the Paris Agreement: Proposal by the President: Draft Decision 5
aerosols 32, 153, 184
Africa 12–13, 49, 132, 248, 253
agriculture 8, 12–13, 49, 69, 94–95, 101–102, 171, 182–183, 216, 221
air pollution 95, 147–152, 180–196, 220
Alaska 93, 95, 169, 182–183, 215, 221, 262–264
alkalization 13
allergens 181
American Clean Energy and Security Act of 2009 173–176
American Indians *see* Native Americans
Antarctica 2, 32, 130, 154, 157, 277–281
Arctic 169, 215, 245–246, 249–250, 262–264, 280
Arrhenius, Svante 1
Asia 12–13, 132
Associated Press 136
Association of Clean Water Administrators (ACWA) 208
Atmosphere of Pressure: Political Interference in Federal Climate Science 2, 141–146
atmospheric pollution 52
Australia 67, 87, 250
Austria 87

Baliunas, Sallie 131
Bangladesh 102
Belgium 87
Belize 102
Berkeley Earth Surface Temperature Project 201–205
Berlin Mandate 66–67
Boehner, John 237
Brazil 200
Brokaw, Tom 137
Bromochloromethane 25
Brown, Edmund G., Jr. 5, 240–242
Brown, Jerry *see* Brown, Edmund G., Jr.
Bulgaria 87
Buses 185

Bush, George H. W. 2–3, 43
Bush, George W. 4, 97–99, 145, 172, 252; administration 1, 4, 99–103, 132, 145–146, 172, 196; political pressure on federal scientists 141–146, 160
Byrd, Robert 67
Byrd-Hagel Resolution *see* Senate Resolution 98 [Report No. 105-84] (Byrd-Hagel Resolution)

California 4–5, 98, 131, 191, 207–208, 240–242
California Global Warming Solutions Act of 2006 240
Canada 87, 94, 250
cancer 32
cap-and-trade 2, 40–44, 117–126, 173–176, 258–260; China 273
carbon dioxide 3, 5, 8, 13, 63, 69, 71, 86, 97–98, 106, 129–130, 157, 160, 167–169, 180–181, 184, 196, 213, 218, 236, 254–258
carbon sequestration 101
carbon tax 159
carbon tetrachloride 20
cars 185
Carter, Bob 139
Carter, Jimmy 2, 15; administration 1
CBS 134–135
Central American–United States of America Joint Accord (CONCAUSA) 102–103
Chicago Tribune 134
Chile 251
China 5, 88, 98, 132, 157, 184, 200, 234–239, 265, 271–276, 316
Chirac, Jacques 132
chlorine atoms 32
chlorofluorocarbons 2, 19, 31–32, 34, 153
chlorofluoromethanes 8, 153
Christy, John 131
Chu, Steven 5, 161–166
Civil Aviation Organization 70
Clean Air Act 4, 43, 97, 147–152, 173–176, 254–260; *Amendments of 1990* 2, 39–44
Clean Air Scientific Advisory Committee 194

Clean Power Plan *see Federal Plan Requirements for Greenhouse Gas Emissions from Electric Utility Generating Units Constructed on or Before January 8, 2014; Model Trading Rules; Amendments to Framework Regulations*
Clean Water State Revolving Fund 206
Climate Change Conference in Copenhagen, Denmark 5
Climate Change Impacts in the United States 4, 213–222
Climate Change Impacts on the United States: The Potential Consequences of Climate Variability and Change 4, 92–96
Climate Change Research Initiative 101
Climate Change Workgroup of the Advisory Committee on Water Information 211
Climate Ready Water Utilities Program 206
Climate Resilience Evaluation and Awareness Tool (CREAT) 207
Climate Stewardship Act of 2003 4, 104–127
Clinton, William "Bill" 3–4, 44, 88, 90–91; administration 4, 65–67
coal 12–13, 97–98, 158, 167, 213, 261
coastal waters 209–210
coastal wetlands 93, 182
Colorado River 219
Columbia 251
Columbia River 219
Committee on Climate Change Science and Technology Integration 100
CONCAUSA *see* Central American–United States of America Joint Accord
Congress *see* United States Congress
Connecticut 4, 208
Copenhagen Accord 5, 177–179, 197–200
Copenhagen Consensus 132
Copenhagen Green Climate Fund 199–200

329

coral reefs 95, 168, 221
Council of Environmental Quality 15
Croatia 87
Crutzen, Paul 2, 153–155
Cruz, Ted 280–281
Czech Republic 87

de Blasio, Bill 5, 223–233
Debt-for-Nature 102
Declaration of the United Nations Conference on the Human Environment 45
Declaration of the World Climate Conference 17–10
Defense Support of Civil Authorities 245
deforestation 8, 12, 199, 213
Deming, David 129
Denmark 87
Department of Defense *see* United States Department of Defense
Department of State *see* United States Department of State
desertification 13, 46, 49, 52
developing countries 7, 12, 25, 32, 46, 48, 52, 65–67, 88, 90, 198–200, 302–304, 308–310, 317
Discovery Channel 137
disease 13, 220, 245
Doyal, Alister 138
drought 7, 46, 49, 52, 207, 219, 244–245

Earth Summit 1992 3, 88
Eizenstat, Stuart 90
El Salvador 102, 251
Endangerment and Cause or Contribute Findings for Greenhouse Gases Under Section 202(a) of the Clean Air Act: Final Rule 4
energy 7, 49, 182, 275
Environmental Protection Agency (EPA) 4, 44, 147–152, 180–196, 254–261; Healthy Watersheds Program 208; National Water Program 206–212
Estonia 87
European Union 32, 87, 200, 236, 253, 276
extinction 13
ExxonMobil 158

Fact Sheet: The United States and China Issue Joint Presidential Statement on Climate Change with New Domestic Policy Commitments and a Common Vision for an Ambitious Global Climate Agreement in Paris 5, 271–276
Federal Plan Requirements for Greenhouse Gas Emissions from Electric Utility Generating Units Constructed on or Before January 8, 2014; Model Trading Rules; Amendments to Framework Regulations 254–261, 264, 270–271, 273

fertility 13
fertilizers 8, 12
Finland 87
fires *see* wildfires
floods 7, 46, 207, 215, 220, 244–245
food 7, 12–13, 48; scarcity 170
forests 13, 49, 93, 95, 182–183; degradation 199; southeast United States 95; tropical forests 13
fossil fuels 34; *see also* coal; gas; oil
France 87

gas 12, 97, 158, 167, 213
Geophysical Research Letters 130
Georgia 207
Germany 87–88
GLACIER Conference 5, 262–267
Global Change Research Act of 1990 96, 100, 171, 222
Global Change Research Program 96, 172
Global Climate Change Impacts in the United States 4, 167–172
Global Climate Change Policy Book 4, 99–103
Global 2000 Report to the President: Entering the Twenty-First Century 1, 11–15
Global Warming Twenty Years Later: Tipping Points Near 2, 156–160
Gore, Albert "Al," Jr. 3, 90–91, 128, 135–138, 140, 205, 236, 280
Gosar, Paul 270
Government Accountability Project 141–146
Governor Brown Establishes Most Ambitious Greenhouse Gas Reduction Target in North America 5, 240–242
Gray, Vincent 130
Gray, William 131
Great Britain 67, 87–88, 250
Great Lakes 219
Greece 87
Green Climate Fund 274, 288–289, 317
greenhouse effect 35–36
greenhouse gas emissions 3–5, 37, 44–45, 47, 49–50, 64–89, 99–103, 147–152, 167, 173–196, 200, 225–228, 238–239, 254–261, 316–317
greenhouse gas intensity 100
Greenland 157
Group of 77 Plus China 88

H. Res. 593—Congratulating Scientists F. Sherwood Rowland, Mario Molina, and Paul Crutzen for their Work in Atmospheric Chemistry, Particularly Concerning the Formation and Decomposition of Ozone, that Led to the Development of the Montreal Protocol on Substances that Deplete the Ozone Layer 2, 153–155
Habitat destruction 95
Hadley Centre/Climate Research Unit 193, 201, 203
Hagel, Chuck 67
halocarbons 169
halons 19, 32
Hansen, James 2, 35–38, 134–135, 137, 145–146, 156–160
health 7
Herbert, Bob 138
hockey stick 129, 136
Hot & Cold Media Spin: A Challenge to Journalists Who Cover Global Warming 3, 128–140
House of Representatives *see* United States House of Representatives
humanitarian disaster relief 243, 245
Hungary 87
hunger 13
hydrobromofluorocarbons 23
hydrochlorofluorocarbons 21, 63, 71
hydrofluorocarbons 86, 106, 180, 184, 196, 236, 272–274
hydrological data 9

Iceland 87
Illinois 4
An Inconvenient Truth 3, 128, 136–138, 140
Independent Review Board of the President's Committee of Advisors on Science & Technology 96
India 5, 98, 132, 157, 184, 200, 238–239
Inhofe, James 3, 128–140
Inter-American Defense Board 251
Intergovernmental Oceanographic Commission 1, 10
Intergovernmental Panel on Climate Change 62, 71, 73, 130, 173, 180, 240
International Atomic Energy Agency 55
International Council of Scientific Unions 1, 10
International Court of Justice 59
International Maritime Organization 70
Ireland 87
irrigation 12; *see also* water
Italy 87, 102

Japan 32, 87, 102, 200
Johannessen, Ola 134
Joint Meeting to Hear an Address by Pope Francis of the Holy See 5, 268–270
journalism 128–140

Kansas 207–208
Kentucky 260–261
Kerry, John 135, 137, 281
Kerry, Teresa Heinz 135

Klein, Naomi 270
Kyoto Protocol to the United Nations Framework Convention on Climate Change 3–5, 68–91, 103, 179; Bush, George W. 97–99; opposition from United States Senate 65–67

lakes 13, 170; *see also* water
land 12; arable 12; arid 52; semi-arid 52
Latin America 12
Latvia 87
Letter to Members of the Senate on the Kyoto Protocol on Climate Change 4, 97–98
Lieberman, Joe 4, 104–127
Liechtenstein 87
Lindzen, Richard 131
Lithuania 87
Lomborg, Bjorn 132
Los Angeles Times 133, 137
Louisiana 221
Luxembourg 87

Maine 4, 211
malnutrition 13
Mann, Michael 129
Marshall Islands 200, 317
Massachusetts 4
Massachusetts et al. v. Environmental Protection Agency et al. 4, 147–152, 180, 184, 196
McCain, John 4, 104–127
McConnell, Mitch 237, 261
Message to the Senate Transmitting the Montreal Protocol on Substances that Deplete the Ozone Layer 2
meteorological data 9
methane 63, 69, 86, 106, 167, 169, 180, 184, 196, 214, 273–274
methyl bromide 23
methyl chloroform 21
Mexico 94
Michaels, Patrick 131
Middle East 253
Ministerial Declaration of the Second World Climate Conference 46
Minnesota 208
Missouri 207
Mitchell, George 43
Molina, Mario 2, 31–32, 153–155
Monaco 87
Montreal Protocol on Substances that Deplete the Ozone Layer 2, 16–34, 46, 68, 153–155
motorcycles 185
Muller, Richard A. 201–205
Multilateral Fund for the Implementation of the Montreal Protocol 29, 32, 154

NASA 143, 193, 201, 203, 277–279
NASA Goddard Institute for Space Studies 2, 35–38, 277–279
National Academy of Sciences 129, 173

National Aeronautics and Space Administration *see* NASA
National Assessment of the Potential Consequences of Climate Variability 92–96, 167–172
National Center for Atmospheric Research 141
National Climate Assessment 213
National Climate Change Technology Initiative 101
National Energy Policy 4, 101, 103
National Fuel Efficiency Policy 190
National Highway Safety Administration 185
National Oceanic and Atmospheric Administration (NOAA) 193, 201, 203, 246
National Research Council 180
National Security Implications of Climate-Related Risks and a Changing Climate 243–253
National Water Program 206–212
Native Americans 211, 221, 256, 262–263, 266
natural gas *see* gas
Nature Conservancy 208
Netherlands 87
New Jersey 4, 245
New Mexico 4
New York 4, 245; New York City 5, 223–233; Ulster County 208
New York Times 133–135, 138
New Zealand 87, 250, 280
Newsweek 133
nitrogen 210
nitrogen fertilizers *see* fertilizers
nitrogen oxides 43, 63
nitrous oxide 86, 106, 169, 180, 184, 196
North Carolina 210
Norway 87
Novim 201

Obama, Barack 14–5, 44, 172–173, 176–179, 234–237, 262–267, 270–276; administration 163–166, 196, 237, 317; *Federal Plan Requirements for Greenhouse Gas Emissions from Electric Utility Generating Units Constructed on or Before January 8, 2014; Model Trading Rules; Amendments to Framework Regulations* 254–261; National Fuel Efficiency Policy 190
oceans 49, 168–169, 209, 215, 221; acidification 174, 221, 263
oil 12, 158, 167, 185, 213
oil shale 12
Oklahoma 215
One City Built to Last: Transforming New York City's Buildings for a Low-Carbon Future 5, 223–233
Oppenheimer, Michael 137
Oregon 4
Oreskes, Naomi 137–138
ozone 181

ozone-depleting chemicals 13
ozone depletion hypothesis 32
ozone layer 2, 16–34, 153–154

Pakistan 245
Paramount Pictures 136
Paris Agreement 5, 282–318; Brown, Jerry 240–242; Cruz, Ted 280–281; Obama, Barack 265–267; United States and China Cooperation 234–237, 271–276; United States Senate 238–239
pastoral practices 8
Peabody Coal 158
Peiser, Benny 138
Pelley, Scott 134–135
Peoples Climate March 232, 316
perfluorocarbons 63–64, 71, 86, 106, 180, 184, 196
pesticides 12, 14
Pielke, Roger, Sr. 131
Poland 87
Pope Francis of the Holy See 5, 268–270
population 11, 13; displacement 174
Portugal 87
poverty 11, 14, 244
power plants 254–261
Presidential Task Force on Global Resources and Environment 15
Public Policy Research 139

Reagan, Ronald 2, 33; administration 2, 34
Remarks by the President at the GLACIER Conference—Anchorage, Alaska 5, 262–267
Remarks by the President at the Morning Plenary Session of the United Nations Climate Change Conference 5, 177–179
Reuters 138
Revkin, Andrew 135
Rhode Island 4
Rivers 170
Rocky Mountains 92, 95, 158
Romania 87
Rowland, Frank Sherwood 2, 31–32, 153–155
Russia 5, 87, 184, 200, 238–239
Russian Federation *see* Russia

Safe Climate Act 2, 173–176
salination 13
salmon 209
sea level rise 171, 182, 207, 215, 217, 220, 245, 253; *see also* oceans
sea walls 94
Secretary Chu's Remarks at the Harvard University Commencement—As Prepared for Delivery 5
seed banks 14
Senate *see* United States Senate
Senate Bill 66: To Prohibit Any Regulation Regarding Carbon Dioxide or Other Greenhouse

Gas Emissions Reduction in the United States Until China, India, and Russia Implement Similar Reductions 5, 238–239
Senate Resolution 98 [Report No. 105-84] (Byrd-Hagel Resolution) 3, 5, 65–67, 88
60 Minutes 134
skin Singer, S. Fred 131
Slovakia 87
Slovenia 87
soils 13
solar panels 2
solar power 229–230
Soon, Willie 131
South America 12, 132
South Korea 132
Spain 87
Spencer, Roy 131
Speth, Gus 15
Statement on the Kyoto Protocol on Climate Change 90–91
Subsidiary Body for Scientific and Technological Advice 70–71, 73, 81–82
sugar maples 93
sulphur dioxide 39–44
sulphur hexafluoride 64, 71, 86, 106, 180, 184, 196
Superstorm Sandy 245
Supreme Court *see* United States Supreme Court
Sweden 87
Switzerland 87
Syria 245

tar sands 12
Taylor, George 131
Taylor, Mitchell 138
technology transfer 29–30
Texas 207, 215
Thailand 102, 250
thermonuclear war *see* warfare
Time 133, 135
transfer of technology *see* technology transfer
transportation 183–185, 273, 275
trichloroethane 21
Trinidad & Tobago 251
tropical forests *see* forests
trucks 185

Ukraine 87
ultraviolet radiation 32

Union of Concerned Scientists 141–146
United Kingdom of Great Britain and Northern Ireland *see* Great Britain
United Nations 5, 55; Climate Change Conference 2009 177–179; *see also* Intergovernmental Panel on Climate Change
United Nations' Environment Programme 32, 46
United Nations Framework Convention on Climate Change 2–3, 46–67, 88, 102, 197–200, 224; *Kyoto Protocol to the United Nations Framework Convention on Climate Change* 68–91; *Paris Agreement* 282–318
United States Academy of Sciences 32, 154
United States Congress 14, 96, 260–261, 267; *see also* United States House of Representatives; *see also* United States Senate
United States Department of Defense 243–253; National Security Strategy 244; North American Aerospace Defense Command / United States Northern Command 247; Quadrennial Defense Review 244, 252; United States Africa Command 247, 248; United States Central Command 248–249; United States European Command 247, 249; United States Pacific Command 247; United States Southern Command 246–247, 250–252
United States Department of State 15
United States Global Research Program 4, 96, 167, 171, 194, 213, 222
United States Greenhouse Gas Emissions and Sinks 107
United States House of Representatives 2, 153–155; Committee on Science, Space, and Technology of the United States 201–205
United States National Academy of Sciences 13
United States Senate 3–5, 33–34, 65–67, 88, 238–239, 317; *Climate Stewardship Act of 2003* 4, 104–127; Committee on Appropriations 243; Environment and Public Works Committee 132, 140; Science Subcommittee of the Senate Commerce Committee 280–281; United States Senate Commission on Energy and Natural Resources 35–38
United States Supreme Court 4, 147–152, 187–188, 196, 261
uranium 12
U.S.-China Joint Announcement on Climate Change 5, 234–237

Vermont 4
Vienna Convention for the Protection of the Ozone Layer 16, 32, 46
Virginia 207

warfare 8
Washington 4
Washington, D.C. 207
Washington Post 133
waste management 13, 49
water 7, 12–13, 95, 170, 182, 216, 220–221, 253, 275; National Water Program 206–212; quality 210, 220–221; western United States 93–94; *see also* irrigation
waterlogging 13
WaterSense 207
watersheds 207–208
Waxman, Henry 43, 176
West Virginia 208, 260
wetlands 207–208
wildfires 207, 220, 263
wood 12
World Climate Conference 17–10
World Climate Program 10
World Climate Programme 9–10
World Climate Research Programme 10
World Meteorological Association 1, 10
World Meteorological Organization 7–10, 46

Xi, Jinping 234–237, 271–276

Zwally, Jay 277–279

www.ingramcontent.com/pod-product-compliance
Ingram Content Group UK Ltd.
Pitfield, Milton Keynes, MK11 3LW, UK
UKHW050543150426
5217IPUK00026B/2051